International Series in Operations Research & Management Science

Volume 246

Series Editor

Camille C. Price
Stephen F. Austin State University, TX, USA

Associate Series Editor

Joe Zhu
Worcester Polytechnic Institute, MA, USA

Founding Series Editor

Frederick S. Hillier
Stanford University, CA, USA

More information about this series at http://www.springer.com/series/6161

John H. Drew • Diane L. Evans • Andrew G. Glen
Lawrence M. Leemis

Computational Probability

Algorithms and Applications in the Mathematical Sciences

Second Edition

 Springer

John H. Drew
Department of Mathematics
The College of William and Mary
Williamsburg, VA, USA

Diane L. Evans
Department of Mathematics
Rose-Hulman Institute of Technology
Terre Haute, IN, USA

Andrew G. Glen
Department of Mathematics
and Computer Science
Colorado College
Colorado Springs, CO, USA

Lawrence M. Leemis
Department of Mathematics
The College of William and Mary
Williamsburg, VA, USA

ISSN 0884-8289 ISSN 2214-7934 (electronic)
International Series in Operations Research & Management Science
ISBN 978-3-319-82790-2 ISBN 978-3-319-43323-3 (eBook)
DOI 10.1007/978-3-319-43323-3

Printed on acid-free paper

This Springer imprint is published by Springer Nature
The registered company is Springer International Publishing AG
The registered company address is: Gewerbestrasse 11, 6330 Cham, Switzerland

Preface

For decades, statisticians have enjoyed the use of "statistical packages" which read in a (potentially) large data set, calculate numerical summaries such as the sample mean and sample variance, calculate more sophisticated statistical quantities such as confidence interval bounds and p-values associated with hypothesis tests, and generate statistical graphics such as histograms, box plots, and multidimensional plots. But pity the poor probabilist, who through all those decades had only paper and pencil for symbolic calculations. The purpose of this monograph is to address the plight of the probabilist by providing algorithms to perform calculations associated with random variables. We refer to a collection of data structures and algorithms that automate probability calculations as "computational probability." The data structures and algorithms introduced here have been implemented in a language known as APPL (A Probability Programming Language). Several illustrations of problems from the mathematical sciences that can be solved by implementing these algorithms in a computer algebra system are presented in the final chapters of this monograph.

The algorithms for manipulating random variables (e.g., adding, multiplying, transforming, ordering) symbolically result in an entire class of new problems that can now be addressed. APPL is able to perform exact probability calculations for problems that would otherwise be deemed intractable. The work is quite distinct from traditional probability analysis in that a computer algebra system, in this case Maple, is used as a computing platform.

The use of a computer algebra system to solve problems in operations research and probability is increasing. Other researchers also sense the benefits of incorporating a computer algebra system into fields with probabilistic applications, for example, Parlar's *Interactive Operations Research with Maple* [98], Karian and Tanis's second edition of *Probability and Statistics: Explorations with Maple* [58], Rose and Smith's *Mathematical Statistics and Mathematica* [105], and Hasting's second edition of *Introduction to the Mathematics of Operations Research with Mathematica* [46].

This monograph significantly differs from the four titles listed above in two ways. First, the four titles listed above are all textbooks, rather than research monographs. They contain exercises and examples geared toward students, rather than researchers. Second, the emphasis in most of these texts is much broader than the emphasis being proposed here. For example, Parlar and Hasting consider all of OR/MS, rather than the probabilistic side of OR/MS proposed here in much more depth. Also, Karian and Tanis emphasize Monte Carlo solutions to probability and statistics problems, as opposed to the exact solutions given in APPL.

The monograph begins with an introductory chapter that contains short examples involving the elementary use of APPL. Chapter 2 reviews the Maple data structures and functions necessary to implement APPL. This is followed by a discussion of the development of the data structures and algorithms (Chaps. 3–6 for continuous random variables and Chaps. 7–9 for discrete random variables) used in APPL. The monograph concludes with Chaps. 10–15 introducing a sampling of various applications in the mathematical sciences. The two most likely audiences for the monograph are researchers in the mathematical sciences with an interest in applied probability and instructors using the monograph for a special topics course in computational probability taught in a mathematics, statistics, operations research, management science, or industrial engineering department. The intended audience for this monograph includes researchers, MS students, PhD students, and advanced practitioners in stochastic operations research, management science, and applied probability.

An indication of the utility of APPL is that the research efforts of the authors and other colleagues have produced many related refereed journal publications, many conference presentations, the ICS Computing Prize with INFORMS, a government patent, the INFORMS Undergraduate Research Prize, a GEM Scholarship based on APPL research, and multiple improvements to pedagogical methods in numerous colleges and universities around the world. We believe that the potential of this field of computational probability in research and education is unlimited. It is our hope that this monograph encourages people to join us in attaining future accomplishments in this field.

We are grateful to Camille Price, Gary Folven, Carolyn Ford, and Matt Amboy from Springer for their support of this project. We also thank Professor Ludolf Meester from TU Delft for class testing and debugging portions of the APPL code. We thank our coauthors Matt Duggan, Kerry Connell, Jeff Mallozzi, and Bruce Schmeiser for their collaboration on the applications given in Chap. 11. We also thank the editors, anonymous referees, and colleagues who have been helpful in the presentation of the material, including John Backes, Donald Barr, Don Campbell, Jacques Carette, Gianfranco Ciardo, Mike Crawford, Mark Eaton, Jerry Ellis, Bob Foote, James Fritz, Winfried Grassmann, Greg Gruver, Carl Harris, Rex Kincaid, Hank Krieger, Marina Kondratovitch, Sid Lawrence, Barry Nelson, David Nicol, Raghu Pasupathy, Steve Roberts, Sheldon Ross, Bruce Schmeiser, Jim Scott, Bob Shumaker,

Paul Stockmeyer, Bill Treadwell, Michael Trosset, Mark Vinson, and Bert Zwart. We thank techie Robert Marmorstein for his LaTeX support, Chris Weld for proofreading the second edition, and Barbara Boyer for her help with the index keywords for the first edition of this monograph. The authors gratefully acknowledge support from the Clare Boothe Luce Foundation, the National Science Foundation (for providing funding for an Educational Innovation Grant CDA9712718 "Undergraduate Modeling and Simulation Analysis," for scholarship funds provided under the CSEMS Grant 0123022 "Effective Transitions Through Academe to Industry for Computer Scientists and Mathematicians," for support by an NSF CSUMS grant DMS-0703532), the Army Research Office, Grant 67906-MA, and The College of William & Mary for research leave to support this endeavor.

Finally, our thanks goes to all of our coauthors who have made such significant contributions to this volume and to APPL language. Nearly all of the material in this monograph has originally appeared in journal articles: Chap. 1 is from Glen et al. [38]; Chaps. 2 and 3 contain original material by Diane Evans; Chap. 4 is from Glen et al. [37]; Chap. 5 is from Yang et al. [122]; Chap. 6 is from Glen et al. [39]; Chap. 7 is from Glen et al. [38]; Chap. 8 is from Evans and Leemis [29]; Chap. 9 is from Evans et al. [30]; Chap. 10 is from Leemis [73], Evans et al. [30], and Marks et al. [82]; Chap. 11 is from Webb and Leemis [115]; Chap. 12 is from Duggan et al. [24] and Drew et al. [23]; Chap. 13 is from Kaczynski et al. [56]; Chap. 14 is original work by Matt Robinson; and Chap. 15 is from Leemis [73], Leemis et al. [74], and Glen et al. [38].

Williamsburg, VA, USA John H. Drew
Terre Haute, IN, USA Diane L. Evans
Colorado Springs, CO, USA Andrew G. Glen
Williamsburg, VA, USA Lawrence M. Leemis
June 2016

Contents

Part I

Introduction

Civilization advances by extending the number of important operations which we can perform without thinking about them.

—Alfred North Whitehead

1

Computational Probability

The purpose of this chapter is to lure you into reading the rest of the monograph. We present four examples of probability questions that would be unpleasant to solve by hand, but are solvable with computational probability using APPL (A Probability Programming Language). We define the field of *computational probability* as the development of data structures and algorithms to automate the derivation of existing and new results in probability and statistics. Section 12.3, for example, contains the derivation of the distribution of a well-known test statistic that requires 99,500 carefully crafted integrations.

1.1 Four Simple Examples of the Use of APPL

The first example is one that a probability student would typically solve approximately by using the central limit theorem or Monte Carlo simulation. We show that APPL can compute the desired result exactly.

Example 1.1. Let X_1, X_2, \ldots, X_{10} be independent and identically distributed (iid) random variables that are uniformly distributed between 0 and 1. Find the probability that their sum lies between 4 and 6, i.e.,

$$\Pr \left(4 < \sum_{i=1}^{10} X_i < 6 \right).$$

The central limit theorem yields only one digit of accuracy for this particular problem. Monte Carlo requires custom programming and the result is typically stated as an interval around the true value. Also, the number of replications required is a quadratic function of the desired accuracy: each additional digit of accuracy requires a 100-fold increase in the number of Monte Carlo replications. On the other hand, the APPL statements to solve this problem are

© Springer International Publishing AG 2017
J.H. Drew et al., *Computational Probability*, International Series
in Operations Research & Management Science 246,
DOI 10.1007/978-3-319-43323-3_1

```
> X := UniformRV(0, 1);
> Y := ConvolutionIID(X, 10);
> CDF(Y, 6) - CDF(Y, 4);
```

which yields

$$\Pr\left(4 < \sum_{i=1}^{10} X_i < 6\right) = \frac{655177}{907200} \cong 0.7222.$$

The first line of APPL code defines X as a $U(0, 1)$ random variable. The second line defines the random variable Y as the sum of 10 iid random variables, each having the same distribution as X. Finally, the last line evaluates the cumulative distribution function (CDF) of Y at 6 less the CDF of Y at 4, yielding the exact result as the fraction shown above.

The second example concerns finding the probability density function (PDF) of the product of two triangular random variables.

Example 1.2. Let X and Y be independent random variables with triangular distributions with parameters $(1, 2, 4)$ and $(1, 2, 3)$, respectively. (The three parameters are the minimum, mode, and maximum, respectively.) Find and plot the PDF of $V = XY$.

The APPL code to solve this problem is:

```
> X := TriangularRV(1, 2, 4);
> Y := TriangularRV(1, 2, 3);
> V := Product(X, Y);
> PlotDist(V);
```

which returns the PDF of V as

$$f_V(v) = \begin{cases} -\frac{4}{3}v + \frac{2}{3}\ln v + \frac{2v}{3}\ln v + \frac{4}{3} & 1 < v \le 2 \\ -8 + \frac{14}{3}\ln 2 + \frac{7v}{3}\ln 2 + \frac{10}{3}v \\ \qquad -4\ln v - \frac{5v}{3}\ln v & 2 < v \le 3 \\ -4 + \frac{14}{3}\ln 2 + \frac{7v}{3}\ln 2 + 2v \\ \qquad -2\ln v - v\ln v - 2\ln 3 - \frac{2v}{3}\ln 3 & 3 < v \le 4 \\ \frac{44}{3} - 14\ln 2 - \frac{7v}{3}\ln 2 - \frac{8}{3}v \\ \qquad -2\ln 3 + \frac{22}{3}\ln v - \frac{2v}{3}\ln 3 + \frac{4v}{3}\ln v & 4 < v \le 6 \\ \frac{8}{3} - 8\ln 2 - \frac{4v}{3}\ln 2 - \frac{2}{3}v \\ \qquad +\frac{4}{3}\ln v + \frac{v}{3}\ln v + 4\ln 3 + \frac{v}{3}\ln 3 & 6 < v \le 8 \\ -8 + 8\ln 2 + \frac{2v}{3}\ln 2 + \frac{2}{3}v \\ \qquad +4\ln 3 - 4\ln v + \frac{v}{3}\ln 3 - \frac{v}{3}\ln v & 8 < v < 12. \end{cases}$$

The first two APPL statements define X and Y as triangular random variables. The third statement uses the APPL `Product` procedure to compute the PDF of the product. The last statement plots the PDF of V, which is shown in Fig. 1.1.

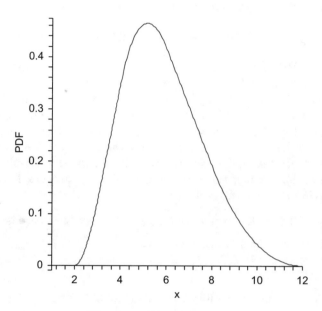

Fig. 1.1. PDF of $V = XY$

The first two examples are complex enough to require about an hour to work out by hand, assuming that no mathematical or calculation errors occurred. In both examples, the result was available with just a few lines of code and a few seconds of computer time. We next turn to a problem involving discrete random variables.

Example 1.3. A bag contains 15 billiard balls, numbered 1 to 15. If 7 balls are drawn from the bag at random, find the probability that the median number drawn is 5 when (*a*) sampling is performed without replacement; (*b*) sampling is performed with replacement.

In both of the solutions given below, the discrete random variable Y denotes the median number drawn in the sample of size 7. In part (*a*), the support of Y is $4, 5, \ldots, 12$ due to the sampling without replacement, but in part (*b*), the support of Y is $1, 2, \ldots, 15$ due to the fact that each ball drawn is placed back into the bag after being sampled.

(a) The APPL code in the sampling without replacement case is

```
> X := UniformDiscreteRV(1, 15);
> Y := OrderStat(X, 7, 4, "wo");
> PDF(Y, 5);
```

which returns the probability that the median is 5 as

$$\Pr(Y = 5) = \frac{32}{429} \cong 0.07459.$$

The first APPL statement establishes the distribution X has the uniform discrete, or rectangular, distribution, which implies that each of the balls in the bag is equally likely to be drawn. The second statement makes a call to the APPL procedure OrderStat with four arguments: the population distribution, the number of draws from the population, the order statistic of interest, and the optional "wo" argument to indicate that sampling is without replacement (the default is with replacement). Finally, the third statement evaluates the PDF of Y at 5, giving the desired result.

(b) The APPL code in the sampling with replacement case is

```
> X := UniformDiscreteRV(1, 15);
> Y := OrderStat(X, 7, 4);
> PDF(Y, 5);
```

which returns the probability that the median is 5 as

$$\Pr(Y = 5) = \frac{2949971}{34171875} \cong 0.08633.$$

The APPL code is identical to the without replacement case except that the optional fourth argument in OrderStat has been defaulted.

Although these two quantities of interest could have been calculated by hand, the calculations would have been much more tedious if some of the billiard balls were more likely to be drawn than others. Such a change, however, would not pose a difficulty for APPL. Also, although it has not been exploited in the two calls to OrderStat, the procedure has computed the entire distribution of Y, so expected values and other subsequent computations on Y could be performed.

The three examples given thus far have concerned well-known distributions (i.e., uniform, triangular) parameterized by constants. The fourth and final example, drawn from a popular mathematical statistics text, highlights a nonstandard distribution with symbolic parameters. Unlike most of the earlier examples, this example can be worked by hand, then the solution can be checked using APPL. Furthermore, a student can change any of the arguments in the problem (e.g., the sample size or critical value) to see if the results of the change match their intuition.

Example 1.4. Let X_1 and X_2 be iid observations drawn from a population with PDF

$$f(x) = \theta x^{\theta-1} \qquad 0 < x < 1,$$

where $\theta > 0$. Test H_0: $\theta = 1$ versus H_1: $\theta > 1$ using the test statistic $X_1 X_2$ and the critical region $C = \{(X_1, X_2) \mid X_1 X_2 \geq 3/4\}$. Find the power function and significance level α for the test [52, page 270].

The APPL code to compute the power function is

```
> n := 2;
> c := 3 / 4;
> assume(theta > 0);
> X := [[x -> theta * x ^ (theta - 1)], [0, 1],
        ["Continuous", "PDF"]];
> T := ProductIID(X, n);
> power := SF(T, c);
```

which yields

$$\Pr(\text{rejecting } H_0 \mid \theta) = 1 - (3/4)^\theta + \theta(3/4)^\theta \ln(3/4)$$

for $\theta > 0$. The sample size n is defined in the first APPL statement. The critical value c is defined next. The Maple `assume` procedure defines the parameter space $\theta > 0$. The fact that the population distribution is non-standard requires the random variable X to be defined using the "list-of-sublists" data structure used in APPL (described subsequently in Chaps. 3 and 7). The first sublist gives the functional form of the PDF, the second sublist gives the support, and the third sublist indicates that the random variable being defined is continuous and that the function in the first sublist is a PDF. Unlike the previous three examples, all of which had numeric values for the parameters in the probability distributions, this PDF involves the symbolic parameter θ. The `ProductIID` procedure computes the product of the two iid random variables having the distribution of X, which results in the creation of the random variable T. Finally, the power function is defined using the survivor function (`SF`) procedure to compute $\Pr(T \geq c = 3/4)$ for various values of $\theta > 0$.

To compute the significance level of the test, the additional Maple statement

```
> alpha := subs(theta = 1, power);
```

is required, yielding

$$\alpha = 1/4 + (3/4) \ln(3/4) \cong 0.0342.$$

Plotting the power function requires the additional Maple statement

```
> plot(power, theta = 0 .. 20);
```

Fig. 1.2. The power function

The power function is shown in Fig. 1.2. Obviously, this example can be generalized for different sample sizes, population distributions, and critical values with only minor modifications.

The careful reader has probably noted the distinction between APPL statements and Maple statements from the previous example. If APPL is to be a serious research tool, then it is necessary to learn certain aspects of Maple, such as the **subs** and **plot** procedures used in the fourth example. A brief review of those aspects of Maple that are germane to APPL are given in Chap. 2.

The next section shows that there is a non-traditional way of thinking that is required in order to mix probability and computing.

1.2 A Different Way of Thinking

A few weeks into the introductory calculus-based probability class, the properties associated with the PDF of a continuous random variable

$$\int_{-\infty}^{\infty} f(x)\, dx = 1 \qquad \text{and} \qquad f(x) \geq 0, -\infty < x < \infty$$

are introduced. Probability definitions and problems often need to be recast in order to make them amenable to processing by a computer algebra system. The first property is fairly easy to verify in a computer algebra system [if $f(x)$ can be integrated symbolically, do so and show that the area under the curve is 1, else, integrate numerically and show that the area under the curve is within epsilon of 1]. The second property, however, is quite difficult to show. One might imagine checking the PDF on a very fine grid over the support of the random variable, but this is unappealing for two reasons. First, the grid should be very fine, resulting in considerable CPU time. Second, and more importantly, if $f(x)$ drops below the x-axis between grid points, a PDF will be accepted with $f(x) < 0$.

A better way to verify the PDF properties is to recast them as the equivalent pair

$$\int_{-\infty}^{\infty} f(x)\, dx = 1 \qquad \text{and} \qquad \int_{-\infty}^{\infty} |f(x)|\, dx = 1,$$

which can be easily verified in a computer algebra system. An APPL procedure named `VerifyPDF` verifies this latter pair of properties for a continuous random variable rather than the more intuitive pair typically introduced in an introductory probability class.

As a second example of a more algorithmic approach to thinking about probability, consider the following problem. If $X \sim U(0,1)$, find the distribution of $Y = g(X) = X^2$. The "transformation technique" is commonly used to find the distribution of Y when $g(X)$ is a 1–1 transformation from the support of X, which we call A, to the support of Y, which we call B. The general formula for the PDF of Y is

$$f_Y(y) = f_X\big(g^{-1}(y)\big) \left| \frac{dg^{-1}(y)}{dy} \right| \qquad y \in B.$$

Since the PDF of X is $f_X(x) = 1$ for $0 < x < 1$, and the inverse transformation is $g^{-1}(y) = \sqrt{y}$, the PDF of Y is

$$f_Y(y) = f_X\big(\sqrt{y}\big) \left| \frac{1}{2\sqrt{y}} \right| \qquad 0 < y < 1$$

or

$$f_Y(y) = \frac{1}{2\sqrt{y}} \qquad 0 < y < 1.$$

This is exactly the way that the transformation technique should be implemented in an algorithm. In fact, the APPL code to find the distribution of Y is

```
> X := UniformRV(0, 1);
> g := [[x -> x ^ 2], [0, 1]];
> Y := Transform(X, g);
```

which returns the PDF of Y as

$$f_Y(y) = \frac{1}{2\sqrt{y}} \qquad 0 < y < 1$$

as expected. Unfortunately, the algorithm behind the `Transform` procedure is not quite as simple as this example would suggest. For example, when Maple finds the inverse function $g^{-1}(y)$ it returns not the single inverse suggested above, but rather two inverses:

$$g_1^{-1}(y) = +\sqrt{y} \qquad \text{and} \qquad g_2^{-1}(y) = -\sqrt{y}.$$

How do we determine which one is the correct inverse? Here is the thinking: First, we determine the midpoint of the support of X, which we denote by x_0. In the current example, $x_0 = 1/2$. Next, we calculate the value of the transformation at x_0, which in the current example is $g(x_0) = 1/4$. Finally, we loop through all of the inverses that are returned by the computer algebra system inserting $g(x_0)$ as an argument, and determine which of the inverses returns x_0. In the current example, $g_1^{-1}(g(x_0)) = x_0$, so the first inverse is the right one and the second inverse should be discarded. Adjustments in this procedure must be made for distributions with infinite support, such as the exponential distribution and the normal distribution.

Likewise, what if the distribution of X is defined in a piecewise fashion, e.g., the triangular distribution? In this case an outside loop must be added to the `Transform` procedure in order to do the appropriate bookkeeping so that all of the density gets transformed from the distribution of X to the distribution of Y appropriately.

Furthermore, how does the `Transform` procedure handle portions of the transformation that are 2–1, or, more generally, k–1? The algorithm behind the `Transform` procedure tracks each piecewise segment and determines the associated mapping onto the support of Y.

These two examples provide a window into the thinking that is required to develop algorithms for performing probability calculations. APPL requires many conditions [e.g., if $g_1^{-1}(g(x_0)) = x_0$, then choose the right inverse] and loops [e.g., loop through all inverses] to perform various manipulations of random variables. Automating probability calculations in this manner is precisely the vision we have for computational probability. APPL and similar languages can be used to explore new probability theory or calculate known but mathematically intractable probability measures.

1.3 Overview

We end this chapter with an overview of the organization of the monograph. As mentioned earlier, Chap. 2 contains a brief review of Maple syntax, data structures, and programming constructs that were used to write the procedures that comprise APPL. We survey only a small portion of the Maple language.

The second part of the monograph, Chaps. 3–6, considers continuous random variables. The data structure that we have used for defining a continuous random variable is defined in Chap. 3. Chapter 4 through Chap. 6 contain examples of algorithms that have been devised for manipulating continuous random variables. Chapter 4 considers transformations of univariate continuous random variables; Chap. 5 considers transformations of bivariate continuous random variables; Chap. 6 considers products of continuous random variables.

The third part of the monograph, Chaps. 7–9, considers discrete random variables. The data structure that we have used for defining a discrete random variable is defined in Chap. 7. Chapters 8 and 9 contain examples of algorithms that have been devised for manipulating discrete random variables. Chapter 8 considers the distribution of sums of discrete random variables; Chap. 9 considers the distribution of order statistics drawn from discrete distributions.

The fourth part of the monograph, Chaps. 10–15, considers applications of APPL in computational probability. Chapter 10 contains applications in reliability and survival analysis, including system design, lower confidence bounds on system reliability, and bootstrapping. Chapter 11 contains applications of APPL to ARMA (autoregressive moving average) models in time series analysis. Chapter 12 contains APPL applications in discrete-event simulation, including random number testing, input modeling, and goodness-of-fit testing. Chapter 13 contains APPL applications in queueing theory, more specifically, determining the probability distribution of initial customer sojourn times. Chapter 14 contains APPL applications in Bayesian statistics, including the calculation of posterior distributions and Bayesian credible sets. Finally, Chap. 15 contains miscellaneous applications, such as determining the exact distribution of the time to complete a stochastic activity network, probabilistic analysis of Benford's law, and the generation of values in statistical tables.

Maple for APPL

Maple is a computer algebra system and programming language that can be used for numerical computations, solving equations, manipulating symbolic expressions, plotting, and programming, just to name a few of the basics. APPL is, simply, a set of supplementary Maple commands and procedures that augments the existing computer algebra system. In effect, APPL takes the capabilities of Maple and turns it into a computer algebra system for computational probability. This chapter contains guidelines for using Maple, and discusses the Maple commands that are used in APPL programming. After reading this chapter, an APPL user will have the knowledge necessary to modify the APPL code to meet his or her particular needs. We will start with a discussion of basic numeric computation, then advance to defining variables, symbolic computations, functions, data types, solving equations, calculus and graphing. Then we will discuss the programming features of Maple that facilitate building the APPL language: loops, conditions, and procedures.

2.1 Numerical Computations

Numerical computations in Maple give it the functionality of a hand-held calculator. The command prompt is >. To execute an arithmetic expression in Maple, the expression must be terminated with a semicolon or colon. The # symbol is used for commenting in a Maple worksheet. Below are several examples of numerical computations with their corresponding outputs.

© Springer International Publishing AG 2017
J.H. Drew et al., *Computational Probability*, International Series
in Operations Research & Management Science 246,
DOI 10.1007/978-3-319-43323-3_2

```
> 2 + 2;
```

$$4$$

```
> 1 + 1 / 2;
```

$$\frac{3}{2}$$

```
> 1 + 0.5;
```

$$1.5$$

```
> sqrt(2);                    # sqrt() takes the square root
```

$$\sqrt{2}$$

```
> Pi;
```

$$\pi$$

```
> evalf(Pi);
```

$$3.141592654$$

```
> 2 * 2.5:
> % + 1 / 2;
```

$$5.500000000$$

From these few examples, it is important to note the following:

- Spaces between symbols are optional, so 2 + 2; and 2+2; are equivalent. We include spaces between operators for readability, consistent with good programming practice.
- Maple performs exact calculations with rational numbers and approximate calculations with decimals. To Maple, the rational number 3/2 and the floating-point approximation 1.5 are different objects. Using both decimals and rational numbers in a statement produces a decimal output.
- Maple interprets irrational numbers as exact quantities. Maple also recognizes standard mathematical constants, such as π and e, and works with them as exact quantities.
- The evalf command converts an exact numerical expression to a floating-point number. By default, Maple calculates the result using ten digits of accuracy, but any number of digits can be specified. The optional second argument of evalf controls the number of floating-point digits for that particular calculation.
- For particularly precise numerical expressions, a call to evalhf evaluates an expression to a numerical value using the hardware floating-point precision of the underlying system. The evaluation is done in double precision. The evalhf function computes only with real floating-point arguments.

- If a statement ends with a colon, instead of a semicolon, then Maple suppresses the output, although it will store the result (if assigned to a variable) internally.
- The ditto operator % refers to your last calculated result, even if that result is not on the line preceding the %.

2.2 Variables

It is convenient to assign variable names to expressions that are referred to one or more times in a Maple session. Maple's syntax for assigning a variable name is name := *expression*. Almost any expression, including numbers, equations, sets, lists, and plots, can be given a name, but it is helpful to choose a name that describes the expression. The restart command makes Maple act (almost) as if just started, and it clears the values of all Maple variables. Some guidelines for variable names are:

- Maple is case sensitive, so the names X and x denote unique variables.
- A variable name can contain alphanumeric characters and underscores, but it *cannot* start with a number.
- Once a variable is assigned to an expression, it remains that expression until changed or cleared.
- A variable name can be cleared by assigning variable := 'variable'; or executing the statement unassign('variable');
- Maple has some predefined and reserved names, such as Sum, sqrt, and length, that are not available for variable assignment. Maple will not allow an expression to be assigned to a predefined variable name.
- When you close a Maple session, variable names assigned during that session will be forgotten. When a Maple worksheet is re-opened, variable names must be reactivated.
- The restart command at the top of the worksheet, followed by the sequence of keystrokes Alt, E, E, W (Edit, Execute, Worksheet) restarts the memory of the variables and then executes the entire worksheet in order. Often one will be in the middle of a series of commands, fix an error and need to re-execute all commands in order once again.

Below are several examples of defining variable names.

```
> restart:
> EventA := 0.3:
> EventB := 0.3:
> EventC := 0.4:
> S := EventA + EventB + EventC;
```

$$S := 1.0$$

```
> Sum := EventA + EventB;
Error, attempting to assign to 'Sum' which is protected

> prob := p * (1 - p) ^ 2:
> p := 1 / 4:
> prob;
```

$$\frac{9}{64}$$

```
> unassign('p'):              # You could also write p := 'p'
> newprob := 3 * p ^ 2 * (1 - p);
```

$$newprob := 3p^2(1-p)$$

Once a Maple expression is given a name, it can be evaluated at different values using subs() or eval(). The command subs(p = 1 / 2, newprob) or eval(newprob, p = 1 / 2) yields the value $\frac{3}{8}$. Since newprob is a variable, and not a function, Maple does not understand the syntax newprob(1 / 3), which the user may incorrectly try to use to determine the value of newprob for $p = 1/3$. If an expression is intended to actually define a function, then the function must be formally defined, and a technique for doing so is in Sect. 2.4.

Sometimes assumptions must be made on variables in order to set variable properties or relationships. A common use of the assume function is to assume a constant is positive, i.e., assume(p > 0). Making such assumptions allow Maple routines to use this information to simplify expressions, for example, $\sqrt{p^2}$. When an assumption is made about a variable, thereafter the variable is displayed with an appended tilde ~ to indicate that it carries assumptions. The additionally function adds additional assumptions without removing previous assumptions. For example, we could further restrict p to be less than one with the command additionally(p < 1).

2.3 Symbolic Computations

One of Maple's main strengths is its ability to manipulate symbolic expressions. Symbols can be treated in the same way that numbers were in the previous section and can do much more. Entering a constant or variable followed by a variable, e.g., 2x or ab, does not imply multiplication in Maple. In fact, ab would be treated as a new two-letter variable instead of the product of two single-letter variables. You may *not* omit the multiplication symbol ($*$) in expressions with more than one factor. Below are a few examples of Maple's symbolic abilities using three commands, combine, expand, and simplify, that appear in the APPL code.

```
> exp(-t ^ 2) * exp(-s ^ 2);
```

$$e^{(-t^2)}e^{(-s^2)}$$

```
> combine(%);
```

$$e^{(-t^2 - s^2)}$$

```
> mgf_1 := (1 / 3) * exp(t) + (2 / 3) * exp(2 * t):
> mgf_2 := (1 / 4) + (3 / 4) * exp(-t):
> mgf_1 * mgf_2;
```

$$\left(\frac{1}{3}e^t + \frac{2}{3}e^{(2t)}\right)\left(\frac{1}{4} + \frac{3}{4}e^{-t}\right)$$

```
> expand(%);
```

$$\frac{7}{12}e^t + \frac{1}{4} + \frac{1}{6}(e^t)^2$$

```
> mgf_1 / mgf_2;
```

$$\frac{\frac{1}{3}e^t + \frac{2}{3}e^{(2t)}}{\frac{1}{4} + \frac{3}{4}e^{(-t)}}$$

```
> simplify(%);
```

$$\frac{4}{3}\frac{e^{(2t)}(1 + 2e^t)}{e^t + 3}$$

2.4 Functions

Functions are defined in Maple by using the arrow notation -> or the unapply()
command. The assignment operator := associates a function name with a
function definition. Two equivalent ways of defining a function f are

```
> f := x -> exp(-2) * 2 ^ x / x!:
> f := unapply(exp(-2) * 2 ^ x / x!, x);
```

$$f := x \rightarrow \frac{e^{(-2)}2^x}{x!}$$

This notation allows a function to be evaluated in the "usual" way, i.e., f(0),
when it appears in Maple expressions. Such functions are an integral part of
APPL, and are used to create PDFs, CDFs, and transformations of random
variables. Unassigning a function is done in the same way that a variable is
unassigned to a value, f := 'f'. As shown in the examples below, piecewise
functions may also be defined in Maple, and we certainly take advantage of
this in APPL, e.g., the triangular distribution.

```
> g := unapply(exp(-lambda) * lambda ^ x / x!, lambda, x);
```

$$g := (\lambda, x) \to \frac{e^{(-\lambda)}\lambda^x}{x!}$$

> g(2, 0);

$$e^{(-2)}$$

> h := x -> piecewise(x <= 1, 0, x > 1, 2):
> h(x);

$$\begin{cases} 0 & x \le 1 \\ 2 & x > 1 \end{cases}$$

In addition to facilitating the creation of custom functions, Maple has built-in mathematical functions, e.g., trigonometric functions such as tan, inverse trigonometric functions such as arccos, absolute value (abs), the exponential function (exp), the gamma function (GAMMA), and binomial coefficients (binomial). As illustrated earlier in the chapter, the exponential function in Maple is written as exp(x). In order to compute e^3, enter exp(3), *not* e ^ 3. The binomial function binomial(n, x) determines binomial coefficients for selecting x items from n items without replacement. Examples of calling Maple functions that might arise in the APPL code are given below.

> abs(ln(0.4) + arcsin(0.65));

$$0.2087062952$$

> binomial(4, 2);

$$6$$

> GAMMA(5);

$$24$$

> evalf(erf(2));

$$0.9953222650$$

Maple has packages of functions available with specialized commands to perform tasks from an extensive variety of disciplines. The commands in a package can be activated using the with command. For example, with(networks) adds the networks package, which supplies tools for constructing, drawing, and analyzing combinatorial networks. Likewise, with(plots) enables more options with plotting functions.

When a package's functions are accessed with with, occasionally a previous definition or a function, either built-in or defined by the user, will be overridden. For example, trace is by default defined to be the program-tracing function. However, after invoking with(linalg), trace is redefined to mean the trace of a matrix. The with function prints out a warning message whenever it overrides a previous definition or function.

2.5 Data Types

This section examines some basic types of Maple objects, including lists, sets, arrays, and strings.

A Maple *list* can be created by enclosing any number of expressions (separated by commas) in brackets. Maple preserves the order and repetition in a list. Thus, [a,b,c] and [b,c,a] are different lists. The nops command determines the total number of elements in a list. The membership of an element in a list can be checked with member(), which returns a *true* or *false* Boolean value.

```
> probs := [0.3, 0.5, 0.2]:
> nops(probs);
```

$$3$$

```
> member(0.3, probs);
```

$$true$$

Because order is preserved in a list, a particular element can be extracted from a list without searching for it. Use L[i] to access the ith element of a list L. The op command applied to a list removes the brackets of the list, and sort preserves the list structure while sorting the elements into ascending order.

```
> probs := [0.3, 0.5, 0.2]:
> op(probs);
```

$$0.3, 0.5, 0.2$$

```
> sort(probs);
```

$$[0.2, 0.3, 0.5]$$

In APPL, we utilize the functionality of Maple's list structure to represent a random variable as a "list-of-sublists." A list-of-sublists is just a list, each of whose elements is itself a list. For example, we create a random variable as a list of three sublists, where the first sublist contains the functional form that describes the probability distribution, the second sublist contains its support, and the third sublist uses strings to describe the type of the random variable, i.e., continuous or discrete, and the type of function, e.g., the PDF, CDF or HF. When APPL receives a random variable for manipulation, it first checks the type of the random variable before continuing; that is, is the random variable continuous or discrete? Depending on the answer, APPL treats that random variable accordingly, for example, integrating to determine expected value versus summing. Additional information about APPL's list-of-sublists data structure for random variables is contained in Chaps. 3 and 7. Elements in a list-of-sublists can be extracted using indices. For example,

```
> x := [[1, 2, 3], [4, 5, 6], [7, 8]];
```

sets x to a list of three elements, each of which is a list. Sublists or individual elements can be extracted as follows:

```
> x[2];
```

$$[4, 5, 6]$$

```
> x[2, 1];
```

$$4$$

```
> x[2][1];
```

$$4$$

A Maple *set* is constructed by enclosing any number of Maple objects (separated by commas) in braces. The braces identify the object as a set. Maple does not preserve order or repetition in a set. Thus the sets {a,b,c}, {b,c,a}, and {c,c,b,a} are identical. The **member** command validates membership in a set, and to choose an item from a set, the subscript notation [i] is used, where i identifies the position of the desired element in the set. Also, the **nops** function counts the number of elements in a set.

The **seq** command is used to generate sequences, which can be contained in a list or a set if desired. For example, a sequence of binomial coefficients can be obtained by using **seq** as follows:

```
> seq(binomial(4, i), i = 0 .. 4);
```

$$1, 4, 6, 4, 1$$

The Maple *array* data structure is an extension of the Maple list data structure. Each element is still associated with an index (which can be negative or zero), but an array is not restricted to one dimension.

A *string* is a Maple object created by enclosing a sequence of characters in double quotes. We use strings in APPL in the third sublist of the list-of-sublists that define a random variable to help identify its functional form, such as "Continuous", "Discrete", "PDF", "CDF", etc.

2.6 Solving Equations

Maple can find analytic solutions for a large class of algebraic equations. The **solve** command is a general-purpose equation solver. It takes a set of one or more equations and attempts to solve it exactly for the specified set of unknowns. The **fsolve** command is the numeric equivalent of **solve**; **fsolve** calculates the solution(s) to equations using a variation on Newton's

method, producing approximate (floating-point) solutions. For a general equation, `fsolve` searches for a single root and often returns the first root it finds, which may not be the desired root. For a polynomial, it looks for all real roots. Fortunately, there is a way to make `fsolve` look for a specific solution by specifying an interval over which `fsolve` may search, as illustrated in the examples to follow. Some special notes about solving equations are:

- If the variable(s) being solved for are not specified, Maple solves for all variables.
- If no right-hand side to an equation is given (for example, `solve(2 * x - 8)`), the right-hand side is assumed to be 0.
- The `fsolve` command often only returns one real solution.
- If there are multiple solutions to an equation, they can be extracted by setting the solution set to a variable, then using subscripts `[1]`, `[2]`, etc.

Here are several examples of solving one or more equations at a time.

```
> solve(x ^ 3 = 2 * x, x);
```

$$0, \sqrt{2}, -\sqrt{2}$$

```
> fsolve(x ^ 3 = 2 * x, x);
```

$$-1.414213562, 0., 1.414213562$$

```
> eqns := {x + 2 * y = 3, y + 1 / x = 1}:
> solns := solve(eqns, {x, y});
```

$$solns := \{y = 2, x = -1\}, \left\{x = 2, y = \frac{1}{2}\right\}$$

```
> solns[2];
```

$$\left\{x = 2, y = \frac{1}{2}\right\}$$

```
> solve(cos(x) - x = 0);
```

$$\text{RootOf}(_Z - \cos(_Z))$$

```
> fsolve(cos(x) - x = 0);
```

$$0.7390851332$$

```
> express1 := (1 / x) + (1 / x ^ 2) - (3 / x ^ 3):
> solve(express1);
```

$$-\frac{1}{2} + \frac{\sqrt{13}}{2}, -\frac{1}{2} - \frac{\sqrt{13}}{2}$$

```
> fsolve(express1);
```

$$1.30277563$$

```
> fsolve(express1, x = -3 .. 0);
```

$$-2.302775638$$

Maple occasionally returns solutions in terms of the `RootOf` function, which is a place holder for representing all the roots of an equation in one variable. The prefix _Z on the variable in a `RootOf` solution indicates that it has integer values. The `fsolve` command is able to approximate these roots (at most one at a time for non-polynomial equations, and a specific root if a search interval is specified). Below is also an indication of the range of a variable using two periods in a row. The range delimiter "$..$" denotes endpoints of intervals in plots, integrations, and other situations.

Two other commands of interest for obtaining either the left-hand side or right-hand side of an expression are `lhs` and `rhs`, respectively. Especially useful to APPL is the ability of these commands to extract the first and last term of a sequence. This functionality allows APPL to read off the bounds on the support of a probability function, which is displayed in the second sublist for a random variable. Both uses of `lhs` and `rhs` are displayed below.

```
> eqn1 := x + y = z + 3:
> lhs(eqn1);
```

$$x + y$$

```
> rhs(eqn1);
```

$$z + 3$$

```
> support := 1 .. infinity:
> lhs(support);
```

$$1$$

```
> rhs(support);
```

$$\infty$$

2.7 Graphing

Another capability of Maple is graphing. The explicit function $y = f(x)$ and its domain must be specified for a Maple plot. Plotting the standard normal distribution on the interval $-3 \leq x \leq 3$, for example, can be done with the Maple statements shown below. We will use the function form `->` to define the PDF of the standard normal distribution. The plot is displayed in Fig. 2.1.

```
> f := x -> 1 / sqrt(2 * Pi) * exp(-x ^ 2 / 2):
> plot(f(x), x = -3 .. 3);
```

In order to plot more than one function on a single graph, the functions must be contained in brackets or braces. The use of brackets means that order is important and Maple should graph them in the given order; the use of braces means that order is not important and Maple may graph them in any order. Also, it is sometimes useful to specify the range of the y-values in order to display important features of a graph. Figure 2.2 contains a plot of two normal PDFs. The functions are defined with two different variables, but used as arguments in plot with a third variable, as one would expect.

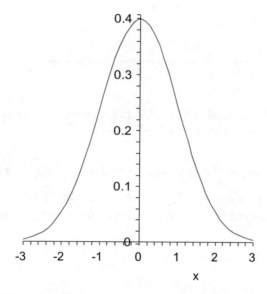

Fig. 2.1. Plot of a standard normal PDF between $x = -3$ and $x = 3$

```
> f1 := t -> 1 / sqrt(2 * Pi) * exp(-t ^ 2 / 2):
> f2 := z -> 1 / (2 * sqrt(2 * Pi)) * exp(-(z - 1) ^ 2 / 8):
> plot({f1(x), f2(x)}, x = -7 .. 9, color = [red, blue]);
```

2.8 Calculus

Maple provides many powerful tools for solving calculus problems. This section describes Maple's ability to determine limits, symbolically compute derivatives and definite integrals, and determine summations. The need for calculus operations in APPL is significant because many probability theorems rely on calculus to create new distribution functions. A few comments about determining limits, derivatives, and definite integrals in Maple follow:

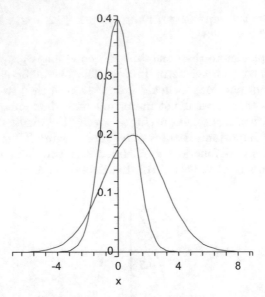

Fig. 2.2. Plot of a standard normal PDF and a normal PDF with mean $\mu = 1$ and standard deviation $\sigma = 2$ between $x = -7$ and $x = 9$

- Maple will compute both two-sided limits, e.g., $\lim_{x \to 5} \left(\frac{x+2}{x-1} \right)$, and one-sided limits, e.g., $\lim_{x \uparrow 1} \left(\frac{x+2}{x-1} \right)$, with the `limit` procedure.
- Maple can determine limits at ∞, such as the limit of an exponential CDF at infinity.

Examples of the `limit` command are

```
> limit((3 ^ (x + h) - 3 ^ x) / h, h = 0);
```

$$3^x \ln(3)$$

```
> assume(lambda > 0):
> limit(1 - exp(-lambda * x), x = infinity);
```

$$1$$

The `diff` procedure computes the derivative of an expression. Maple differentiates with respect to the variable placed in the second argument of the `diff()` procedure. Clearly, `diff` is used often in APPL for changing from one functional form of a continuous random variable's distribution to another, e.g., deriving a PDF from a CDF. One such example of the `diff()` command follows; however it is important to note that APPL will do such conversions in a more automated manner. Also note the use of the `unapply` command to turn the final expression into a function.

```
> F := x -> 1 - exp(-2 * x);
```

$$F := x \rightarrow 1 - e^{(-2x)}$$

```
> f := unapply(diff(F(x), x), x);
```

$$f := x \rightarrow 2\, e^{(-2x)}$$

The capabilities of Maple include a robust set of integration operations. Symbolic integrals (often used in APPL) can be calculated as in the following example:

```
> f := y -> 2 * exp(-2 * y):
> F := unapply(int(f(y), y = 0 .. x), x);
```

$$F := x \rightarrow 1 - e^{(-2x)}$$

Maple's ability to integrate also includes the numerical evaluation of integrals for which no closed-form solution is available, such as the case with the standard normal random variable:

```
> g := 1 / sqrt(2 * Pi) * exp(-x ^ 2 / 2):
> int(g, x = -infinity .. 1.96);
```

$$0.9750021049$$

However, if we integrate the standard normal PDF over exact values, rather than floating point values, Maple responds with \mathtt{erf}. For example, integrating a standard normal PDF from $x = -1$ to 1 yields the following:

```
> int(g, x = -1 .. 1);
```

$$\mathrm{erf}\left(\frac{\sqrt{2}}{2}\right)$$

The function $\mathtt{erf(x)}$ is the error function and it is defined for real and complex x by $\mathrm{erf}(x) = \frac{2}{\pi}\int_0^x e^{-t^2}\,dt$. Maple can numerically evaluate an $\mathtt{erf(x)}$ output with the \mathtt{evalf} command.

```
> evalf(int(g, x = -1 .. 1));
```

$$0.6826894920$$

Maple can also compute finite sums and infinite series, which are necessary capabilities for working with discrete random variables. The next few examples demonstrate the use of sum. The first example verifies that a geometric distribution with $p = 1/4$ has a PDF that sums to one.

```
> geometric := (1 / 4) * (3 / 4) ^ (x - 1);
```

$$geometric := \frac{\left(\dfrac{3}{4}\right)^{(x-1)}}{4}$$

```
> sum(geometric, x = 1 .. infinity);
```

$$1$$

The second example determines the expected value of a binomial distribution with parameters n and p.

```
> assume(p > 0):
> additionally(p < 1):
> pdf := binomial(n, x) * p ^ x * (1 - p) ^ (n - x):
> simplify(sum(x * pdf, x = 0 .. n));
```

$$np\sim$$

2.9 Loops and Conditions

Like all programming languages, loops in Maple allow the user to execute a sequence of statements repeatedly, either for a prescribed number of times or until a condition is satisfied (or not satisfied). This is a useful tool in APPL when working with random variables. Loops are used in APPL to do many things: cycle through segments of PDFs, verify inverse transformations, and error-checking, to name a few. For example, if a discrete random variable's PDF consists of a finite number of numeric probabilities in the first sublist in a list-of-sublists, we require these probabilities to be non-negative. Using a for loop allows APPL to cycle through the items in this sublist (from item one to nops(sublist)) until the list is complete or a negative value is identified. Similarly, if numeric probabilities in the first sublist are supposed to be equally-likely probabilities, then APPL can repeatedly check each probability value while the probability values remain the same. The expression in the while clause is a Boolean expression which must evaluate to *true* or *false* upon each cycle.

Conditional statements in APPL have a variety of purposes including error checking, branching in logic, and conditional plotting. If an APPL procedure requires input arguments from the user, then an if statement can be used to verify that the input is valid. If a procedure does not receive the correct type of arguments, e.g., lists, an error message is sent back to the APPL user. Also, in the third sublist is the "form" of the random variable. The first element of the third sublist tells APPL whether the random variable is continuous or discrete. If the random variable is continuous, then continuous mathematics (e.g., integration) is used to verify the validity of its PDF or compute its mean, for example. If discrete random variables are indicated by the field, then discrete mathematical techniques (e.g., summation) are used.

Boolean expressions in the conditional statement are most often formed by using <, <=, >, >=, =, <>, or, and, or not. For example, if the "form" of the random variable is *not* one of the approved formats (e.g., PDF, CDF, SF, HF, CHF, IDF), then an error message is reported to the user.

A few examples of small loops and conditionals are presented here to amplify the above explanations:

```
> probs := [0.1, 0.05, 0.2, 0.25, 0.15, 0.10, 0.15]:
> totprob := 0:
> for i from 1 to nops(probs) do
>    totprob := totprob + probs[i]:
> end do;            # 'od' can also be used to end a 'do' loop
```

$$totprob := 0.1$$
$$totprob := 0.15$$
$$totprob := 0.35$$
$$totprob := 0.60$$
$$totprob := 0.75$$
$$totprob := 0.85$$
$$totprob := 1.00$$

```
> die := rand(1 .. 6):          # random integer from 1 to 6
> Roll1 := die():
> Roll2 := die():
> if (Roll1 >= Roll2) then
>    print("Roll 1 is greater than or equal to Roll 2");
> else
>    print("Roll 1 is less than Roll 2");
> end if;    # 'fi' can also be used to end an 'if' conditional
```

To conclude this section, a while loop is created in which the user is encouraged to experiment with various values for the probability prob.

```
> i := 1:
> totprob := 0:
> prob := 1 / 2:
> while (totprob <= 1) and (i <= 1000) do
>    totprob := prob + totprob:
>    prob := prob ^ 2 + 0.001:
>    i := i + 1:
>    print(i):
>    print(totprob):
> end do:
```

When a **break** statement is used in a procedure, a break is executed and the result is to exit from the innermost repetition (for/while/do) statement within which it occurs.

2.10 Procedures

The capability to create new procedures in Maple is a key element to enabling APPL programming. In fact, APPL is a text file of new Maple procedures that is imported into a worksheet session with the **read** command. When a file is read into Maple, each line in the file is treated as if it had been typed in order. The APPL procedures are available during the Maple session, but the commands are not displayed on the Maple worksheet. When the user reads in the APPL commands, it's as if they used the **with** command to add extra capability. The syntax for reading APPL into Maple from the ASCII source code file named **APPL.txt** is **read('APPL.txt')**.

The syntax for creating a procedure in Maple is

$$\texttt{proc}(\langle argseq \rangle) \, [\text{local} \, \langle nseq \rangle \, ; \text{global} \, \langle nseq \rangle \, ; \,]$$

$$\langle statement \ sequence \rangle$$

end;

where [] indicates optional parameters. Then

- $\langle argseq \rangle$ is a sequence of variable names, separated by commas. These variables are the arguments to the procedure.
- $\langle nseq \rangle$ is a sequence of variable names, separated by commas.
- local $\langle nseq \rangle$ is a list of variables local to the procedure. Any assignments of these variables will only have a scope of the procedure. The values of these variables will be unassigned when the procedure starts, regardless of the value of the variable outside the procedure.
- global $\langle nseq \rangle$ is a list of global variables used by the procedure. Any assignments made in the procedure will be global in scope and the initial values of the variables will be as they are in the existing Maple session.
- $\langle statement \ sequence \rangle$ is the body of the procedure and may consist of any valid sequence of Maple statements.

The **proc** command is usually used in conjunction with the **RETURN** keyword. The syntax is

$$\texttt{RETURN} \, (expression) \, ;$$

This command will terminate the execution of the procedure and return the value of *expression*.

Since APPL's random variables often have several arguments, it's helpful that Maple provides an easy way to do data typing in procedures. The arguments passed to a procedure may be given a specific data type. If any of these variables is then assigned an incorrect type, Maple generates an error message. The syntax for declaring a variable to be of a given data type is the declaration is

$$variable \ name :: data \ type$$

where Maple accepts a wide range of data types. The types most frequently encountered in APPL are `list` and `listlist` (because of the list-of-sublists structure of APPL random variables), `constant`, `posint`, `int`, and `array`. Another type used frequently in APPL is `symbol`. Infinity (∞) is of type `symbol`, and this is important information for APPL procedures to know when evaluating random variables over supports that are not finite. The type of an expression can be determined in Maple with the command `type`. The syntax is

 type (*expression, type*)

which returns a Boolean value, either *true* or *false*.

Another useful error-checking command that can used with procedures is `nargs`, which is the number of arguments passed to a procedure. We use `if` statements in APPL to determine if optional arguments are passed to a procedure and how the procedure should proceed given this number of arguments. For example, the `OrderStat` procedure in APPL will perform differently based on whether or not it is given an optional fourth argument that specifies whether sampling is done without replacement.

To summarize the last few sections on programming in Maple, the following excerpt from the APPL source code is noteworthy. The procedure is called `ReduceList` and it is a small sub-procedure in APPL that looks for redundant support entries in a random variable list-of-sublist's second sublist. One sees how the `proc` command begins the procedure as well as some argument checking, local variable declarations, a `for` loop and some conditional branching. The `RETURN` and `end` commands complete the procedure.

```
#
# ReduceList is a procedure that eliminates floating point
# redundancies (e.g., 3 vs. 3.0) from a sorted Maple list.
#
ReduceList := proc(LST :: list)
local i, size, delt, deltamin, ListIn:
deltamin := 0.0000001:
ListIn := LST:
size := nops(ListIn):
for i from (size - 1) by -1 to 1 do
  if (ListIn[i] <> -infinity and ListIn[i + 1] <> infinity) then
    delt := evalf(ListIn[i + 1]) - evalf(ListIn[i]):
    if (delt < deltamin) then
      if (whattype(ListIn[i]) <> float) then
        ListIn := subsop((i + 1) = NULL, ListIn):
      else
        ListIn := subsop(i = NULL, ListIn):
      fi:
    fi:
  fi:
```

```
od:
RETURN(ListIn):
end:
```

This concludes our brief introduction to the Maple computer algebra system. Additional help on a specific Maple topic can be accessed from the Maple worksheet by entering ?topic or help(topic) at the prompt. Also, help can be obtained directly by going to the "Help" menu in Maple, scrolling down to "Topic Search," and entering the topic of interest in the "Topic" line. We now turn to the main topic of the monograph, defining random variables in APPL and the associated algorithms to manipulate them. We begin with continuous random variables.

Part II

Algorithms for Continuous Random Variables

All models are wrong; some models are useful.

—George E.P. Box

3

Data Structures and Simple Algorithms

This chapter and the three that follow it concern continuous random variables. We have chosen to present continuous random variables first because they are defined with a somewhat simpler data structure than that for discrete random variables. The development described here gives a probabilist the ability to automate the instantiation and processing of continuous random variables— key elements of computational probability.

3.1 Data Structures

There are dozens of ways that one could represent the distribution of a random variable in Maple. Whatever choice is made, the syntax associated with the data structure should be succinct, intuitive, and easily manipulated. This implies there should be a single data structure that allows a user to define the

- probability density function (PDF) $f_X(x)$,
- cumulative distribution function (CDF) $F_X(x) = \int_{-\infty}^{x} f_X(s)\, ds$,
- survivor function (SF) $S_X(x) = 1 - F_X(x)$,
- hazard function (HF) $h_X(x) = f_X(x)/S_X(x)$,
- cumulative hazard function (CHF) $H_X(x) = \int_{-\infty}^{x} h_X(s)\, ds$, and
- inverse distribution function (IDF) $F_X^{-1}(u)$

for a continuous random variable X. It should also provide a seamless transition from one representation to another. The common data structure that we used in the development of the APPL software package is referred to as the "list-of-sublists." Specifically, any functional representation of a random variable is presented in a list that contains three sublists, each with a specific purpose. The first sublist contains the ordered functions that define the

© Springer International Publishing AG 2017
J.H. Drew et al., *Computational Probability*, International Series
in Operations Research & Management Science 246,
DOI 10.1007/978-3-319-43323-3_3

functional forms of the distribution. The PDF representation of the triangular distribution, for example, would have the two linear functions that comprise the two segments of its PDF for its first sublist. The second sublist is an ordered list of real numbers that delineate the end points of the intervals for the functions in the first sublist. The end point of each interval is assumed to be the starting point of the succeeding interval. The third sublist has two elements. The first element of the third sublist is either the string "Continuous" for continuous distributions or "Discrete" for discrete distributions. The second element of the third sublist shows which of the six functional forms listed earlier in this paragraph is used in the first sublist. The string "PDF", for example, indicates that the first sublist denotes a PDF.

Example 3.1. The following seven random variables give a variety of ways continuous random variables can be defined in APPL.

- The following APPL statement assigns a list-of-sublists that represents the PDF of a $U(0, 1)$ random variable to the list X:

    ```
    > X := [[x -> 1], [0, 1], ["Continuous", "PDF"]];
    ```

- The triangular distribution has a PDF with two segments to its distribution. The following statement defines a triangular(0, 1, 2) random variable named X with a minimum of 0, mode of 1, and maximum of 2, as a list-of-sublists:

    ```
    > X := [[x -> x, x -> 2 - x], [0, 1, 2],
             ["Continuous", "PDF"]];
    ```

- An exponential random variable X with a mean of 1/3 can be defined in terms of its hazard function with the APPL statement

    ```
    > X := [[x -> 3], [0, infinity], ["Continuous", "HF"]];
    ```

- The three random variables defined above all have numeric parameters. APPL is also able to accommodate symbolic parameters. A $N(\theta, 1)$ random variable X can be defined with the statement

    ```
    > X := [[x -> exp(-(x - theta) ^ 2 / 2) / sqrt(2 * Pi)],
             [-infinity, infinity], ["Continuous", "PDF"]];
    ```

- The parameter space can be specified by using the Maple `assume` function. Consider the random variable T with HF

$$h_T(t) = \begin{cases} \lambda & 0 < t < 1 \\ \lambda t & t \geq 1 \end{cases}$$

for $\lambda > 0$. The random variable T can be defined by the statements

```
> assume(lambda > 0);
> T := [[t -> lambda, t -> lambda * t],
        [0, 1, infinity], ["Continuous", "HF"]];
```

- The syntax allows the endpoints of intervals associated with the support of a random variable to be specified symbolically. A $U(a, b)$ random variable X is defined by

```
> assume(a < b);
> X := [[x -> 1 / (b - a)], [a, b],
        ["Continuous", "PDF"]];
```

- There is no error checking to insure that the distribution associated with a random variable is legitimately represented by a particular list-of-sublists. This means that the statement shown below will create a list-of-sublists that is not a legitimate PDF.

```
> X := [[x -> 17], [0, 49], ["Continuous", "PDF"]];
```

An APPL procedure named VerifyPDF will be introduced subsequently that can be used to verify the PDF properties for a random variable.

After repeatedly typing the list-of-sublists for common continuous distributions, we soon realized that the list-of-sublists format could be automated for certain common (e.g., normal, exponential) distributions. We appended the letters RV onto these procedure names so as to avoid any conflicts with existing Maple procedures. The following continuous distributions have a procedure that creates the appropriate list-of-sublists in PDF form for use in APPL: *ArcSin, ArcTan, Beta, Cauchy, Chi, ChiSquare, Erlang, Error, Exponential, ExponentialPower, ExtremeValue, F, Gamma, GeneralizedPareto, Gompertz, HyperbolicSecant, HyperExponential, HypoExponential, IDB, InverseGaussian, InvertedGamma, KS, LaPlace, LogGamma, Logistic, LogLogistic, LogNormal, Lomax, Makeham, Muth, Normal, Pareto, Rayleigh, StandardCauchy, StandardNormal, StandardTriangular, StandardUniform, T, Triangular, Uniform,* and *Weibull.* This list of distributions, along with their parameters, is displayed every time the APPL code is read into Maple. A chi square random variable with six degrees of freedom, denoted by X, for example, can be defined by the APPL statement

```
> X := ChiSquareRV(6);
```

This is much more compact than the standard list-of-sublists approach:

```
> X := [[x -> x ^ 2 * exp(-x / 2) / 16], [0, infinity],
        ["Continuous", "PDF"]];
```

The `ChiSquareRV` procedure (or function, to use the more standard parlance) relieves the user of typing the list-of-sublists every time a chi square random variable is required. The APPL source code that defines a chi square random variable with **n** degrees of freedom, complete with error trapping, is given by:

```
ChiSquareRV := proc(n :: {posint, symbol})
local ListOfLists, x:
global n1:
n1 := n:
if (nargs <> 1) then
   print('ERROR: This procedure requires 1 argument'):
   RETURN():
fi:
if type(n1, infinity) then
   print('ERROR: The shape parameter must be finite'):
   RETURN():
fi:
if type(n1, symbol) then
   assume(n1, posint):
   about(n1):
fi:
ListOfLists := [[unapply(x ^ (n1 / 2 - 1) * exp(-x / 2) /
   (2 ^ (n1 / 2) * GAMMA(n1 / 2)), x)],
   [0, infinity], ["Continuous", "PDF"]]:
RETURN(ListOfLists):
end:
```

We use Maple's :: in the procedure call just after the argument for Maple error checking to ensure that the argument is either a positive integer or a symbol. If the argument is a symbol, Maple appends a tilde (\sim) to the symbol as a reminder to the user that an assumption (in this case that it is a positive integer) has been placed on the symbol. The introduction of global variables for parameters allows any assumptions about the parameters to be passed from APPL code to the worksheet one is using. For example, the APPL statement

```
> X := ChiSquareRV(6);
```

returns a list-of-sublists with 6 substituted for **n1** in the PDF. On the other hand, the APPL statement

```
> X := ChiSquareRV(m);
```

returns a list of lists with $m\sim$ substituted for **n1** in the PDF. This allows for certain symbolic operations to be done subsequently on **X**, such as verifying (symbolically) that the area under the PDF is one. The first two **if** statements in `ChiSquareRV` check that there is just one argument passed to the procedure and that the number of degrees of freedom is not passed as **infinity**. The third **if** statement places an assumption on the parameter if it is passed

as a symbol and the Maple `about` procedure prints information about the assumption to the screen. Finally, the Maple `unapply` procedure transforms the expression for the PDF into the `x ->` format consistent with the earlier examples, and the PDF is returned as a list-of-sublists. The PDF for the chi square distribution contains the gamma function, which accounts for the use of the Maple `GAMMA` procedure.

This concludes our description of the data structure that is used to represent a continuous random variable. We now briefly outline some simple algorithms that can be applied to this data structure.

3.2 Simple Algorithms

The algorithms described in this section conduct the following elementary operations on data structures that represent continuous random variables:

- convert any functional representation (PDF, CDF, SF, HF, CHF, and IDF) of a random variable into another functional representation using the common data structure,
- plot any of the six functional forms of any fully-specified distribution, e.g., the CDF,
- determine common summary characteristics of random variables, such as the mean and variance,
- calculate the PDF of the rth order statistic from a sample of n iid random variables.

Changing Functional Forms of a Random Variable. The functional form of a distribution can be converted to and from any of its six well-known forms: PDF, CDF, SF, HF, CHF, and IDF. The routines return functional forms when they are called with a single argument and scalars when a second optional argument is included. The names of the conversion procedures are `PDF`, `CDF`, `SF`, `HF`, `CHF`, and `IDF`, and can be applied to both discrete and continuous random variables. The first example finds a CDF from a PDF for a continuous random variable.

Example 3.2. Let $X \sim U(0, 4)$. Find the CDF of X.

The APPL code required to define the distribution of X (which will be in PDF form) and determine the CDF is

```
> X := UniformRV(0, 4);
> CDF(X);
```

which returns the list-of-sublists

$$[[x \to x/4], [0, 4], [\text{"Continuous"}, \text{"CDF"}]]$$

as expected.

Since APPL uses Maple's symbolic processing capability, it will typically return the appropriate function without error whenever the function can be expressed analytically. As the next example shows, however, there are many elementary examples where Maple is not able to integrate functions analytically.

Example 3.3. Let $Z \sim N(0, 1)$. Find the CDF of Z and $\Pr(Z < 1.96)$.

Analogous to the last example, the APPL statements required to define the distribution of Z and determine its CDF are

```
> Z := NormalRV(0, 1);
> CDF(Z);
```

Since the CDF for a standard normal random variable is not closed form, APPL returns

$$\left[\left[x \to \frac{1}{2}\operatorname{erf}\left(\frac{1}{2}x\sqrt{2}\right) + \frac{1}{2}\right], [-\infty, \infty], [\text{"Continuous"}, \text{"CDF"}]\right]$$

where **erf** is the Maple error function defined by $\operatorname{erf}(x) = \frac{2}{\pi}\int_0^x e^{-s^2}ds$. To find $\Pr(Z < 1.96)$ requires an additional APPL statement which uses the second optional argument in the call to CDF:

```
> prob := evalf(CDF(Z, 1.96));
```

which sets the Maple variable **prob** to 0.9750021048. The Maple **evalf** function evaluates the value of CDF(Z, 1.96) as a floating-point (decimal) value. A second optional argument can be included in the **evalf** function to control the number of digits displayed.

The previous examples have considered numeric parameters. The six conversion procedures work on distributions with symbolic parameters as well.

Example 3.4. Find the survivor function (SF) of a Weibull(λ, κ) random variable and find the 99th percentile of a Weibull random variable with $\lambda = 2$ and $\kappa = 3$.

For the case of symbolic parameters, the following statements define a Weibull random variable in PDF form and convert it to a Weibull random variable in SF form:

```
> X := WeibullRV(lambda, kappa);
> X := SF(X);
```

which sets X to the list-of-sublists format shown below:

$$\left[\left[x \to e^{(-x^{\kappa^-}\lambda^{-\kappa^-})}\right], [0, \infty], [\text{"Continuous"}, \text{"SF"}]\right]$$

As indicated earlier, the tildes after the parameters indicate that assumptions have been made concerning the parameters (i.e., $\lambda > 0$ and

$\kappa > 0$) in the `WeibullRV` procedure. Finding a percentile of a distribution requires the `IDF` procedure. The 99th percentile of a Weibull$(2, 3)$ distribution can be found with the APPL statement

```
> IDF(WeibullRV(2, 3), 0.99);
```

which returns 0.8318631745. If 0.99 is replaced by a random number, a Weibull$(2, 3)$ random variate is generated. Maple automatically senses that 0.99 is floating point. The nearly-identical APPL statement

```
> IDF(WeibullRV(2, 3), 99 / 100);
```

returns the expression $2^{-2/3}\left(\ln(10)\right)^{1/3}$, rather than the floating point approximation, for the 99th percentile.

The conversion procedures handle distributions defined in a piecewise fashion in the same way as those defined over a single interval.

Example 3.5. Find the SF for the random variable T with hazard function
$$h_T(t) = \begin{cases} \lambda & 0 < t < 1 \\ \lambda t & t \geq 1 \end{cases}$$

for $\lambda > 0$.

The following APPL statements are used to find the survivor function:

```
> assume(lambda > 0);
> T := [[t -> lambda, t -> lambda * t],
         [0, 1, infinity], ["Continuous", "HF"]];
> SF(T);
```

which returns

$$S_T(t) = \begin{cases} e^{-\lambda t} & 0 < t < 1 \\ e^{-\lambda(t^2+1)/2} & t \geq 1 \end{cases}$$

as expected.

The conversion procedures can be nested; for example, the APPL statements

```
> X := ExponentialRV(1);
> X := PDF(CDF(HF(SF(CHF(IDF(X))))));
```

do absolutely nothing to the list-of-sublists representation for X because all transformations can be performed analytically.

Most of the code for the six conversion routines consists of error checking, for example, to assure that

- there are one or two arguments passed,
- the distribution of interest is in the list-of-sublists format and there are three sublists,
- the list-of-sublists corresponds to one of the six forms.

Once the appropriate if statements have concluded that we are converting, for example, from a CDF to a PDF, which is done by simply differentiating, the real "work" in the PDF procedure consists of just a single loop shown in the code segment

```
for i from 1 to nsegments do
   ffx := unapply(diff(X[1][i](x), x), x):
   fX := [op(fX), op([unapply(simplify(ffx(x)), x)])]:
od:
```

The Maple variable nsegments is a count of the number of piecewise segments associated with a PDF (e.g., nsegments is 2 for a triangular distribution). The first statement in the loop differentiates the ith piecewise function in the CDF, then uses the Maple unapply procedure to place it in the x -> format for the list-of-sublists. The second statement in the loop simplifies the segment of the PDF using the Maple simplify procedure, then concatenates the function into the list named fX, which is eventually returned as the first sublist in a list-of-sublists containing the PDF.

Plotting Distributions. APPL contains the PlotDist procedure that plots any distribution represented in the list-of-sublists format.

Example 3.6. The statements used to plot the PDF for the triangular(1, 2, 4) distribution are:

```
> X := TriangularRV(1, 2, 4);
> PlotDist(X);
```

The Maple plot drawn by the APPL PlotDist procedure is shown in Fig. 3.1.

APPL is not able to plot "generic" distributions with symbolic parameters. Attempting such a plot will produce an "empty plot" error message.

Calculating Expected Values. Expected values can be calculated with the ExpectedValue procedure. The procedures Mean, Variance, Skewness, Kurtosis, MGF, and CF all produce these special cases of expectation by invoking the ExpectedValue procedure.

Example 3.7. Find the expected value, kurtosis, and moment generating function of a normal random variable X.

The second argument in the NormalRV procedure is σ, unlike the customary notation for a normal random variable. The APPL code to compute the mean, kurtosis, and moment generating function is

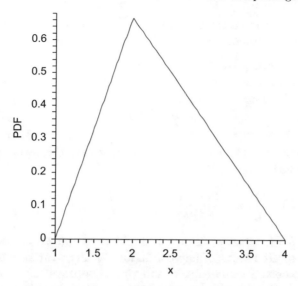

Fig. 3.1. Triangular PDF plot

```
> X := NormalRV(mu, sigma);
> Mean(X);
> Kurtosis(X);
> MGF(X);
```

which returns μ for the mean, 3 for the kurtosis, and

$$M_X(t) = e^{\mu t + \sigma^2 t^2 / 2} \qquad -\infty < t < \infty$$

for the moment generating function.

The `ExpectedValue` procedure can be used to find the expected value of a function of a random variable.

Example 3.8. Find the expected value of the square root of Y, where $Y \sim \text{exponential}(\lambda)$.

In this case, the APPL procedure `ExpectedValue` is used with a second argument that defines the function of the random variable of interest. The APPL statements

```
> Y := ExponentialRV(lambda);
> ExpectedValue(Y, y -> sqrt(y));
```

return $\sqrt{\pi}/(2\sqrt{\lambda})$.

The `ExpectedValue` procedure calculates the expected value of a function of a random variable by calculating the appropriate integrals over each segment of the support of the distribution, then summing the results.

Example 3.9. To find the mean of a triangular(a, b, c) random variable X, the APPL statements

```
> X := TriangularRV(a, b, c);
> Mean(X);
```

return

$$\frac{a+b+c}{3},$$

which is computed as a sum of the integrals over the two segments of the distribution as

$$E[X] = \int_a^b x\,\frac{2(x-a)}{(c-a)(b-a)}\,dx + \int_b^c x\,\frac{2(c-x)}{(c-a)(c-b)}\,dx.$$

As Example 3.9 illustrates, the essence of the computation associated with finding the expected value of any continuous random variable is to initialize the desired expected value to zero and run a loop over the number of segments, incrementing (or decrementing) the expected value by the appropriate integral.

Computing the Distribution of Order Statistics. The OrderStat procedure returns the PDF of the rth of n order statistics drawn from a population having the same distribution as the random variable X. For a continuous random variable X having support on $a < x < b$ with PDF $f(x)$, CDF $F(x)$, and SF $S(x)$, the PDF of the rth of n order statistics drawn from this distribution is given by Hogg et al. [52, page 240] as

$$f_r(x_{(r)}) = \frac{n!}{(r-1)!(n-r)!} F(x_{(r)})^{r-1} f(x_{(r)}) S(x_{(r)})^{n-r} \qquad a < x_{(r)} < b.$$

This well-known result has been implemented in an APPL procedure named OrderStat(X, n, r), where X is a list-of-sublists containing the population distribution, n is the sample size, and r is the order statistic of interest.

Example 3.10. Find the distribution of the minimum of eight iid exponential(1) random variables.

The distribution of the minimum is found with the following APPL statements:

```
> X := ExponentialRV(1);
> Y := OrderStat(X, 8, 1);
```

Not surprisingly, these statements place the exponential distribution with a rate of eight into the variable Y. The list-of-sublists for Y is

$$[\,[\,x \to 8\,e^{-8x}\,],[0,\infty\,],[\,\text{``Continuous''},\text{``PDF''}\,]\,]$$

For many distributions, the `OrderStat` procedure can accommodate distributions with symbolic parameters.

Example 3.11. Find the distribution of the median of a sample of size $n = 3$ drawn from an exponential(λ) population.

In this example, an exponential distribution with a symbolic argument is entered as an argument for `OrderStat`.

```
> X := ExponentialRV(lambda);
> OrderStat(X, 3, 2);
```

The result is the symbolic PDF of the sample median:

$$[\,[\,x \to 6\,\lambda\tilde{}\,(1 - \mathrm{e}^{-\lambda\tilde{}\,x}\,)\,\mathrm{e}^{-2\,\lambda\tilde{}\,x}\,],[\,0,\infty\,],[\,\text{``Continuous''}, \text{``PDF''}\,]\,]$$

All of the standard programming constructs available in Maple, such as loops and conditions, can be used alongside APPL statements, as illustrated in the next example.

Example 3.12. To plot a progression of the five PDFs of the order statistics for a sample of $n = 5$ observations drawn from an exponential(1) population, use the APPL code

```
> X := ExponentialRV(1);
> n := 5;
> for i from 1 to n do
>    PlotDist(OrderStat(X, n, i), 0, 6);
> od;
```

The result is five PDFs plotted sequentially. The optional second and third arguments to `PlotDist` define the lower and upper plotting limit. This sequence could be of use to an instructor explaining the progressive nature of order statistics to first-year probability students. The last of the five plots is given in Fig. 3.2.

This chapter has presented data structures for representing continuous random variables and illustrated some of the basic procedures comprising the core of APPL. The procedures shown in the APPL tree diagram in Fig. 3.3 provide a roadmap for the rest of the monograph. The next three chapters describe three of APPL's nontrivial procedures that can be applied to continuous random variables: `Transform` (transformations of univariate random variables), `BiTransform` (transformations of bivariate random variables), and `Product` (products of two independent random variables).

Fig. 3.2. PDF of the largest of five unit exponential random variables

Fig. 3.3. The APPL tree

Transformations of Random Variables

This chapter presents a generalized version of the univariate change-of-variable technique for transforming continuous random variables. Extending a theorem from Casella and Berger [16] for many–to–1 transformations, we consider more general univariate transformations. Specifically, the transformation can range from 1–to–1 to many–to–1 on various subsets of the support of the random variable of interest. We also present an implementation of the theorem in APPL and present four examples.

In its simplest application, the change-of-variable technique is used to determine the distribution of a continuous random variable Y given the distribution of a continuous random variable X and a 1–to–1 transformation from the support of X to the support of Y. As the conditions on the transformation $Y = g(X)$ become more general, the technique requires more detail in its description and is less likely to appear in introductory probability and statistics texts. Casella and Berger [16, page 53] discuss transforming random variables using the change-of-variable technique when the entire transformation is many–to–1, except for a finite number of points, i.e., the cardinality of the set $g^{-1}(y)$ is the same for almost all y in the support of Y. Hogg, et al. [52, pages 128–129] extend this many–to–1 technique to n-dimensional random variables. We are concerned with a more general univariate case in which the transformations are "piecewise many–to–1," where "many" may vary based on the subinterval of the support of Y under consideration. We state and prove a theorem for this case and present APPL code to implement the result. Although our theorem is a straightforward generalization of Casella and Berger's theorem, there are a number of details that have to be addressed in order to produce an algorithm for finding the PDF of Y.

Consider the following simple example. Let

$$f_X(x) = \frac{1}{3} \qquad -1 < x < 2,$$

© Springer International Publishing AG 2017

J.H. Drew et al., *Computational Probability*, International Series in Operations Research & Management Science 246, DOI 10.1007/978-3-319-43323-3_4

be the PDF for the continuous random variable X. Consider the transformation $Y = g(X) = X^2$. This transformation is a 2–to–1 transformation on the interval $X \in (-1, 1)$ (except at $X = 0$) and it is 1–to–1 on the interval $X \in [1, 2)$; see Figure 4.1 in Sect. 4.3. Some introductory probability texts use this as an example (e.g., Larsen and Marx [68, page 174]), but fail to state a general theorem that treats such piecewise many–to–1 transformations. The difficulty lies in identifying an appropriate partition of the support of X and a corresponding partition for the support of Y, and then determining which of these subsets of the support of X correspond to each of the subsets of the support of Y. A further complication is encountered when the transformation itself is either discontinuous or non-differentiable at certain points. For example, consider the random variable X, where $f_X(x) = (x + 1)/18$ for $-1 < x < 5$, and the transformation (see Figure 4.3 in Sect. 4.3)

$$Y = g(X) = \begin{cases} X^2 & -1 < X < \frac{3}{2} \\ X & \frac{3}{2} < X < 5. \end{cases}$$

In this example, the transformation is discontinuous as well as "piecewise many–to–1." Our theorem and resulting APPL implementation will determine the PDF of Y for such transformations.

In Section 4.2, we will present our transformation theorem, modeled after Casella and Berger's Theorem 2.1.8 [16, page 53]. Our theorem has been presented more generally in earlier textbooks. Barr and Zehna [8, page 225] consider multivariate many–to–one transformations. Rohatgi [104, pages 73–74] and Port [102, page 462] consider the piecewise many–to–one case in the univariate and multivariate settings, respectively. Our theorem is strictly univariate, but permits implementation in APPL. It determines the distribution of $Y = g(X)$ for any univariate random variable X of the continuous type with few restrictions on the transformation $g(X)$. Our theorem will be stated in a somewhat elaborate form for the purpose of facilitating its implementation.

4.1 Theorem

Before we present the theorem, we overview the rationale for the notation. We assume that the support of X, denoted by \mathcal{X}, consists of a finite union of open intervals. The points $x_1 < x_2 < \cdots < x_{n+1}$ generate n consecutive subintervals and are determined as follows. The first subinterval of \mathcal{X} begins with x_1 and the last subinterval of \mathcal{X} ends with x_{n+1}. The remaining x_i's correspond to other endpoints of the intervals of \mathcal{X} or to the locations in \mathcal{X} where the function g is discontinuous or non-differentiable, or g' is zero. Let $g_i(x)$ denote the restriction of $g(x)$ to (x_i, x_{i+1}); by design, g_i is monotone and therefore invertible. Let $X^\star = \{x_1, x_2, \ldots, x_{n+1}\}$; note that since X is continuous, $\Pr(X \in X^\star) = 0$.

We introduce Y^\star, a set of points on the y-axis, to partition the support of Y into subintervals. The range of each g_i, denoted by (m_i, M_i), either contains or is disjoint from each Y subinterval. The set Y^\star is designed so that the final PDF of Y may be specified by designating its value on each Y subinterval. The set Y^\star is defined using one-sided limits because the transformations g_1, g_2, \ldots, g_n are not necessarily defined at the points in X^\star. The set I_j consists of integers i with the property that the range of g_i is either equal to or properly contains the jth Y subinterval. The PDF for each Y subinterval depends on the PDF of X and the associated transformations g_i.

Theorem 4.1. Let X be a random variable of the continuous type with PDF $f_X(x)$ and with support \mathcal{X}, where \mathcal{X} consists of a finite union of open intervals. Let $g(x)$ be a real-valued function whose domain includes \mathcal{X}. Let $-\infty \leq x_1 < x_2 < \cdots < x_n < x_{n+1} \leq +\infty$ be a sequence of extended real numbers which satisfy the following conditions:

1. The sequence includes the endpoints of the intervals whose union is \mathcal{X}.
2. The PDF $f_X(x)$ is continuous on each open interval $A_i = (x_i, x_{i+1})$ for $i = 1, 2, \ldots, n$.
3. If $f_X(x)$ is not identically zero on A_i, then the function $g_i(x)$, which is the restriction of $g(x)$ to A_i, is monotone on A_i and has a nonzero derivative at each point in A_i, for $i = 1, 2, \ldots, n$.

Let $X^\star = \{x_1, x_2, \ldots, x_{n+1}\}$.
Let $\alpha = \{i \mid f_X(x) \text{ is not identically zero on } A_i\}$.
Let $m_i = \min\left\{\lim_{x \downarrow x_i} g(x), \lim_{x \uparrow x_{i+1}} g(x)\right\}$ for $i = 1, 2, \ldots, n$.
Let $M_i = \max\left\{\lim_{x \downarrow x_i} g(x), \lim_{x \uparrow x_{i+1}} g(x)\right\}$ for $i = 1, 2, \ldots, n$.
Let $Y^\star = \cup_{i \in \alpha} \{m_i, M_i\}$.
Let $m = \|Y^\star\| - 1$, where $\|\cdot\|$ denotes cardinality.
Order the elements of y_j of Y^\star so that $y_1 < y_2 < \cdots < y_{m+1}$.
Let $I_j = \{i \mid m_i \leq y_j \text{ and } y_{j+1} \leq M_i\}$, for $j = 1, 2, \ldots, m$.
Then for $y \in (y_j, y_{j+1})$,

$$f_Y(y) = \sum_{i \in I_j} f_X\left(g_i^{-1}(y)\right) \left|\frac{dg_i^{-1}(y)}{dy}\right|$$

for $j = 1, 2, \ldots, m$.

Proof. Without loss of generality, consider the jth Y subinterval (y_j, y_{j+1}). Also suppose that a and b are any points that lie inside (y_j, y_{j+1}) such that $a < b$. Furthermore, let $M_i = \max\{g_i^{-1}(a), g_i^{-1}(b)\}$ and $m_i = \min\{g_i^{-1}(a), g_i^{-1}(b)\}$ for $i \in I_j$. Then $Y \in (a, b)$ if and only if X lies in (m_i, M_i) for some i in I_j, which implies that

$$\Pr(a < Y < b) = \sum_{i \in I_j} \Pr\left(m_i < X < M_i\right).$$

Since Y is a continuous random variable,

$$\Pr(a < Y < b) = \sum_{i \in I_j} \int_{m_i}^{M_i} f_X(x)\, dx$$

$$= \sum_{i \in I_j} \int_a^b f_X\left(g_i^{-1}(y)\right) \left|\frac{dg_i^{-1}(y)}{dy}\right|\, dy$$

$$= \int_a^b \sum_{i \in I_j} f_X\left(g_i^{-1}(y)\right) \left|\frac{dg_i^{-1}(y)}{dy}\right|\, dy$$

where we performed the change-of-variable $y = g_i(x)$, or equivalently $x = g_i^{-1}(y)$, and have used the absolute value to treat both increasing and decreasing transformations (see Miller and Miller's proof of their Theorem 7.1 [89, pages 240–242]). Referring to their theorem, we see that the integrand

$$\sum_{i \in I_j} f_X\left(g_i^{-1}(y)\right) \left|\frac{dg_i^{-1}(y)}{dy}\right|$$

is the PDF for Y on the subinterval (y_j, y_{j+1}). □

The theorem could have been stated more compactly. It is given in the form presented here in order to facilitate data structures and algorithms that will be implemented in the APPL `Transform` procedure.

4.2 Implementation in APPL

It could be quite tedious to implement this theorem by hand for large m and n. Since transformations of random variables are commonplace in applied probability, we wanted to include an implementation of the theorem in APPL's suite of procedures. Two significant implementation issues emerged. First, Maple may produce several candidates for g_i^{-1}, e.g., when $g_i(x) = x^2$. As alluded to earlier, Maple returns $g_i^{-1}(y) = -\sqrt{y}$ and $g_i^{-1}(y) = \sqrt{y}$. The correct inverse is selected by requiring that $g_i^{-1}(g_i(c_i)) = c_i$, where c_i is any point in the ith X subinterval. When the ith X subinterval has finite endpoints, c_i can be chosen to be the midpoint of the subinterval. In the cases in which $x_1 = -\infty$ or $x_{n+1} = \infty$ [e.g., $X \sim N(\mu, \sigma^2)$], c_i must be selected more carefully. The algorithm for determining c_i is:

1. If $x_i = -\infty$ and $x_{i+1} = +\infty$, then $c_i = 0$.
2. If $x_i = -\infty$ and $x_{i+1} \neq +\infty$, then $c_i = x_{i+1} - 1$.
3. If $x_i \neq -\infty$ and $x_{i+1} = +\infty$, then $c_i = x_i + 1$.
4. For all other cases (i.e., x_i and x_{i+1} both finite), $c_i = \dfrac{x_i + x_{i+1}}{2}$.

The second implementation issue involves redundancies in lists. Maple doesn't recognize that the integer 3 and the floating-point value 3.0, for example, are redundant in a list. The ReduceList procedure finds all adjacent points that differ by a small prescribed threshold δ and discards redundant floating-point type elements.

The Transform(X, g) procedure has two arguments: X and g. The first argument is a random variable in the list-of-sublists format described in Chapter 3. The data structure for g is a "list-of-two-sublists" that is similar in nature to the first two sublists of X. Here we characterize g by listing its monotone components and the endpoints of the corresponding domains.

The following are some additional minor implementation issues that arose during the coding of the algorithm:

- The user must supply $x_1, x_2, \ldots, x_{n+1}$. This ordered list consists of the endpoints of the open intervals that constitute \mathcal{X} and all locations in \mathcal{X} where $g(x)$ is discontinuous or non-differentiable, or $g'(x)$ is zero. A preprocessor to determine elements of X^\star corresponding to $g'(x) = 0$ could be added to the algorithm if desired. Thus if the support of X is $(-1, 2)$, and $Y = g(X) = X^2$, this preprocessor would include $x = 0$ in X^\star yielding $X^\star = \{-1, 0, 2\}$.
- The partition points $x_1, x_2, \ldots, x_{n+1}$ must be chosen so that $f_X(x)$ is not defined in a piecewise fashion on any X subinterval (e.g., if X has a triangular distribution, the mode must be a partition point).
- Adding extra x-values in \mathcal{X} to X^\star that are not maxima, minima, or saddle points of $g(x)$ or discontinuities of $f_X(x)$ or $g(x)$ will not affect the correctness of the algorithm's implementation of the theorem, although the implementation will be slower.
- Many transformations such as $g(x) = x + e^x$ do not have an inverse that can be expressed in a closed-form. Numerical methods could be used in conjunction with the algorithm to find individual values of $f_Y(y)$. We have not implemented this in our code. If $g(x) = x + e^x$, for example, then

$$\left| \frac{dg^{-1}(y)}{dy} \right| = \left. \left| \frac{1}{1 + e^x} \right| \right|_{x=g^{-1}(y)}.$$

- Periodic transformations can pose a problem in the implementation of the theorem. Maple uses the usual principal inverses for $\sin x$, $\cos x$, and $\tan x$, but the restricted ranges of these principal inverses are often inappropriate. When the algorithm tries to identify which inverse is appropriate, it is only given one choice, the principal inverse, but the domain of X might not coincide with the range of the principal inverse. Example 4.4 illustrates this problem and a circumvention.
- No error-trapping has been done to insure that $f_X(x)$ is a legitimate PDF. Also, there is limited error-trapping on $g(x)$ in that the procedure gives an error message when the inverse cannot be found.
- Symbolic parameters are allowed in $f_X(x)$ but not in $g(x)$.

4.3 Examples

This section contains some illustrative examples of using the algorithm described in the previous section to determine the distribution of $Y = g(X)$. Before using the program, we tested its performance on several common transformations. Given $X \sim N(0,1)$ and $Y = g(X) = X^2$, the program correctly returned the PDF for a chi square random variable with one degree of freedom. Given $X \sim N(\mu, \sigma^2)$ and $Y = g(X) = (X - \mu)/\sigma$, the program correctly returned the PDF for a standard normal random variable. Now we will consider more complex examples that illustrate the theorem's implementation.

Example 4.1. Consider the first example introduced earlier: given the random variable $X \sim U(-1, 2)$, find the PDF of $Y = g(X) = X^2$.

Given X^\star, the algorithm determines the relevant partitions of the supports of X and Y. Then it determines which X subintervals map onto which Y subintervals. The subintervals are apparent in Fig. 4.1. Note the set X^\star is displayed on the horizontal axis, Y^\star on the vertical axis, both marked with the \times symbol. The transformation is partitioned into monotone segments (with identical or disjoint ranges) delineated by \bullet. The assignment of data structures for f and g and the call to `Transform` are as follows:

```
> X := UniformRV(-1, 2);
> g := [[x -> x ^ 2, x -> x ^ 2], [-infinity, 0, infinity]];
> Y := Transform(X, g);
```

Fig. 4.1. The transformation $Y = g(X) = X^2$ for $-1 < X < 2$

The `Transform` procedure determines that the transformation is 2–to–1 on $-1 < x < 1$ (excluding $x = 0$) and 1–to–1 on $1 \leq x < 2$. Since $Y = X^2$ has two inverses, the procedure determines which inverse to apply to which X subinterval. The resulting PDF for Y is

$$f_Y(y) = \begin{cases} 1/(3\sqrt{y}) & 0 < y < 1 \\ 1/(6\sqrt{y}) & 1 < y < 4. \end{cases}$$

The corresponding returned value of `Y` is a list-of-sublists that specifies the PDF of Y:

$$\left[\left[y \to \frac{1}{3\sqrt{y}}, y \to \frac{1}{6\sqrt{y}} \right], [0, 1, 4], [\text{"Continuous"}, \text{"PDF"}] \right]$$

The previous example could have been completely automated by letting Maple determine the extreme point, as opposed to specifying g separately on each of its monotone segments. The next example will use a nondifferentiable function to indicate why we require the user to break a function into its monotone segments.

Example 4.2. Consider the random variable $X \sim U(0, 7)$. Find the PDF of $Y = g(X) = \big|\,|X - 3| - 1\big|$.

A graphical representation of this transformation is shown in Fig. 4.2, with X^\star, Y^\star, and the monotone segments of g marked as before. This transformation is more complex than Example 4.1 in that it is 4–to–1 for $0 < y < 1$, 2–to–1 for $1 < y < 2$, and 1–to–1 for $2 < y < 3$. The APPL code

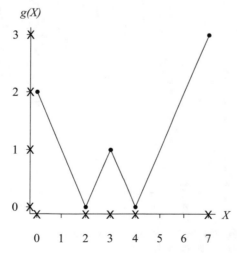

Fig. 4.2. The transformation $Y = g(X) = \big|\,|X - 3| - 1\big|$ for $0 < X < 7$

```
> X := UniformRV(0, 7);
> g := [[x -> 2 - x, x -> x - 2, x -> 4 - x, x -> x - 4],
        [0, 2, 3, 4, 7]];
> Y := Transform(X, g);
```

yields the correct PDF for Y:

$$f_Y(y) = \begin{cases} 4/7 & 0 < y < 1 \\ 2/7 & 1 < y < 2 \\ 1/7 & 2 < y < 3. \end{cases}$$

Example 4.3. Consider the second example from beginning of the chapter: if the random variable X has PDF

$$f_X(x) = (x+1)/18 \qquad -1 < x < 5,$$

find the distribution of

$$Y = g(X) = \begin{cases} X^2 & -1 < X < \frac{3}{2} \\ X & \frac{3}{2} < X < 5. \end{cases}$$

The appropriate partition for the transformation with $n = 3$ and $m = 4$ is depicted in Fig. 4.3. The APPL code for calculating the distribution of Y is

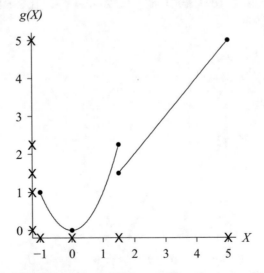

Fig. 4.3. The transformation $Y = g(X)$ has a discontinuity and is variously 2–to–1 and 1–to–1 on different subintervals

```
> X := [[x -> (x + 1) / 18], [-1, 5],
        ["Continuous", "PDF"]];
> g := [[x -> x ^ 2, x -> x ^ 2, x -> x],
        [-1, 0, 3 / 2, 5]];
> Y := Transform(X, g);
```

The `Transform` procedure returns the following PDF for Y as a list-of-sublists:

$$
f_Y(y) = \begin{cases}
\frac{1}{18\sqrt{y}} & 0 < y < 1 \\[2mm]
\frac{\sqrt{y}+1}{36\sqrt{y}} & 1 < y < 3/2 \\[2mm]
\frac{(2y+3)\sqrt{y}+1}{36\sqrt{y}} & 3/2 < y < 9/4 \\[2mm]
\frac{y+1}{18} & 9/4 < y < 5.
\end{cases}
$$

Example 4.4. As a final example, we will consider the problem that Casella and Berger [16] discussed, without providing $f_Y(y)$, as a prelude to their transformation theorem. Letting X be a uniform random variable on $(0, 2\pi)$, find the distribution of $Y = g(X) = \sin^2(X)$.

As Fig. 4.4 shows, this transformation is 4–to–1 for the single subinterval for Y, which is $(0,1)$. Furthermore, since $n = 4$, $X^\star = \{0, \frac{\pi}{2}, \pi, \frac{3\pi}{2}, 2\pi\}$. The PDF of Y is

$$
f_Y(y) = \frac{1}{\pi\sqrt{y - y^2}} \qquad 0 < y < 1,
$$

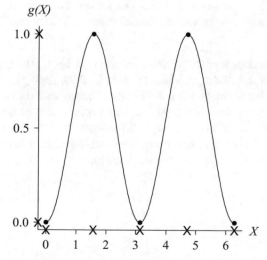

Fig. 4.4. The transformation $Y = g(X) = \sin^2(X)$ for $0 < X < 2\pi$

which is commonly known as the arcsin distribution (Johnson, et al. [55, pages 253–255]). To overcome the principal inverse difficulty alluded to in the previous section, we devised the following equivalent problem: let the random variable $X \sim U(-\frac{\pi}{2}, \frac{\pi}{2})$ and let $Y = g(X) = \sin^2(X)$. In this case the domain of X will be the same as the range of g^{-1} which results from using the standard arcsin function. This solution yielded a random variable with a notable feature. The distribution's HF, $h_Y(y) = f_Y(y)/S_Y(y)$, might be of interest to a reliability engineer. For this distribution,

$$h_Y(y) = \frac{2}{\sqrt{y - y^2}\left[\pi - 2\arcsin(2y - 1)\right]} \qquad 0 < y < 1.$$

By plotting this HF, we see it has the rare property of a closed-form, "bathtub"-shaped HF. Furthermore, we can now apply the transformation $W = g(Y) = \lambda Y$ in order to derive a random variable W that is a one-parameter random variable with a closed-form, bathtub-shaped HF. To our knowledge, although the arcsin distribution has been discussed in the literature, we have yet to find mention of the fact that it has a bathtub-shaped HF. Most distributions with bathtub-shaped HFs must be analyzed numerically because their HFs are not closed form. One of the useful features of the `Transform` procedure is that it can help the practitioner gain insight into what transformations could result in a useful model. In this example, it becomes apparent that the transformation "crowds" the uniformly distributed X random variable into a distribution for Y that is "heavy" on either end of its support. The result is a random variable with a bathtub-shaped HF. The program enables the model designer to confirm, or gain insight into the univariate transformations that could result in a useful model.

The `Transform` procedure, originally devised by Glen, et al. [37], is one of the fundamental building blocks of APPL. Although the presentation in this chapter has been solely for continuous random variables, `Transform` has also been extended to find the distribution of a function of a discrete random variable. Applications of `Transform` will be postponed to the later chapters of this monograph. The next chapter considers a generalization of the `Transform` procedure to allow for bivariate transformations.

5

Bivariate Transformations of Random Variables

This chapter extends the work in the previous chapter in order to automate the bivariate change-of-variables technique for bivariate continuous random variables with arbitrary distributions. The algorithm from the previous chapter for univariate change-of-variables was originally devised by Glen et al. [37]. The bivariate transformation procedure presented in this chapter handles 1-to-1, k-to-1, and piecewise k-to-1 transformations for both independent and dependent random variables. We also present other procedures that operate on bivariate random variables (e.g., calculating correlation and marginal distributions).

The most relevant software we found to automate bivariate transformations is Mathematica's statistical package, MathStatica by Rose and Smith [105]. Their examples appear to be limited to 1-to-1 transformations and independent random variables X and Y. We present an addendum to the APPL language that handles 1-to-1 transformations, k-to-1 transformations, and piecewise k-to-1 transformations for both dependent and independent continuous random variables X and Y.

The bivariate transformation technique calculates the distribution of $U = g(X, Y)$ from the joint distribution of two continuous random variables X and Y with joint PDF $f_{X,Y}(x, y)$ defined on the support set \mathcal{A} in the xy-plane. In the simplest case, an auxiliary dummy function $V = h(X, Y)$ can be found so that the two functions $U = g(X, Y)$ and $V = h(X, Y)$ define a 1-to-1 transformation from the support set \mathcal{A} in the xy-plane onto a support set \mathcal{B} in the uv-plane. Denote the inverse transformation from \mathcal{B} to \mathcal{A} by $X = r(U, V)$ and $Y = s(U, V)$. The formula from Hogg et al. [52]

$$f_{U,V}(u, v) = f_{X,Y}\big(r(u, v), s(u, v)\big)|J| \qquad \text{for } (u, v) \in \mathcal{B}$$

J.H. Drew et al., *Computational Probability*, International Series in Operations Research & Management Science 246, DOI 10.1007/978-3-319-43323-3_5

gives the joint PDF of U and V, where $J = \begin{vmatrix} \frac{dx}{du} & \frac{dx}{dv} \\ \frac{dy}{du} & \frac{dy}{dv} \end{vmatrix}$ is the Jacobian. The marginal PDF of U is then calculated by integrating v out of the joint PDF

$$f_U(u) = \int_{-\infty}^{\infty} f_{U,V}(u,v)\,dv = \int_{\mathcal{D}(u)} f_{U,V}(u,v)\,dv \qquad \text{for } u \in \mathcal{U},$$

where $\mathcal{D}(u) = \{v \mid (u,v) \in \mathcal{B}\}$ and \mathcal{U} is the support of U.

Next, we present a bivariate transformation algorithm used to handle such 1-to-1 transformations, and, more generally, k-to-1 transformations and piecewise k-to-1 transformations for the joint continuous random variables X and Y. The algorithm is modeled after the theorem for univariate transformations that was described in the previous chapter.

5.1 Algorithm

The purpose of the algorithm is to find the PDF of $U = g(X, Y)$, where X and Y are bivariate continuous random variables with joint PDF $f_{X,Y}(x, y)$. Let the support of X and Y in the xy-plane be denoted by \mathcal{A}, which is partitioned into a finite number of disjoint components. Denote the ith component of \mathcal{A} by \mathcal{A}_i and the associated PDF on \mathcal{A}_i by $f_i(x, y)$ for $i = 1, 2, \ldots, n$. The purpose of this partitioning might be to accommodate a distribution that is defined piecewise, a transformation that is defined piecewise, a transformation that is many-to-one, or any combination of these situations. Since the \mathcal{A}_i might include some of their boundary points on which the transformation could be badly behaved, we consider the interior of each \mathcal{A}_i, denoted by \mathcal{A}_i°, which is an open set. Let $(U, V) = w(X, Y) = \big(g(X, Y), h(X, Y)\big)$ be a function from \mathbb{R}^2 to \mathbb{R}^2 whose domain includes $\cup_{i=1}^n \mathcal{A}_i^\circ$. The following assumptions must hold:

1. The PDF $f_i(x, y)$ is continuous for all $(x, y) \in \mathcal{A}_i^\circ$ for $i = 1, 2, \ldots, n$.
2. The function w is a 1-to-1 transformation from \mathcal{A}_i° onto a set \mathcal{B}_i in the uv-plane, for $i = 1, 2, \ldots, n$. Denoting the inverse of this transformation from \mathcal{A}_i° onto \mathcal{B}_i by $(X, Y) = w_i^{-1}(U, V) = \big(r_i(U, V), s_i(U, V)\big)$, we require that the Jacobian

$$J_i = \begin{vmatrix} \frac{\partial r_i(u,v)}{\partial u} & \frac{\partial r_i(u,v)}{\partial v} \\ \frac{\partial s_i(u,v)}{\partial u} & \frac{\partial s_i(u,v)}{\partial v} \end{vmatrix}$$

is non-vanishing on \mathcal{B}_i for $i = 1, 2, \ldots, n$, and that the partial derivatives are everywhere defined and continuous.

Since the \mathcal{B}_i's are not necessarily pairwise disjoint, we find the contribution to the joint PDF of U and V over the component \mathcal{B}_i that results from \mathcal{A}_i°, is, from Hogg et al. [52],

$$f_{i,U,V}(u,v) = f_i\big(r_i(u,v), s_i(u,v)\big)|J_i| \qquad \text{for } (u,v) \in \mathcal{B}_i$$

for $i = 1, 2, \ldots, n$. Then $f_{U,V}(u, v) = \sum f_{i,U,V}(u, v)$, where the sum is taken over those i's for which $(u, v) \in \mathcal{B}_i$. To calculate the marginal PDF $f_U(u)$, we define the following additional notation. For those values of u that can occur for points (u, v) in \mathcal{B}_i, denote the contribution of the component \mathcal{A}_i to the marginal PDF of U by

$$f_{U_i}(u) = \int_{-\infty}^{\infty} f_{i,U,V}(u, v)\, dv = \int_{\mathcal{D}_i(u)} f_{i,U,V}(u, v)\, dv,$$

where $\mathcal{D}_i(u) = \{v \mid (u, v) \in \mathcal{B}_i\}$. In order to automate the integration over $\mathcal{D}_i(u)$, we have made the additional assumption that, for each value of u realizable in the interior of \mathcal{B}_i, $\mathcal{D}_i(u)$ is a single line segment (i.e., each vertical line segment spanning \mathcal{B}_i is entirely contained in \mathcal{B}_i) and that the endpoints of the closure of $\mathcal{D}_i(u)$ are determined by two different constraints. For some transformations defined on \mathcal{A}, these assumptions can only be satisfied by a judicious partitioning of \mathcal{A}. Because the boundary of \mathcal{B}_i typically consists of many distinct constraint curves, and the limits of integration for v depend on which constraint curves are relevant for a particular value of u, the function $f_{U_i}(u)$ is defined in a piecewise fashion with m_i pieces, where m_i is a positive integer. Then proceed as follows to find the PDF of U.

- For $j = 1, 2, \ldots, m_i$, denote the jth piece of $f_{U_i}(u)$ by $f_{U_{ij}}(u)$, which is defined on an interval denoted by $u_{ij} < u < u_{i(j+1)}$.
- Let $U^* = \cup_{i=1}^{n} \{u_{ij} \mid j = 1, 2, \ldots, m_i + 1\}$.
- Order the elements of U^* without repeats and relabel them using the notation u_i^* so that $u_1^* < u_2^* < \cdots < u_{l+1}^*$, where $l = \|U^*\| - 1$, and $\|\cdot\|$ denotes cardinality.
- Let $I_k = \{(i, j) \mid u_{ij} \le u_k^* \text{ and } u_{k+1}^* \le u_{i(j+1)}\}$ for $k = 1, 2, \ldots, l$.

Then for $u \in (u_k^*, u_{k+1}^*)$, the PDF of U is given by

$$f_U(u) = \sum_{(i,j) \in I_k} f_{U_{ij}}(u)$$

for $k = 1, 2, \ldots, l$.

5.2 Data Structure

In order to implement the algorithm, we will use a data structure for the distribution of the bivariate random variable that expands on the list-of-sublists format used in APPL and described in Chap. 3. The distribution of the bivariate random variable is presented in a list-of-three-sublists format. The first sublist contains the ordered two-variable PDF expressions $f_i(x, y)$ of the distribution, corresponding to the components \mathcal{A}_i into which the support of X and Y has been partitioned. The second sublist describes each component of \mathcal{A}_i by using a list of constraints for that component. The third sublist contains

the strings "`Continuous`" and "`PDF`" to specify the type of distribution and meaning of the function in the first sublist. For proper implementation of our algorithms, each component described in the second sublist must be a simply connected set (i.e., contains no holes) whose boundary can be formed by consecutive segments of curves that satisfy equations of the form $p(x, y) = 0$. In addition, the constraints listed in the second sublist for each component must satisfy the following conditions:

- The constraints must be entered in adjacent order; clockwise or counter-clockwise is acceptable.
- The constraints must completely enclose a region.
- The constraints must be entered as strict inequalities.
- Except for constraints of the form $x < a$ or $x > a$, each constraint, when written as an equality, must pass the vertical line test, i.e., only a single value of y corresponds to each value of x. For example, use $y < \sqrt{1 - x^2}$ or $y > -\sqrt{1 - x^2}$, rather than $x^2 + y^2 < 1$.
- If the component is unbounded, a less than ∞ or greater than $-\infty$ constraint must be included.

This data structure will be illustrated in Sect. 5.4.

5.3 Implementation

The bivariate transformation procedure `BiTransform(Z, g, h)` takes up to three arguments as input. The first two arguments are required and the third is optional. The first required argument is the joint distribution of X and Y, which is given in the list-of-three-sublists format. If $U = g(X, Y)$ is a k-to-1 transformation or defined in a piecewise fashion, or both, then the user must partition the support of X and Y such that $g(X, Y)$ is 1-to-1 and is not defined in a piecewise fashion on each component of the partition. The second required argument is the transformation of interest $U = g(X, Y)$, given in the Maple function format `[(x, y) → g(x, y)]`. If only one transformation is provided, then `g(x, y)` is applied to all components of the support of X and Y. Otherwise the user must supply a number of transformations equal to the number of components, where the first transformation corresponds to the first component, the second transformation corresponds to the second component, and so on. The third optional argument is the transformation $V = h(X, Y)$, which is described using the same Maple function format and conventions used when describing U. The default transformation is the dummy transformation $V = h(X, Y) = Y$ for all values in \mathcal{A}.

We give detail in this paragraph concerning the automatic determination of the inverse. Each component \mathcal{A}_i of the support of X and Y is associated with a corresponding transformation $U = g_i(X, Y)$ and $V = h_i(X, Y)$ for

$i = 1, 2, \ldots, n$. The algorithm adapts to the case in which the inverse transformation $(X, Y) = \big(r_i(U, V), s_i(U, V)\big)$ found by the Maple `solve` procedure returns multiple solutions, and the correct inverse must be chosen. For example, the 2-to-1 function $(U, V) = (X^2, Y)$ yields inverses $(X, Y) = (\sqrt{U}, V)$ and $(X, Y) = (-\sqrt{U}, V)$. The `BiTransform` procedure finds the correct inverse by selecting the inverse that satisfies $x_i = r_i\big(g_i(x_i, y_i), h_i(x_i, y_i)\big)$ and $y_i = s_i\big(g_i(x_i, y_i), h_i(x_i, y_i)\big)$ where $(x_i, y_i) \in \mathcal{A}_i^{\circ}$. An algorithm for determining such a point (x_i, y_i) in \mathcal{A}_i follows. Let W_i denote the set of x-values for all intersections of adjacent constraint equalities that define \mathcal{A}_i. Let t_i be the minimum of W_i and T_i be the maximum of W_i. If W_i contains more than two elements, let \hat{t}_i be the second smallest element of W_i and \hat{T}_i be the second largest element of W_i.

1. If $t_i = -\infty$ and $T_i = \infty$, then $x_i = 0$ and y_i is the average of the y-values for the two constraint equalities associated with $x_i = 0$ that maximize the difference of the y-values.
2. If $t_i = -\infty$ and T_i is finite, then $x_i = (T_i + \hat{T}_i)/2$ and y_i is the average of the two y-values that correspond to x_i on the two constraint equalities associated with x_i.
3. If t_i is finite, then $x_i = (t_i + \hat{t}_i)/2$ and y_i is the average of the two y-values that correspond to x_i on the two constraint equalities associated with x_i.

Each constraint defining \mathcal{A}_i is an inequality of the form $p(x, y) < 0$, where p is a real-valued continuous function. The corresponding constraint for \mathcal{B}_i is found by substituting the appropriate inverse transformations determined by the algorithm described in the previous paragraph to achieve the inequality $p\big(r_i(u, v), s_i(u, v)\big) < 0$. One difficulty arises when the number of constraints for \mathcal{A}_i is greater than the number of constraints for \mathcal{B}_i. In other words, at least one of the corresponding constraints for \mathcal{B}_i has been deleted as redundant. The following example illustrates this problem.

Consider the bivariate random variables X and Y with support on the open unit square $0 < x < 1, 0 < y < 1$. The transformation $U = XY$ with dummy transformation $V = Y$ is a 1-to-1 transformation from \mathcal{A} to \mathcal{B} (shown in Fig. 5.1 where the boundaries of \mathcal{A} and \mathcal{B} are not a part of the support) with inverse $X = U/V$, $Y = V$. Substituting $X = U/V$ and $Y = V$ into the linear constraints $x > 0, y > 0, x < 1, y < 1$ gives the linear constraints $u > 0, v > 0, u < v, v < 1$. The second constraint, $v > 0$, is redundant and should not be included in the description of the support of UV. The support of X and Y is bounded by four linear constraints whose corresponding equalities intersect at the adjacent points $(0, 0), (1, 0), (1, 1), (0, 1)$. These points on the boundary of \mathcal{A} are mapped to $(0, 0), (0, 0), (1, 1), (0, 1)$, respectively on the boundary of \mathcal{B}. The `BiTransform` procedure checks for equality between adjacent intersections for \mathcal{B} and deletes redundant constraints. In this case, the first and second intersections for \mathcal{A} map to the same point, $(0, 0)$, for \mathcal{B}, so $v > 0$ is deleted.

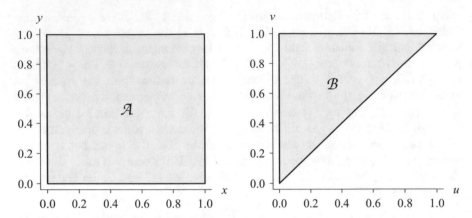

Fig. 5.1. Transformation from \mathcal{A} to \mathcal{B} given by $(U, V) = (XY, Y)$. The constraint equality $y = 0$ of \mathcal{A} is mapped to a point on the boundary of \mathcal{B}

There are cases where Maple functions don't operate as desired. The most troublesome of these functions is `solve`. In particular, the input `solve({x = a, y = infinity})` returns NULL for $a = 0$, but returns $\{x = a,\ y = \infty\}$ for all $a \neq 0$. The `solve` procedure is used primarily to find the intersection points between adjacent constraints associated with the support. Since joint distributions constrained to the first quadrant are fairly common, if-statements in the code handle the $a = 0$ case. However, it is impossible to add conditional statements for all problems of this type (i.e., `solve({x = y * y, y = infinity})` also returns NULL). Thus, the capability of our transformation procedure, `BiTransform`, is limited in handling distributions with infinite support. The scope of problems that can be solved by `BiTransform` is limited to those whose adjacent constraint equalities can be solved analytically by the `solve` function.

5.4 Examples

This section illustrates the use of the bivariate transformation procedure `BiTransform` for a variety of distributions and transformations. Example 5.1 illustrates the use of `BiTransform` on independent random variables. Example 5.2 illustrates transforming dependent random variables. Example 5.3 illustrates piecewise transformations. Example 5.4 illustrates k-to-1 transformations. Example 5.5 illustrates piecewise k-to-1 transformations. Example 5.6 illustrates a method for working around a support that is not a simply connected set.

Example 5.1 (Independent X and Y). The joint PDF of the independent continuous random variables $X \sim U(0,1)$ and $Y \sim U(0,1)$ is

$$f_{X,Y}(x,y) = f_X(x)f_Y(y) = 1 \qquad 0 < x < 1, 0 < y < 1.$$

Find the distribution of $U = XY$.

The distribution of $U = XY$ is found using the transformation technique as follows:

$$u = g(x,y) = xy \qquad x = r(u,v) = u/v$$
$$v = h(x,y) = y \qquad y = s(u,v) = v,$$

$$J = \begin{vmatrix} 1/v & -u/v^2 \\ 0 & 1 \end{vmatrix} = 1/v,$$

$$f_{U,V}(u,v) = 1 \cdot \left| \frac{1}{v} \right| = \frac{1}{v} \qquad 0 < u < v,$$

$$f_U(u) = \int_u^1 \frac{1}{v} dv$$
$$= [\ln v]_u^1$$
$$= -\ln u \qquad 0 < u < 1.$$

The commands to assign data structures and apply the bivariate transformation technique are

```
> XY := [[(x, y) -> 1], [[y > 0, x > 0, y < 1, x < 1]],
        ["Continuous", "PDF"]];
> g := [(x, y) -> x * y];
> h := [(x, y) -> y];
> U := BiTransform(XY, g, h);
```

which return

$$[[u \to -\ln(u)], [0, 1], [\text{``Continuous''}, \text{``PDF''}]]$$

as the PDF of U as expected. Leaving out the third line of code and the third argument in BiTransform results in identical output.

The next example illustrates a bivariate transformation with a similar support as the previous example, but with dependent continuous random variables X and Y.

Example 5.2 (Dependent X and Y). The joint PDF of dependent continuous random variables X and Y is

$$f_{X,Y}(x,y) = 2 \qquad x > 0, \ y > 0, \ x + y < 1.$$

Find the distribution of $U = X + Y$.

The distribution of $U = X + Y$ is found using the transformation technique as follows:

$$u = g(x, y) = x + y \qquad\qquad x = r(u, v) = \frac{u + v}{2}$$
$$v = h(x, y) = x - y \qquad\qquad y = s(u, v) = \frac{u - v}{2},$$

$$J = \begin{vmatrix} \frac{1}{2} & \frac{1}{2} \\ \frac{1}{2} & -\frac{1}{2} \end{vmatrix} = -\frac{1}{2},$$

$$f_{U,V}(u, v) = 2 \cdot \left| -\frac{1}{2} \right| = 1 \qquad\qquad 0 < u < 1, \ -u < v < u,$$

$$f_U(u) = \int_{-u}^{u} 1 \, dv = 2u \qquad\qquad 0 < u < 1.$$

The commands to assign data structures and apply the bivariate transformation technique are

```
> XY := [[(x, y) -> 2], [[x > 0, y > 0, x + y < 1]],
         ["Continuous", "PDF"]];
> g := [(x, y) -> x + y];
> h := [(x, y) -> x - y];
> U := BiTransform(XY, g, h);
```

which correctly return

$$[[u \to 2\,u]\,,\ [0,\ 1],\ [\text{``Continuous''},\ \text{``PDF''}]]$$

as the PDF of U.

The next example considers a bivariate transformation that is defined in a piecewise fashion.

Example 5.3 (Piecewise Transformation). The joint PDF of the independent continuous random variables $X \sim U(0, 1)$ and $Y \sim U(0, 1)$ is

$$f_{X,Y}(x, y) = f_X(x) f_Y(y) = 1 \qquad\qquad 0 < x < 1, 0 < y < 1.$$

Find the probability distribution of U, which is defined by the piecewise transformation with $U = X + Y$ for $0 < x < 1/2, 0 < y < 1$ and $U = XY$ for $1/2 < x < 1, 0 < y < 1$.

The transformation is illustrated in Fig. 5.2, using $V = X - Y$ on $0 < X < 1/2, 0 < Y < 1$ and $V = Y$ on $1/2 < X < 1, 0 < Y < 1$. For \mathcal{A}_1 with $0 < x < 1/2, 0 < y < 1$,

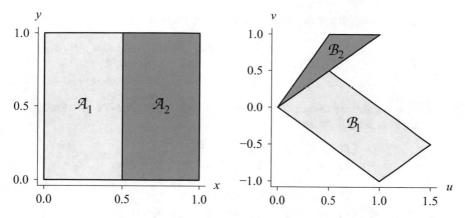

Fig. 5.2. Piecewise transformation from \mathcal{A}_1 to \mathcal{B}_1 and \mathcal{A}_2 to \mathcal{B}_2

$$u = g_1(x,y) = x+y \qquad x = r_1(u,v) = \frac{u+v}{2}$$
$$v = h_1(x,y) = x-y \qquad y = s_1(u,v) = \frac{u-v}{2},$$

$$J_1 = \begin{vmatrix} \frac{1}{2} & \frac{1}{2} \\ \frac{1}{2} & -\frac{1}{2} \end{vmatrix} = -\frac{1}{2},$$

$$f_{1,U,V}(u,v) = 1 \cdot \left| -\frac{1}{2} \right| = 1/2 \qquad 0 < u+v < 1, 0 < u-v < 2,$$

$$f_{U_1}(u) = \begin{cases} \displaystyle\int_{-u}^{u} 1/2\,dv & 0 < u < 1/2 \\ \displaystyle\int_{-u}^{1-u} 1/2\,dv & 1/2 < u < 1 \\ \displaystyle\int_{u-2}^{1-u} 1/2\,dv & 1 < u < 3/2 \end{cases}$$
$$= \begin{cases} u & 0 < u < 1/2 \\ 1/2 & 1/2 < u < 1 \\ (3-2u)/2 & 1 < u < 3/2. \end{cases}$$

For \mathcal{A}_2 with $u = g_2(x,y) = xy$, $v = h_2(x,y) = y$, we have $J_2 = 1/v$ from Example 5.1 and

$$f_{2,U,V}(u,v) = 1 \cdot \left| \frac{1}{v} \right| = \frac{1}{v} \qquad v/2 < u < v,\ 0 < v < 1,$$

$$f_{U_2}(u) = \begin{cases} \displaystyle\int_{u}^{2u} 1/v\,dv = \ln(2) & 0 < u < 1/2 \\ \displaystyle\int_{u}^{1} 1/v\,dv = -\ln(u) & 1/2 < u < 1. \end{cases}$$

Using the algorithm for bivariate transformations

$$f_U(u) = f_{U_1}(u) + f_{U_2}(u) = \begin{cases} u + \ln(2) & 0 < u < 1/2 \\ 1/2 - \ln(u) & 1/2 < u < 1 \\ (3 - 2u)/2 & 1 < u < 3/2. \end{cases}$$

The commands to assign data structures and apply the bivariate transformation technique are

```
> XY := [[(x, y) -> 1, (x, y) -> 1],
         [[y > 0, x < 0, y < 1, x < 1 / 2],
          [y > 0, x > 1 / 2, y < 1, x < 1]],
         ["Continuous", "PDF"]];
> g := [(x, y) -> x + y, (x, y)-> x * y];
> h := [(x, y) -> x - y, (x, y)-> y];
> U := BiTransform(XY, g, h);
```

which return the correct PDF of U in the APPL list-of-sublists format as

$$[[u \to u + \ln(2), u \to -\ln(u) + 1/2, u \to -u + 3/2], [0, 1/2, 1, 3/2],$$
$$["Continuous", "PDF"]]$$

The previous three examples have all involved 1-to-1 transformations. The next example considers a 2-to-1 bivariate transformation of the continuous random variables X and Y.

Example 5.4 (2-to-1 Transformation). The joint PDF of the independent continuous random variables X and Y is

$$f_{X,Y}(x,y) = 1/4 \qquad -1 < x < 1, -1 < y < 1.$$

Find the distribution of $U = X^2 - Y$.

The distribution of U under the 2-to-1 transformation $U = X^2 - Y$ and $V = Y$ is found using the algorithm described in Sect. 5.2. Partition the support of X and Y such that the transformation is 1-to-1 in each component. Let \mathcal{A}_1 be the component bound by $-1 < x < 0, -1 < y < 1$ and \mathcal{A}_2 be the component bound by $0 < x < 1, -1 < y < 1$. Both components map onto \mathcal{B} under the transformation $U = X^2 - Y$ and $V = Y$, as illustrated in Fig. 5.3. For \mathcal{A}_1,

$$\begin{aligned} u = g_1(x,y) = x^2 - y \qquad & x = r_1(u,v) = -(u+v)^{1/2} \\ v = h_1(x,y) = y \qquad & y = s_1(u,v) = v, \end{aligned}$$

$$J_1 = \begin{vmatrix} -\frac{1}{2}(u+v)^{-1/2} & -\frac{1}{2}(u+v)^{-1/2} \\ 0 & 1 \end{vmatrix} = -\frac{1}{2}(u+v)^{-1/2},$$

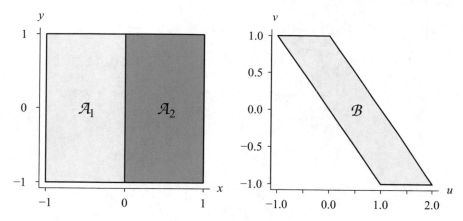

Fig. 5.3. 2-to-1 transformation from $\mathcal{A} = \mathcal{A}_1 \cup \mathcal{A}_2$ to \mathcal{B}

$$f_{1,U,V}(u,v) = \frac{1}{4} \cdot \left| -\frac{1}{2}(u+v)^{-1/2} \right| = \frac{1}{8}(u+v)^{-1/2}$$

for $-v < u < 1-v$, $-1 < v < 1$,

$$f_{U_1}(u) = \begin{cases} \int_{-u}^{1} \frac{1}{8}(u+v)^{-1/2} \, dv & -1 < u < 0 \\ \int_{-u}^{1-u} \frac{1}{8}(u+v)^{-1/2} \, dv & 0 < u < 1 \\ \int_{-1}^{1-u} \frac{1}{8}(u+v)^{-1/2} \, dv & 1 < u < 2 \end{cases}$$

$$= \begin{cases} \sqrt{u+1}/4 & -1 < u < 0 \\ 1/4 & 0 < u < 1 \\ -\sqrt{u-1}/4 + 1/4 & 1 < u < 2. \end{cases}$$

For \mathcal{A}_2,

$$\begin{aligned} u = g_2(x,y) = x^2 - y && x = r_2(u,v) = (u+v)^{1/2} \\ v = h_2(x,y) = y && y = s_2(u,v) = v, \end{aligned}$$

$$J_2 = \begin{vmatrix} \frac{1}{2}(u+v)^{-1/2} & \frac{1}{2}(u+v)^{-1/2} \\ 0 & 1 \end{vmatrix} = \frac{1}{2}(u+v)^{-1/2},$$

$$f_{2,U,V}(u,v) = \frac{1}{4} \cdot \left| \frac{1}{2}(u+v)^{-1/2} \right| = \frac{1}{8}(u+v)^{-1/2}$$

for $-v < u < 1-v$, $-1 < v < 1$,

$$f_{U_2}(u) = \begin{cases} \displaystyle\int_{-u}^{1} \frac{1}{8}(u+v)^{-1/2}\,dv & -1 < u < 0 \\[2ex] \displaystyle\int_{-u}^{1-u} \frac{1}{8}(u+v)^{-1/2}\,dv & 0 < u < 1 \\[2ex] \displaystyle\int_{-1}^{1-u} \frac{1}{8}(u+v)^{-1/2}\,dv & 1 < u < 2 \end{cases}$$

$$= \begin{cases} \sqrt{u+1}/4 & -1 < u < 0 \\ 1/4 & 0 < u < 1 \\ -\sqrt{u-1}/4 + 1/4 & 1 < u < 2. \end{cases}$$

Using the algorithm for bivariate transformations,

$$f_U(u) = f_{U_1}(u) + f_{U_2}(u) = \begin{cases} \sqrt{u+1}/2 & -1 < u < 0 \\ 1/2 & 0 < u < 1 \\ -\sqrt{u-1}/2 + 1/2 & 1 < u < 2. \end{cases}$$

The commands to assign data structures and apply the bivariate transformation technique are

```
> XY := [[(x, y) -> 1 / 4, (x, y) -> 1 / 4],
         [[y > -1, x > -1, y < 1, x < 0],
          [y > -1, x > 0, y < 1, x < 1]],
         ["Continuous", "PDF"]];
> g := [(x, y) -> x ^ 2 - y];
> h := [(x, y) -> y];
> U := BiTransform(XY, g, h);
```

which return the correct PDF of U in the APPL list-of-sublists format as

$$[[u \to \sqrt{u+1}/2, u \to 1/2, u \to -\sqrt{u-1}/2 + 1/2], [-1, 0, 1, 2],$$
$$["Continuous", "PDF"]]$$

The two previous examples have considered a piecewise transformation and a 2-to-1 transformation. The next example combines these two situations to analyze a bivariate piecewise 2-to-1 transformation.

Example 5.5 (Piecewise 2-to-1 Transformation). The joint PDF of independent continuous random variables X and Y is

$$f_{X,Y}(x, y) = 1/6 \qquad -2 < x < 1, -1 < y < 1.$$

Find the probability distribution of $U = X^2 - Y$.

The transformation $U = X^2 - Y$ and $V = Y$ is 2-to-1 on $-1 < x < 1$, $-1 < y < 1$ and 1-to-1 on $-2 < x < -1$, $-1 < y < 1$. Partition the support of X and Y such that the transformation $U = X^2 - Y$ and

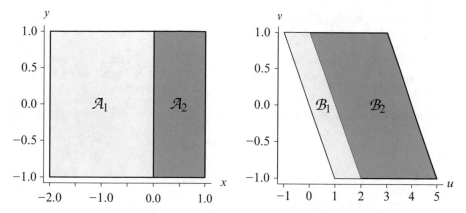

Fig. 5.4. Piecewise 2-to-1 transformation. Note that \mathcal{A}_1 maps onto $\mathcal{B}_1 \cup \mathcal{B}_2$ while \mathcal{A}_2 only maps to \mathcal{B}_2

$V = Y$ is 1-to-1 in each component. Let \mathcal{A}_1 be the component bound by $-2 < x < 0, -1 < y < 1$ and \mathcal{A}_2 be the component bound by $0 < x < 1, -1 < y < 1$, as illustrated in Fig. 5.4. The marginal PDF of U is

$$
f_U(u) = \begin{cases}
\sqrt{u+1}/3 & -1 < u < 0 \\
\sqrt{u+1}/6 + 1/6 & 0 < u < 1 \\
-\sqrt{u-1}/3 + 1/6 + \sqrt{u+1}/6 & 1 < u < 2 \\
-\sqrt{u-1}/6 + \sqrt{u+1}/6 & 2 < u < 3 \\
-\sqrt{u-1}/6 + 1/3 & 3 < u < 5.
\end{cases}
$$

The commands to assign data structures and apply the bivariate transformation technique are

```
> XY := [[(x, y) -> 1 / 6, (x, y) -> 1 / 6],
         [[y > -1, x > -2, y < 1, x < 0],
          [y > -1, x > 0, y < 1, x < 1]],
         ["Continuous", "PDF"]];
> g := [(x, y) -> x ^ 2 - y, (x, y) -> x ^ 2 - y];
> h := [(x, y) -> y, (x, y) -> y];
> U := BiTransform(XY, g, h);
```

Notice that g and h each contain two functions instead of one as in the previous example. Declaring the same transformation for each component achieves the same result as declaring the transformation once. The correct PDF of U is returned in the APPL list-of-sublists format as

$[[u \to \sqrt{u+1}/3, u \to \sqrt{u+1}/6 + 1/6, u \to -\sqrt{u-1}/3 + 1/6 + \sqrt{u+1}/6,$
$u \to -\sqrt{u-1}/6 + \sqrt{u+1}/6, \sqrt{u+1}, u \to -\sqrt{u-1}/6 + 1/3], [-1, 0, 1, 2, 3, 5],$
$[\text{"Continuous"}, \text{"PDF"}]].$

The final example of using the `BiTransform` procedure illustrates its use in the case of a non-simply connected support.

Example 5.6 (Non-simply Connected Support for U and V). Let X and Y be uniformly distributed on $(1, 2) \times (-\pi, \pi)$. Find the PDF of $U = X \cos Y$.

Consider the bivariate transformation

$$U = X \cos Y, \ V = X \sin Y.$$

The joint density of U and V is

$$f_{U,V}(u, v) = \frac{1}{2\pi\sqrt{u^2 + v^2}} \qquad 1 < u^2 + v^2 < 4$$

and the marginal PDF of U is

$$f_U(u) = \begin{cases} \pi^{-1} \sinh^{-1} \sqrt{4u^{-2} - 1} & 1 < |u| < 2 \\ \pi^{-1}\left(\sinh^{-1} \sqrt{4u^{-2} - 1} - \sinh^{-2} \sqrt{u^{-1} - 1}\right) & |u| < 1. \end{cases}$$

The commands to assign data structures and apply the bivariate transformation technique are

```
> X := UniformRV(1, 2);
> Y := UniformRV(-Pi, Pi);
> XY := JointPDF(X, Y);
> g := [(x, y) -> x * cos(y)];
> h := [(x, y) -> x * sin(y)];
> BiTransform(XY, g, h);
```

which displays the joint density of (U, V) as

$$\left[\left[(u, v) \mapsto 1/2 \frac{1}{\pi\sqrt{|v^2 + u^2|}}\right], \left[\left[1 < \sqrt{v^2 + u^2}, \sqrt{v^2 + u^2} < 2,\right.\right.\right.$$

$$\left.\left.\arctan\left(\frac{v}{\sqrt{v^2 + u^2}}, \frac{u}{\sqrt{v^2 + u^2}}\right) < \pi\right]\right], [\text{"Continuous", "PDF"}]\right]$$

The marginal PDF is calculated by manually dividing the support of (U, V) into six regions using $u = -1$, $v = 0$ and $u = 1$, as illustrated in Fig. 5.5. The Maple commands

```
> UV := [[(u, v) -> 1 / (2 * pi * sqrt(u ^ 2 + v ^ 2)),
          (u, v) -> 1 / (2 * pi * sqrt(u ^ 2 + v ^ 2)),
          (u, v) -> 1 / (2 * pi * sqrt(u ^ 2 + v ^ 2)),
          (u, v) -> 1 / (2 * pi * sqrt(u ^ 2 + v ^ 2)),
          (u, v) -> 1 / (2 * pi * sqrt(u ^ 2 + v ^ 2)),
          (u, v) -> 1 / (2 * pi * sqrt(u ^ 2 + v ^ 2))],
```

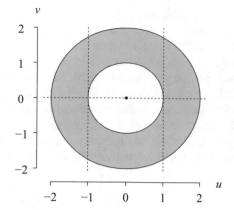

Fig. 5.5. Support of U and V

```
[[v < sqrt(4 - u ^ 2), u < -1, v > 0],
 [v < sqrt(4 - u ^ 2), u > -1,
  v > sqrt(1 - u ^ 2), u < 1],
 [v < sqrt(4 - u ^ 2), u > 1, v > 0],
 [v > -sqrt(4 - u ^ 2), u < -1, v < 0],
 [v > -sqrt(4 - u ^ 2), u > -1,
  v < -sqrt(1 - u ^ 2), u < 1],
 [v > -sqrt(4 - u ^ 2), u > 1, v < 0]],
 ["Continuous", "PDF"]];
> MarginalPDF(UV, u);
```

correctly return

$$\left[\left[u \mapsto \frac{1}{2\pi} \ln\left(\sqrt{-u^2+4}+2\right) - \frac{1}{2\pi} \ln\left(-\sqrt{-u^2+4}+2\right),\right.\right.$$

$$u \mapsto -\frac{1}{2\pi} \ln\left(\sqrt{-u^2+1}+1\right) - \frac{1}{2\pi} \ln\left(\sqrt{-u^2+4}+2\right) +$$

$$\frac{1}{2\pi} \ln\left(-\sqrt{-u^2+4}+2\right) - \frac{1}{2\pi} \ln\left(-\sqrt{-u^2+1}+1\right),$$

$$\left.u \mapsto -\frac{1}{2\pi} \ln\left(\sqrt{-u^2+4}+2\right) + \frac{1}{2\pi} \ln\left(-\sqrt{-u^2+4}+2\right)\right],$$

$$\left.[-2.0, -1.0, 1.0, 2.0], [\text{"Continuous"}, \text{"PDF"}]\right]$$

as the marginal PDF of U. [Recall that $\sinh^{-1}(x) = \ln\left(x + \sqrt{x^2+1}\right)$.]

This completes the six examples that illustrate the use of the **BiTransform** procedure. Calculating the distribution of a transformation of two random

variables by hand can be tedious, especially when working with piecewise joint distributions or piecewise transformations. An automated method for computing bivariate transformations saves the user time and prevents calculation errors. The algorithm and `BiTransform` procedure we developed allows for 1-to-1, k-to-1, and piecewise k-to-1 transformations for both independent and dependent continuous random variables. In the future, we hope to see computer-based algorithms for bivariate discrete transformations, as well as multivariate transformations involving three or more random variables. The algorithm described in this chapter has been implemented in approximately 1300 lines of code and is available at `www.math.wm.edu/~leemis/BiVarAPPL.txt`.

6

Products of Random Variables

This chapter describes an algorithm for computing the PDF of the product of two independent continuous random variables. This algorithm has been implemented in the `Product` procedure in APPL. The algorithms behind the `Transform` and `BiTransform` procedures from the two previous chapters differ fundamentally from the algorithm behind the `Product` procedure in that the transformation algorithms are more general whereas determining the distribution of the product of two random variables is more specific. Some examples given in the chapter demonstrate the algorithm's application.

Rohatgi's well-known result [104, page 141] for determining the distribution of the product of two random variables is straightforward to derive, but difficult to implement: If X and Y are continuous random variables with joint PDF $f_{X,Y}(x,y)$, then the PDF of $V = XY$ is

$$f_V(v) = \int_{-\infty}^{\infty} f_{X,Y}\left(x, \frac{v}{x}\right) \frac{1}{|x|}\, dx.$$

Larsen and Marx [69, page 225] state the formula for the special case in which X and Y are independent. The implementation of this easily-stated result, however, is not straightforward for general X and Y. Springer [109] presents a chapter on finding distributions of products of random variables, relying mostly on Laplace and Mellin transformation techniques, as implementation of Rohatgi's result is often too cumbersome. Difficulties occur as a result of both the myriad variations to the limits of integration and the propensity of the PDF of V to be defined in a piecewise manner. In this chapter, we consider the cases when X and Y are independent and may have PDFs defined in a piecewise fashion. This simplification implies that the joint PDF of X and Y will be defined on a rectangular product space, as shown in Hogg et al. [52, page 109].

© Springer International Publishing AG 2017
J.H. Drew et al., *Computational Probability*, International Series in Operations Research & Management Science 246,
DOI 10.1007/978-3-319-43323-3_6

6.1 Theorem

Considering a special case of Rohatgi's result will illustrate some of the issues associated with a general algorithm for determining the PDF of the product of two independent random variables. For completeness, the theorem below is proven from first principles (using the transformation technique) even though it could be stated as a special case of Rohatgi's result. Assume that the random variable X has support on the positive interval (a, b) and the random variable Y has support on the positive interval (c, d). Also, the product space of the two random variables is assumed to fall entirely in the first quadrant. Theorems and proofs for other rectangular support regions are similar. The algorithm described in Sect. 6.5 includes all possible scenarios.

Theorem 6.1. Let X be a random variable of the continuous type with PDF $f_X(x)$ that is defined and positive on the interval (a, b), where $0 < a < b < \infty$. Similarly, let Y be a random variable of the continuous type with PDF $f_Y(y)$ which is defined and positive on the interval (c, d), where $0 < c < d < \infty$. The PDF of $V = XY$ is

$$
f_V(v) = \begin{cases}
\int_a^{v/c} f_Y\left(\frac{v}{x}\right) f_X(x)\frac{1}{x}\, dx & ac < v < ad \\
\int_{v/d}^{v/c} f_Y\left(\frac{v}{x}\right) f_X(x)\frac{1}{x}\, dx & ad < v < bc \\
\int_{v/d}^{b} f_Y\left(\frac{v}{x}\right) f_X(x)\frac{1}{x}\, dx & bc < v < bd
\end{cases}
$$

when $ad < bc$,

$$
f_V(v) = \begin{cases}
\int_a^{v/c} f_Y\left(\frac{v}{x}\right) f_X(x)\frac{1}{x}\, dx & ac < v < ad \\
\int_{v/d}^{b} f_Y\left(\frac{v}{x}\right) f_X(x)\frac{1}{x}\, dx & ad < v < bd
\end{cases}
$$

when $ad = bc$, and

$$
f_V(v) = \begin{cases}
\int_a^{v/c} f_Y\left(\frac{v}{x}\right) f_X(x)\frac{1}{x}\, dx & ac < v < bc \\
\int_a^{b} f_Y\left(\frac{v}{x}\right) f_X(x)\frac{1}{x}\, dx & bc < v < ad \\
\int_{v/d}^{b} f_Y\left(\frac{v}{x}\right) f_X(x)\frac{1}{x}\, dx & ad < v < bd
\end{cases}
$$

when $ad > bc$.

Proof. Only the case of $ad < bc$ is considered. The other cases are proven analogously. Using the transformation technique [52, page 87], the bivariate transformation $Z = X$ and $V = XY$ constitutes a 1-to-1 mapping from $\mathcal{A} = \{(x, y)|a < x < b, c < y < d\}$ to $\mathcal{B} = \{(z, v)|a < z < b, cz < v < dz\}$. Let u denote the transformation and w the inverse transformation. The transformation, inverse, and Jacobian are

$$
\begin{aligned}
z &= u_1(x, y) = x & \qquad x &= w_1(z, v) = z \\
v &= u_2(x, y) = xy & \qquad y &= w_2(z, v) = v/z
\end{aligned}
$$

$$J = \begin{vmatrix} 1 & 0 \\ -v/z^2 & 1/z \end{vmatrix} = 1/z.$$

The joint PDF of Z and V is

$$f_{Z,V}(z,v) = f_X\big(w_1(z,v)\big) f_Y\big(w_2(z,v)\big) |J|, \qquad (z,v) \in \mathcal{B},$$

or, equivalently,

$$f_{Z,V}(z,v) = f_X(z) f_Y(v/z) \frac{1}{|z|}, \qquad (z,v) \in \mathcal{B}.$$

Integrating with respect to z over the appropriate intervals and replacing z with x in the final result yields the marginal PDF of V,

$$f_V(v) = \begin{cases} \int_a^{v/c} f_Y\left(\frac{v}{x}\right) f_X(x) \frac{1}{x}\, dx & ac < v < ad \\ \int_{v/d}^{v/c} f_Y\left(\frac{v}{x}\right) f_X(x) \frac{1}{x}\, dx & ad < v < bc \\ \int_{v/d}^{b} f_Y\left(\frac{v}{x}\right) f_X(x) \frac{1}{x}\, dx & bc < v < bd, \end{cases}$$

as desired. □

The geometry associated with the transformation for the case $ad < bc$ considered in the proof is shown in Figs. 6.1 and 6.2. The transformation maps A_i onto B_i for $i = 1, 2, 3$. Although the transformation technique has been used to prove this theorem, the CDF technique could also have been used. The

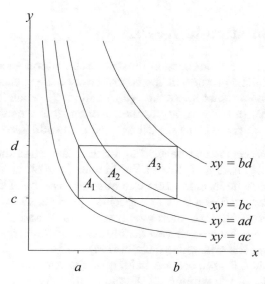

Fig. 6.1. The support of X and Y when $ad < bc$

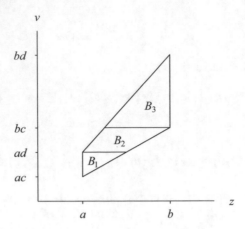

Fig. 6.2. The mapping of $Z = X$ and $V = XY$ when $ad < bc$

theorem also follows directly from Rohatgi's result because $f_{X,Y}(x, v/x) = f_X(x)f_Y(v/x)$, which is nonzero only when $a < x < b$ and $c < v/x < d$. Thus $\max\{a, v/d\} < x < \min\{b, v/c\}$, which yields the limits of integration displayed in the theorem. The geometry of the case $ad < bc$ is shown in Fig. 6.1 in which the regions A_1, A_2, and A_3 correspond to the three different integrals for that case. Each A_i corresponds to a different interval of v for the family of hyperbolas given by $xy = v$. Given v, the interval of integration consists of those values of x for which the curve $xy = v$ intersects the rectangle $\{(x, y) : a < x < b, c < y < d\}$.

6.2 Implementation in APPL

The theorem in the previous section illustrates the importance of considering the magnitudes of the product of the coordinates of the southeast and northwest corners of the product space [i.e., (b, c) and (a, d)] when it lies entirely in the first quadrant. In order to apply the theorem to any continuous random variables X and Y, three generalizations need to be addressed:

1. Analogous theorems must be written for the cases when the product space defined by (a, b) and (c, d) lies wholly in one of the other three quadrants. Rectangles that lie in more than one quadrant can be split into parts.
2. Instead of having PDFs that are specified by a single standard function over their entire support, the random variables X and Y may be defined in a piecewise manner over several intervals, forming many segments of the PDF (e.g., the triangular distribution).
3. The cases when 0 and/or $\pm\infty$ belong to the endpoints of the intervals which constitute the support of X and Y must be considered (e.g., the normal distribution, see Example 6.4).

These generalizations result in 24 (6 for each quadrant) primary cases that must be considered in order to correctly compute the limits of integration of the theorem. Further, some of these 24 cases must be broken into as many as three subcases, as illustrated in the theorem. In the first quadrant (i.e., $a \geq 0$ and $c \geq 0$), for example, we must consider the six primary cases:

1. $c = 0$ and $d = \infty$ (Scenario A, see the algorithm in Sect. 6.5),
2. $c = 0$ and $d < \infty$ (Scenario B),
3. $c > 0$ and $d = \infty$ (Scenario C),
4. $c > 0$ and $d < \infty$ (Scenario D, with Case I: $ad < bc$),
5. $c > 0$ and $d < \infty$ (Scenario D, with Case II: $ad = bc$), and
6. $c > 0$ and $d < \infty$ (Scenario D, with Case III: $ad > bc$).

The paragraphs below provide more details. For quadrants II, III and IV, the limits of integration must also be set based on the appropriate geometry, as in Fig. 6.1.

For random variables that are defined in a piecewise fashion over various intervals, let n be the number of intervals for X and let m be the number of intervals for Y. There are mn rectangular product subspaces and the contribution of each to the value of the PDF of $V = XY$ must be computed. Furthermore, each product subspace can contribute differently to the PDF of V on up to three segments of the support of V (see Theorem 6.1). As a result, the PDF of V tends to become complicated very quickly, with an upper limit of $3mn$ segments to its PDF. Furthermore, each of these segments fits into one of the 24 cases of the integration mentioned above. For example, the product of two $U(0, 1)$ random variables (Scenario B from the list above) yields a random variable V with only one segment (see Example 6.1). But slightly changing X and Y to $X \sim U(1, 2)$ and $Y \sim U(3, 4)$ yields a $V = XY$ defined differently on three segments (see Example 6.2).

The case in which the support of X or Y contains 0 [e.g., $U(-1, 2)$] poses special difficulty because some of the rectangular product spaces will not lie wholly in any one quadrant and cannot be handled by the previously developed techniques. Our solution to this difficulty is to artificially add 0 as one of the endpoints of the intervals for X and Y whenever this case occurs, producing redundant segments, i.e., a segment on each side of zero with the same formula for the PDF. This method allows for implementation when the support area intersects more than one quadrant. Example 6.3 gives an application of this special case, where the area of interest intersects all four quadrants. The algorithm subdivides the total area into four components, each entirely in one quadrant, and calculates the relevant portion of the new PDF for each quadrant.

The algorithm consists of a set-up portion, followed by nested loops that determine the contribution to the PDF of $V = XY$ separately for each of the four quadrants. Section 6.5 contains the set-up portion and the algorithm associated with the first quadrant. The algorithm for the other quadrants is similar.

The set-up phase begins by setting n and m to the number of intervals that form the support of X and Y. Next, 0 is added as an interval delimiter for X and/or Y if the random variable can assume both positive and negative values, and 0 is not already an interval delimiter. Finally, the endpoints of the intervals that form the support of V are determined by taking all products of the endpoints of the X intervals times the endpoints of the Y intervals.

A nested set of loops follows that treats all pairings of X and Y intervals. As shown in Fig. 6.1, the coordinates (a, c) are assigned to the southwest corner of the current rectangle of interest, and the coordinates (b, d) are assigned to the northeast corner of the current rectangle of interest. A test to determine which quadrant contains the current rectangle is made at this point. Adding 0 as an interval delimiter in the set-up phase assures that the current rectangle will be completely contained in just one of the quadrants. Once the quadrant is determined, tests on c and d determine which integrals should be computed and the appropriate limits of integration. Finally, the insertion of 0 sometimes leads to a PDF for V with the same formula on both sides of 0. If this occurs, the procedure simplifies the PDF by removing 0 as an interval endpoint if the function is defined at 0.

When this algorithm is implemented in APPL, Maple occasionally returns a PDF in a less than optimal form. The expression $\ln(-v) - I\pi$, for example, may occur as part of a PDF. Since $\ln(-1) = I\pi$, the expression is equal to $\ln v$. Also the PDF may involve a complicated integral, which Maple fails to recognize is equal to a much simpler standard form (see Example 6.5). We anticipate that future generations of computer algebra systems will overcome these difficulties.

There are three classes of resultant distributions that come from products via this algorithm. Class A is the set of resultant distributions that are completely tractable, such as the product of two $U(0, 1)$ random variables (see Example 6.1). Here the PDF is in closed form, the CDF is calculable, quantiles can be calculated, and plots of distributions are easily produced. The second set, Class B, is the set of resultant distributions that are not in closed form, but are in forms that Maple can still evaluate, such as PDFs that rely on the `erf` function or the `BesselK` functions of Maple. Example 6.4 illustrates one such case, where the non-closed form PDF is expressed in terms of the `BesselK` function. In Class B, it is often possible to plot the PDF and calculate quantiles. Finding a usable form for the CDF, however, might not be possible. The final class of problems, Class C, is the set that only appears as un-evaluated integrals. The convolution of three Weibull random variables is one such example. Here the resulting PDF has little use, as Maple cannot plot the function, and cannot evaluate critical points very effectively.

6.3 Examples

This section contains examples of applying the APPL procedure `Product`, which is our implementation of the algorithm described in the previous section.

Example 6.1. Consider the independent random variables $X \sim U(0,1)$ and $Y \sim U(0,1)$. Find the distribution of $V = XY$.

This is a simple application of the algorithm. The following APPL code defines the distribution of the random variables X and Y and returns the PDF of their product.

```
> X := UniformRV(0, 1);
> Y := UniformRV(0, 1);
> V := Product(X, Y);
```

The resulting PDF for $V = XY$ is

$$f_V(v) = -\ln v \qquad 0 < v < 1,$$

which can be readily verified by hand as follows. The CDF of V is

$$
\begin{aligned}
F_V(v) = \Pr(V \le v) \\
= 1 - \Pr(V > v) \\
= 1 - \Pr(XY > v) \\
= 1 - \int_v^1 \int_{v/x}^1 1 \, dy \, dx \\
= v - v \ln v.
\end{aligned}
$$

Differentiating with respect to v yields the same PDF returned by APPL.

Example 6.1 illustrates the simplest case, where the PDF of V is defined on a single segment. The next example has a rectangular support for X and Y, which results in a piecewise PDF of V.

Example 6.2. Consider the independent random variables $X \sim U(1,2)$ and $Y \sim U(3,4)$. Find the distribution of $V = XY$.

The following APPL code defines the random variables X and Y and returns the PDF of their product.

```
> X := UniformRV(1, 2);
> Y := UniformRV(3, 4);
> V := Product(X, Y);
```

These statements yields the following PDF for $V = XY$:

$$f_V(v) = \begin{cases} \ln v - \ln 3 & 3 < v < 4 \\ \ln 4 - \ln 3 & 4 < v < 6 \\ 3\ln 2 - \ln v & 6 < v < 8. \end{cases}$$

Notice that while the PDF of both X and Y are defined on single segments that have positive interval limits, the PDF of V is defined on three segments. All of the other APPL procedures described earlier can be applied to the random variable V. For instance, the first four moments of V can be calculated via the additional APPL statements

```
> Mean(V);
> Variance(V);
> Skewness(V);
> Kurtosis(V);
```

which return

$$\frac{21}{4} \qquad \frac{175}{144} \qquad \frac{54\sqrt{7}}{875} \qquad \frac{1647621}{765625}.$$

Examples 6.1 and 6.2 have defined X and Y using numeric parameters, with the joint support of X and Y being a bounded product space. The next example considers the distribution of the product of two random variables with symbolic parameters whose joint support is the entire first quadrant.

Example 6.3. Consider the independent random variables $X \sim$ lognormal(μ_1, σ_1) and $Y \sim$ lognormal(μ_2, σ_2). Show that the product of X and Y is also lognormal.

The APPL code to find the distribution of the product is

```
> X := LogNormalRV(mu1, sigma1);
> Y := LogNormalRV(mu2, sigma2);
> V := Product(X, Y);
```

Maple does not return the PDF of V in a form that can be directly identified as a lognormal PDF. But using the **Transform** procedure with transformation $g(V) = \ln V$ on the random variable V yields a random variable that is $N\left(\mu_1 + \mu_2, \sigma_1^2 + \sigma_2^2\right)$. Because of the relationship between the lognormal and the normal distributions, we know that V is lognormal with PDF

$$f_V(v) = \frac{\exp\left(-(\ln v - \mu_1 - \mu_2)^2 / 2(\sigma_1^2 + \sigma_2^2)\right)}{v\sqrt{2\pi(\sigma_1^2 + \sigma_2^2)}} \qquad v > 0.$$

This can be identified as a lognormal$\left(\mu_1 + \mu_2, \sqrt{\sigma_1^2 + \sigma_2^2}\right)$ random variable. Hence the product of two independent lognormal random variables is itself lognormal.

We now move outside the first quadrant to a distribution with a much less tractable distribution for $V = XY$.

Example 6.4. Consider the independent random variables $X \sim N(0, 1)$ and $Y \sim N(0, 1)$. Find the PDF of $V = XY$.

This example illustrates the case of 0 in the support of X and Y and also the case where the support of X and Y includes the endpoints $\pm\infty$. The APPL code to find the distribution of the product is

```
> X := NormalRV(0, 1);
> Y := NormalRV(0, 1);
> V := Product(X, Y);
```

The `Product` procedure yields the following PDF for V:

$$f_V(v) = \begin{cases} \dfrac{\mathrm{BesselK}(0, -v)}{\pi} & -\infty < v < 0 \\[2mm] \dfrac{\mathrm{BesselK}(0, v)}{\pi} & 0 < v < \infty \end{cases}$$

which relies on Maple's `BesselK` function. An APPL plot of this function is given in Fig. 6.3.

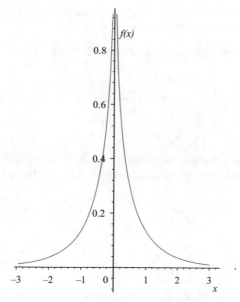

Fig. 6.3. The PDF of $V = XY$ for $X \sim N(0, 1)$ and $Y \sim N(0, 1)$

The following final, more complicated example, involves checking the validity of a well-known random variate generation algorithm.

Example 6.5. Consider the independent uniform random variables $U_1 \sim U(0,1)$ and $U_2 \sim U(0,1)$. The Box–Muller algorithm for generating a single standard normal deviate V can be coded in one line (see Devroye [21]) as

$$V \leftarrow \sqrt{-2 \ln U_1} \cos(2\pi U_2),$$

where U_1 and U_2 are independent random numbers. Use the Transform and Product procedures together to determine the PDF of V.

Due to the principle inverse difficulty with trigonometric functions (see Example 4.4 for an illustration), the transformation must be rewritten as

$$V \leftarrow \sqrt{-2 \ln U_1} \cos(\pi U_2)$$

before using Transform. The APPL code to compute the distribution of V is

```
> U1 := UniformRV(0, 1);
> U2 := UniformRV(0, Pi);
> T1 := Transform(U1, [[x -> sqrt(-2 * ln(x))],[0, 1]]);
> T2 := Transform(U2, [[x -> cos(x)], [0, Pi]]);
> V  := Product(T1, T2);
```

This APPL code yields the following PDF for V:

$$f_V(v) = \begin{cases} \displaystyle\int_{-v}^{\infty} \frac{x e^{-x^2/2}}{\pi \sqrt{x^2 - v^2}} \, dx & -\infty < v < 0 \\ \displaystyle\int_{v}^{\infty} \frac{x e^{-x^2/2}}{\pi \sqrt{x^2 - v^2}} \, dx & 0 < v < \infty. \end{cases}$$

While this form in not easily recognizable as the PDF for the normal distribution, it is mathematically equivalent to the more standard

$$f_V(v) = \frac{1}{\sqrt{2\pi}} e^{-v^2/2} \qquad -\infty < v < \infty.$$

6.4 Extensions

We consider two extensions to the Product procedure: (*a*) adapting the algorithm to automate finding the distribution of sums of random variables by using the Product and Transform procedures together, and (*b*) placing a loop around the Product code to compute the distribution of the product of several iid random variables.

First, by combining the algorithm from Chap. 4 for automating the transformations of random variables, it is possible to create an algorithm for finding the distribution of the sum of independent, univariate random variables. Note that the relationship $W = \ln(e^X e^Y)$ is equivalent to $W = X + Y$, when X and Y are independent random variables. This equivalence allows us to find the distributions of e^X and e^Y, then find the distribution of the product of the two, then find the distribution of the natural logarithm of this product to yield the distribution of the sum of X and Y. So APPL includes the Convolution procedure, which is used to find the distribution of the sum of X and Y. Likewise, there is an ancillary APPL procedure Difference that uses Transform and Product to find the distribution of $X - Y$. Additionally, Transform and Product can be used together to find the distribution of X/Y. For example, the single line of APPL code

```
> Convolution(UniformRV(0, 1), UniformRV(0, 1));
```

calculates the distribution of the sum of two independent $U(0,1)$ random variables in the fashion described above.

Second, applications might arise when it is of interest to find the distribution of the product of several iid random variables. For instance, if W is the product of eight $U(0,1)$ random variables, the distribution of W can be found using the following APPL code:

```
> X := UniformRV(0, 1);
> Y := Product(X, X);
> Z := Product(Y, Y);
> W := Product(Z, Z);
```

The program is clearer, and more easily generalized if a Maple for loop is used:

```
> X := UniformRV(0, 1);
> W := Product(X, X);
> for i from 3 to 8 do
>    W := Product(W, X);
> od;
```

APPL contains a procedure named ProductIID which automates this process, giving a more readable code segment than the two segments given above:

```
> X := UniformRV(0, 1);
> W := ProductIID(X, 8);
```

or, equivalently, but more succinctly,

```
> W := ProductIID(UniformRV(0, 1), 8);
```

Since the Product procedure is fairly short, we end this chapter by presenting the Product algorithm in pseudocode.

6.5 Algorithm

This section gives the details associated with the `Product` algorithm, which computes the PDF of the product of two continuous, independent random variables. Cardinality is denoted by $\|\cdot\|$ and the empty list is denoted by []. Indentation is used to indicate loops and conditions. Comments are set flushright. To allow subscripting, the PDFs $f_X(x)$, $f_Y(y)$, and $f_V(v)$ are replaced by $f(x)$, $g(y)$, and $h(v)$. The conditions prior to the calculation of the integrals are imposed to prevent calculating integrals unnecessarily. For brevity, the pseudocode presented here considers only the calculation of the probability distribution of the product of two random variables whose supports are both positive. Handling the more general cases is analogous.

Procedure `Product`: Find the PDF of the product of X and Y.

Input: The distribution of the random variable X and the distribution of the random variable Y, where X and Y are independent, univariate, continuous random variables. The PDFs are in the list-of-sublists format, i.e., the random variable X is represented by the data structure $[\tilde{f}(x), x^\star,$ ["Continuous", "PDF"]], and the random variable Y is represented by the data structure $[\tilde{g}(y), y^\star,$ ["Continuous", "PDF"]]. The first sublist for X is a list of the piecewise $f(x)$ functions, that is, $\tilde{f}(x) = [f_1(x), f_2(x), \ldots, f_n(x)]$. The second sublist for X is a list of the endpoints of the support of X, that is, $x^\star = [x_1^\star, x_2^\star, \ldots, x_{n+1}^\star]$, where $f(x) = f_i(x)$ for $x_i^\star < x < x_{i+1}^\star$, $i = 1, 2, \ldots, n$. The same format applies to defining the m segments that define the PDF of Y, i.e., $g(y)$.

Output: The PDF of $V = XY$, $h(v)$, in the same list-of-sublists format as the input. The list $\tilde{h}(v) = [h_1(v), h_2(v), \ldots, h_l(v)]$ contains the piecewise elements of the PDF of V and $v^\star = [v_1^\star, v_2^\star, \ldots, v_{l+1}^\star]$ contains the corresponding breakpoints of the support of V. The PDF is determined by a special case of Rohatgi's result for two independent random variables:

$$h(v) = \int_{-\infty}^{\infty} f(x) \cdot g\left(\frac{v}{x}\right) \frac{1}{|x|}\, dx.$$

Algorithm:
$n \leftarrow \|x^\star\| - 1$ n is the number of segments of the PDF of X
$m \leftarrow \|y^\star\| - 1$ m is the number of segments of the PDF of Y
if ($x_1^\star < 0$ and $x_{n+1}^\star > 0$ and $0 \notin x^\star$) then insert 0 into x^\star if necessary
 for $i \leftarrow 1$ to n
 if ($x_i^\star < 0$ and $x_{i+1}^\star > 0$) then
 insert 0 between positions x_i^\star and x_{i+1}^\star
 insert $f_i(x)$ after element i in the list $\tilde{f}(x)$
 $n \leftarrow n + 1$
 break

if ($y_1^\star < 0$ and $y_{m+1}^\star > 0$ and $0 \notin y^\star$) then　　　insert 0 into y^\star if necessary
　　for $i \leftarrow 1$ to m
　　　　if ($y_i^\star < 0$ and $y_{i+1}^\star > 0$) then
　　　　　　insert 0 between positions y_i^\star and y_{i+1}^\star
　　　　　　insert $g_i(y)$ after element i in the list $\tilde{g}(y)$
　　　　　　$m \leftarrow m + 1$
　　　　　　break
$v^\star \leftarrow [\,]$　　　　　　　　　　　　　　　　initialize the support list for V
for $i \leftarrow 1$ to $n + 1$　　　　　　　　　　　build the support list v^\star
　　for $j \leftarrow 1$ to $m + 1$
　　　　$v^\star \leftarrow v^\star \bigcup \left(x_i^\star \cdot y_j^\star \right)$
$v^\star \leftarrow \mathrm{sort}(v^\star)$　　　　　　　　order the endpoints of the support of V
$l \leftarrow \|v^\star\| - 1$　　　　　　　　l is the number of segments of the PDF of V
for $i \leftarrow 1$ to l　　　　　　　　　　initialize PDF segments of V
　　$h_i(v) \leftarrow 0$
for $i \leftarrow 1$ to n　　　　　　　loop through each piecewise segment of X
　　for $j \leftarrow 1$ to m　　　　　loop through each piecewise segment of Y
　　　　$a \leftarrow x_i^\star$　　　　　　　define corners of the rectangular region
　　　　$b \leftarrow x_{i+1}^\star$
　　　　$c \leftarrow y_j^\star$
　　　　$d \leftarrow y_{j+1}^\star$
　　　　if ($a \geq 0$ and $c \geq 0$) then　　　　the region is in the first quadrant
　　　　　　$p(v) \leftarrow \int_a^b f_i(x) \cdot g_j(\frac{v}{x}) \cdot \frac{1}{x}\, dx$　　　　compute required integrals
　　　　　　if ($d < \infty$) then $q(v) \leftarrow \int_{v/d}^b f_i(x) \cdot g_j(\frac{v}{x}) \cdot \frac{1}{x}\, dx$
　　　　　　if ($c > 0$) then $r(v) \leftarrow \int_a^{v/c} f_i(x) \cdot g_j(\frac{v}{x}) \cdot \frac{1}{x}\, dx$
　　　　　　if ($c > 0$ and $d < \infty$ and $ad < bc$) then
　　　　　　　　$s(v) \leftarrow \int_{v/d}^{v/c} f_i(x) \cdot g_j(\frac{v}{x}) \cdot \frac{1}{x}\, dx$
　　　　　　if ($c = 0$ and $d = \infty$) then　　　　Scenario A from Section 6.2
　　　　　　　　for $k \leftarrow 1$ to l
　　　　　　　　　　if ($v_k^\star \geq 0$) then $h_k(v) \leftarrow h_k(v) + p(v)$
　　　　　　if ($c = 0$ and $d < \infty$) then　　　　Scenario B from Section 6.2
　　　　　　　　for $k \leftarrow 1$ to l
　　　　　　　　　　if ($v_k^\star \geq 0$ and $v_{k+1}^\star \leq ad$) then $h_k(v) \leftarrow h_k(v) + p(v)$
　　　　　　　　　　if ($v_k^\star \geq ad$ and $v_{k+1}^\star \leq bd$) then $h_k(v) \leftarrow h_k(v) + q(v)$
　　　　　　if ($c > 0$ and $d = \infty$) then　　　　Scenario C from Section 6.2
　　　　　　　　for $k \leftarrow 1$ to l
　　　　　　　　　　if ($v_k^\star \geq bc$) then $h_k(v) \leftarrow h_k(v) + p(v)$
　　　　　　　　　　if ($v_k^\star \geq ac$ and $v_{k+1}^\star \leq bc$) then $h_k(v) \leftarrow h_k(v) + r(v)$
　　　　　　if ($c > 0$ and $d < \infty$) then　　　　Scenario D from Section 6.2
　　　　　　　　if ($ad < bc$) then　　　　　　　　　　Case I
　　　　　　　　　　for $k \leftarrow 1$ to l
　　　　　　　　　　　　if ($v_k^\star \geq ac$ and $v_{k+1}^\star \leq ad$) then $h_k(v) \leftarrow h_k(v) + r(v)$
　　　　　　　　　　　　if ($v_k^\star \geq ad$ and $v_{k+1}^\star \leq bc$) then $h_k(v) \leftarrow h_k(v) + s(v)$

$$\text{if } (v_k^\star \geq bc \text{ and } v_{k+1}^\star \leq bd) \text{ then } h_k(v) \leftarrow h_k(v) + q(v)$$

if $(ad = bc)$ then Case II

for $k \leftarrow 1$ to l

$$\text{if } (v_k^\star \geq ac \text{ and } v_{k+1}^\star \leq ad) \text{ then } h_k(v) \leftarrow h_k(v) + r(v)$$

$$\text{if } (v_k^\star \geq bc \text{ and } v_{k+1}^\star \leq bd) \text{ then } h_k(v) \leftarrow h_k(v) + q(v)$$

if $(ad > bc)$ then Case III

for $k \leftarrow 1$ to l

$$\text{if } (v_k^\star \geq ac \text{ and } v_{k+1}^\star \leq bc) \text{ then } h_k(v) \leftarrow h_k(v) + r(v)$$

$$\text{if } (v_k^\star \geq bc \text{ and } v_{k+1}^\star \leq ad) \text{ then } h_k(v) \leftarrow h_k(v) + p(v)$$

$$\text{if } (v_k^\star \geq ad \text{ and } v_{k+1}^\star \leq bd) \text{ then } h_k(v) \leftarrow h_k(v) + q(v)$$

analogous steps here for rectangles in quadrants II, III, and IV

remove any redundancies in $f_V(v)$ due to 0 in the support list

return($[\bar{h}(v), v^*, [$"Continuous", "PDF"$]]$)

In conclusion, an algorithm for calculating the PDF of the product of two independent random variables X and Y, as detailed in Glen et al. [39], has been developed and implemented. Special care has been taken to allow for random variables that are defined in a piecewise manner or have 0 as a part of their support. The algorithm can be extended to calculate the PDF of sums of independent random variables and to products of iid random variables. Although its presentation in this chapter has been solely for continuous random variables, the algorithm is also able to find the distribution of the product of two discrete random variables. This concludes our introduction to continuous random variables. The next three chapters are devoted to discrete random variables.

Part III

Algorithms for Discrete Random Variables

Now in building of chaises, I tell you what,
There is always somewhere a weakest spot,–
In hub, tire, felloe, in spring or thill,
In panel, or crossbar, or floor, or sill,
Above or below, or within or without,–
And that's the reason, beyond a doubt,
A chaise breaks down, but doesn't wear out.

—Oliver Wendall Holmes, "The Deacon's Masterpiece"

7

Data Structures and Simple Algorithms

This chapter introduces the data structures necessary to define a discrete random variable in APPL and surveys some simple algorithms associated with discrete random variables. The first section will show that the nature of the support of discrete random variables makes the data structures required much more complicated than for continuous random variables.

7.1 Data Structures

Recall from Chap. 3 that the format for describing a continuous random variable is a Maple list consisting of three sublists:

$$[[f(x)], [support], ["\texttt{Continuous}", "\texttt{XXX}"]]$$

where XXX is PDF, CDF, SF, HF, CHF, or IDF. It might seem like a trivial matter to simply modify the list-of-sublists to

$$[[f(x)], [support], ["\texttt{Discrete}", "\texttt{XXX}"]]$$

but there is a complication. The problem lies in the second sublist—the support. For example, let X denote the number of heads in two tosses of a fair coin. The binomial distribution with $n = 2$ and $p = 1/2$ is an appropriate probability model for X. The APPL statement

```
> X := [[1 / 4, 1 / 2, 1 / 4], [0, 1, 2], ["Discrete", "PDF"]];
```

defines X using the above format for a discrete random variable. But what if $n = 1000$? Keying 1001 support values into the second sublist is impractical. The data structure must provide a way to automate the input when the support contains many values. Furthermore, what if we are interested in finding the distribution of $Y = X^2$ by using the Transform procedure? There also needs to be a mechanism available for easily encoding the support $0, 1, 4, 9, \ldots, n^2$.

© Springer International Publishing AG 2017 89
J.H. Drew et al., *Computational Probability*, International Series
in Operations Research & Management Science 246,
DOI 10.1007/978-3-319-43323-3_7

The acronyms that identify the functional form that defines the distribution of a discrete random variable X in the second element of the third sublist are the same as in the continuous case (log denotes natural logarithm):

- probability density function (PDF) $f_X(x) = \Pr(X = x)$
- cumulative distribution function (CDF) $F_X(x) = \sum_{s \leq x} f_X(s)$,
- survivor function (SF) $S_X(x) = \sum_{s \geq x} f_X(s)$,
- hazard function (HF) $h_X(x) = f_X(x)/S_X(x)$,
- cumulative hazard function (CHF) $H_X(x) = -\log\bigl(S_X(x)\bigr)$, and
- inverse distribution function (IDF) $F_X^{-1}(u)$.

The probability density function is also commonly referred to as the *probability mass function* for discrete random variables.

The common data structure used in APPL is again referred to as a "list-of-sublists." Each APPL random variable is input as a list that contains three sublists, each with a specific purpose. The first sublist contains either a function or a numeric list that defines one of the six functional forms that describe the distribution. For example, the PDF form of the Poisson distribution with mean λ and support $x = 0, 1, 2, \ldots$ has as its first sublist

$$\left[x \to \frac{\lambda^x e^{-\lambda}}{x!} \right].$$

The CDF form of the geometric$(1/4)$ distribution with support $x = 1, 2, \ldots,$ has as its first sublist

$$\left[x \to 1 - \left(\frac{3}{4} \right)^x \right].$$

The SF form (in a numeric Maple list) of the probability of obtaining an x or higher, where $x = 1, 2, \ldots, 6$, on a single roll of a fair 6-sided die is

$$\left[1, \frac{5}{6}, \frac{2}{3}, \frac{1}{2}, \frac{1}{3}, \frac{1}{6} \right].$$

The second sublist contains the random variable's support. Recall from Chap. 3 that for a continuous random variable, this second sublist contains an ordered list of real numbers that delineate the end points of the intervals for the functions in the first sublist. The end point of each interval is automatically the starting point of the subsequent interval. A discrete random variable's support, on the other hand, can be input in one of several different formats. This variation in formats presents greater difficulty than in the continuous case for determining a structure for the second sublist. For example, the Poisson distribution with PDF

$$f(x) = \frac{\lambda^x e^{-\lambda}}{x!} \qquad x = 0, 1, 2, \ldots$$

has as its support the nonnegative integers. This support has a pattern to it; the first value of x at which the PDF is defined is zero and the rest of the support consists of subsequent integers. Many discrete random variables do not have a patterned support. When designing the second sublist structure for discrete random variables, we first had to distinguish between random variables that had some type of pattern to their support versus those that did not. For example, let X be a binomial random variable with parameters $n = 5$ and $p = 0.2$, and associated PDF

$$f(x) = \binom{5}{x} (0.2)^x (0.8)^{5-x} \qquad x = 0, 1, \ldots, 5.$$

The support of the random variable X consists of consecutive integers. For such random variables, only the first and last values of the support are needed to generate the entire support list. We call this support case the *Dot* case, since we can write the support in Maple's range (also called type '..') format: $\min\{\Omega\}$.. $\max\{\Omega\}$, where Ω is the support. Thus, the second sublist for the random variable X is input as [0 .. 5].

Switching to the *NoDot* case, consider the random variable Y with PDF

$$f(y) = \begin{cases} 0.2 & y = 1 \\ 0.5 & y = 7/2 \\ 0.3 & y = 11. \end{cases}$$

Random variables with supports that display no pattern, such as the support of Y, must be entered as a Maple list, where the list elements are separated by commas. We call this support case the *NoDot* case. The support values listed in the second sublist correspond to the distribution's function values in the first sublist. We input the first and second sublists of Y as [0.2, 0.5, 0.3] and [1, 7 / 2, 11], respectively.

After distinguishing between the *Dot* and *NoDot* support cases, there are subcases of these two general cases to consider. First, in the *NoDot* case, the function in the first sublist can be written as a formula or a list of numeric elements. The random variable Y with PDF values [0.2, 0.5, 0.3] is a Maple list of numeric elements (separated by commas). On the other hand, the random variable X with PDF

$$f(x) = \frac{x}{27} \qquad x = 1, 3, 7, 16,$$

is a valid discrete probability mass function whose PDF can be written as a formula in Maple's function notation as $x \rightarrow x/27$. Its first two sublists are input in Maple as [x -> x / 27] and [1, 3, 7, 16]. APPL allows the user to enter a discrete random variable represented in the *NoDot* case in either format, numeric or formulaic. Converting the first sublist from the formulaic *NoDot* format to a "standard" *NoDot* format (where the first sublist is *not* written as a formula) is handled in a procedure called Convert. If we enter X as displayed above and apply the Convert procedure to X with the APPL statements

```
> X := [[x -> x / 27], [1, 3, 7, 16], ["Discrete", "PDF"]];
> X := Convert(X);
```

the random variable X is returned as the list

$$\left[\left[\frac{1}{27}, \frac{1}{9}, \frac{7}{27}, \frac{16}{27}\right], [1, 3, 7, 16], [\text{"Discrete"}, \text{"PDF"}]\right]$$

More examples of the `Convert` procedure are given subsequently.

The *Dot* case also has two subcases: either the random variable has finite or infinite support. Let Ω represent the support of a random variable X. The "standard" *Dot* format for defining the support has the general form

$$[\min\{\Omega\} \; .. \; \max\{\Omega\}, \, k, \, x \to g(x)]$$

where the default value of the support increment k (if not entered) is 1 and the default value of support transformation function $g(x)$ (if not entered) is the identity function.

To introduce the different variations in format, let $X_1 \sim$ geometric$(1/4)$ with PDF

$$f_{X_1}(x) = \frac{1}{4}\left(\frac{3}{4}\right)^{x-1} \qquad x = 1, 2, \ldots .$$

Let $X_2 = 2Y$, where $Y \sim$ binomial$(5, 0.2)$. The PDF of X_2 is

$$f_{X_2}(x) = \binom{5}{x/2}(0.2)^{x/2}(0.8)^{5-x/2} \qquad x = 0, 2, 4, \ldots, 10.$$

Let X_3 be a discrete random variable with PDF

$$f_{X_3}(x) = \frac{x+6}{54} \qquad x = 1, 4, 9, 16.$$

Let X_4 be a discrete random variable with PDF

$$f_{X_4}(x) = \frac{\sqrt{x}}{15} \qquad x = 9, 25, 49.$$

The support of each of these random variables has a pattern that can be accounted for in the *Dot* data structure. X_1's support is incremented by ones, while X_2's support is incremented by twos. The support of X_3 is not only incremented by ones, but is also transformed by the function $g(x) = x^2$. Going a step further, X_4's support is incremented by twos starting with $x = 3$ and then transformed by $g(x) = x^2$. In each of these cases, the minimum and maximum values of the support, the increment of the support, and the type of transformation applied to the support is pertinent information. Thus, in the *Dot* case, the second sublist will consist of either 1, 2, or 3 elements separated by commas containing this information. This general capability has been included in the data structure to accommodate discrete algorithmic procedures, such as `Transform`.

The supports of the random variables X_1, X_2, X_3, and X_4 are input in the second sublist in the APPL list-of-sublists as

$$X_1 : [1 \; .. \; \infty]$$
$$X_2 : [0 \; .. \; 10, \; 2]$$
$$X_3 : [1 \; .. \; 4, \; x \to x^2]$$
$$X_4 : [3 \; .. \; 7, \; 2, \; x \to x^2]$$

In summary, if the random variable X is discrete, its support will match one of the cases displayed in Table 7.1. An example of the APPL list-of-sublists format of a random variable from each of these cases is given in the outline that immediately follows Table 7.1.

Table 7.1. Discrete random variable support categories for various types of support, support increments k, and support transformation functions $g(x)$

Support Ω	Cases	Subcases		Examples
NoDot $\Omega = [x_1, x_2, \dots, x_n]$	Numeric PDF			7.1
	Formulaic PDF			7.2
Dot $\Omega = [\min\{\Omega\} .. \max\{\Omega\},$ $k, \; x \to g(x)]$	Finite support	$k = 1$	$g(x) = x$	7.3
			$g(x) \neq x$	7.4
		$k \neq 1$	$g(x) = x$	7.5
			$g(x) \neq x$	7.6
	Infinite support	$k = 1$	$g(x) = x$	7.7
			$g(x) \neq x$	7.8
		$k \neq 1$	$g(x) = x$	7.9
			$g(x) \neq x$	7.10

1. *NoDot* SUPPORT FORMAT: The random variable's support Ω is a Maple numeric list of elements.

 (a) NUMERIC PDF: The random variable's PDF is a Maple numeric list of elements.

Example 7.1. Let X be a random variable with PDF

$$f(x) = \begin{cases} 0.6 & x = 2.5 \\ 0.3 & x = 3 \\ 0.1 & x = 15. \end{cases}$$

This random variable X is input in APPL as

```
> X := [[0.6, 0.3, 0.1], [2.5, 3, 15],
          ["Discrete", "PDF"]];
```

(b) FORMULAIC PDF: The random variable's PDF is formulaic.

Example 7.2. Let X be a random variable with PDF

$$f(x) = x/8 \qquad x = 1, 3, 4.$$

This random variable X is input in APPL as

```
> X := [[x -> x / 8], [1, 3, 4], ["Discrete", "PDF"]];
```

2. *Dot* SUPPORT FORMAT: The random variable's support Ω is of Maple type "..", i.e., $\min\{\Omega\}$.. $\max\{\Omega\}$.

(a) FINITE SUPPORT: The random variable's support is finite.

(i) The random variable's support is incremented by $k = 1$ and transformed by the identity function, i.e., $g(x) = x$ for all x.

Example 7.3. Let X be a Benford random variable [9] with PDF

$$f(x) = \log_{10}\left(1 + 1/x\right) \qquad x = 1, 2, \ldots, 9.$$

The Benford random variable is input in APPL as

```
> X := [[x -> log[10](1 + 1 / x)], [1 .. 9],
          ["Discrete", "PDF"]];
```

(ii) The random variable's support is incremented by $k = 1$ and transformed by a function other than the identity function, i.e., $g(x) \neq x$ for some or all x.

Example 7.4. Let X be a random variable with PDF

$$f(x) = x/224 \qquad x = 8, 27, 64, 125.$$

This random variable is input in APPL as

```
> X := [[x -> x / 224], [2 .. 5, x -> x ^ 3],
          ["Discrete", "PDF"]];
```

(iii) The random variable's support is incremented by $k \neq 1$ and transformed by $g(x) = x$.

Example 7.5. Let $Y \sim$ binomial(5, p), and let $X = 2Y$. This random variable X is input in APPL as

```
> X :=   [[x -> 120 * p ^ (x / 2) *
                (1 - p) ^ (5 - x / 2) /
                ((5 - x / 2)! * (x / 2)!)],
         [0 .. 10, 2], ["Discrete", "PDF"]];
```

(iv) The random variable's support is incremented by $k \neq 1$ and transformed by $g(x) \neq x$.

Example 7.6. Let X be a random variable with PDF

$$f(x) = 1600/489x \qquad x = 4, 25, 64.$$

This random variable is input in APPL as

```
> X := [[x -> 1600 / (489 * x)],
        [2 .. 8, 3, x -> x ^ 2],
        ["Discrete", "PDF"]];
```

(b) INFINITE SUPPORT: The random variable's support is infinite.

(i) The random variable's support is incremented by $k = 1$ and transformed by $g(x) = x$.

Example 7.7. Let X be a negative binomial random variable with parameters $r = 3$ and $p = 1/3$. The simplified PDF of X is

$$f(x) = \frac{(x-2)(x-1)}{16}\left(\frac{2}{3}\right)^x \qquad x = 3, 4, \ldots .$$

This random variable is input in APPL as

```
> X := [[x -> (2 / 3) ^ x * (x - 2) * (x - 1) / 16],
        [3 .. infinity], ["Discrete", "PDF"]];
```

(ii) The random variable's support is incremented by $k = 1$ and transformed by $g(x) \neq x$.

Example 7.8. Let X be the square of a geometric$\left(\frac{1}{2}\right)$ random variable; i.e.,

$$f(x) = 1/2^{\sqrt{x}} \qquad x = 1, 4, 9, \ldots .$$

This random variable is input in APPL as

```
> X := [[x -> 2 ^ (-sqrt(x))],
        [1 .. infinity, x -> x ^ 2],
        ["Discrete", "PDF"]];
```

(iii) The random variable's support is incremented by $k \neq 1$ and transformed by $g(x) = x$.

Example 7.9. Let $Y \sim \text{geometric}\left(\frac{1}{2}\right)$. Let $X = 2Y$, which has PDF

$$f(x) = \left(\frac{1}{2}\right)^{x/2} \qquad x = 2, 4, 6, \ldots.$$

This random variable is input in APPL as

```
> X := [[x -> (1 / 2) ^ (x / 2)], [2 .. infinity, 2],
        ["Discrete", "PDF"]];
```

(iv) The random variable's support is incremented by $k \neq 1$ and transformed by $g(x) \neq x$.

Example 7.10. Let $Y \sim \text{geometric}\left(\frac{1}{2}\right)$ and $X = 4Y^2$. The PDF of X is

$$f(x) = \left(\frac{1}{2}\right)^{\sqrt{x}/2} \qquad x = 4, 16, 36, \ldots.$$

This random variable is input in APPL as

```
> X := [[x -> (1 / 2) ^ (sqrt(x) / 2)],
        [2 .. infinity, 2, x -> x ^ 2],
        ["Discrete", "PDF"]];
```

Whenever an APPL procedure is invoked to perform an operation on a discrete random variable, the random variable is first converted to its standard format within the procedure. As discussed in the preceding section, there is a standard discrete data structure format for both the *NoDot* and *Dot* cases. The conversion is necessary so that the main body of each procedure can avoid the complexity resulting from multiple formats.

Let X, for example, be the random variable discussed earlier in this chapter with PDF

$$f(x) = \frac{x}{27} \qquad x = 1, 3, 7, 16.$$

Suppose the random variable X is input in APPL as

```
> X := [[x -> x / 27], [1, 3, 7, 16], ["Discrete", "PDF"]];
```

When an APPL procedure, such as `Mean`, receives the random variable X as an argument, it first determines if X has a *Dot* or *NoDot* support format. Since the support $\Omega = [1, 3, 7, 16]$ has a *NoDot* support format, the procedure's formula for computing the mean expects that it is receiving the random variable X in its converted format, which is

$$\left[\left[\frac{1}{27}, \frac{1}{9}, \frac{7}{27}, \frac{16}{27} \right], [1, 3, 7, 16], [\text{"Discrete"}, \text{"PDF"}] \right]$$

In this format, the mean of a random variable with support values x_1, x_2, \ldots with a *NoDot* support format is computed as

$$E[X] = \sum_{i=1}^{|\Omega|} x_i \cdot f(x_i)$$

$$= \sum_{i=1}^{4} X[2][i] \cdot X[1][i]$$

$$= 1 \cdot \frac{1}{27} + 3 \cdot \frac{1}{9} + 7 \cdot \frac{7}{27} + 16 \cdot \frac{16}{27}$$

$$= \frac{35}{3},$$

where `X[2][i]` is the ith element in the second sublist (X's support) and `X[1][i]` is the ith element in the first sublist (X's probability values).

Again, each APPL procedure operates on random variables only in the standard *NoDot* and *Dot* formats. Each APPL procedure internally converts arguments to one of the two pre-defined formats, which allows algorithms to be developed that exploit these formats. Without these standard formats, an APPL procedure would be forced to discern the exact form of each random variable it encountered. The computations required in the procedure would then depend on the unique structure of each random variable, and as indicated in Table 7.1, there are ten acceptable formats for a discrete random variable in APPL. The `Convert` procedure provides each APPL procedure with a standard structural format for a discrete random variable.

Some random variables with finite support, such as the random variable Y with PDF

$$f(y) = \frac{y}{15} \qquad y = 3, 5, 7,$$

can be input in APPL in several different formats, which include both a *NoDot* and *Dot* format. The APPL statements

```
> Y := [[1 / 5, 1 / 3, 7 / 15], [3, 5, 7], ["Discrete", "PDF"]];
> Y := [[y -> y / 15], [3, 5, 7], ["Discrete", "PDF"]];
> Y := [[y -> y / 15], [3 .. 7, 2], ["Discrete", "PDF"]];
```

define the same random variable Y in APPL. The first two lines define Y in its *NoDot* formats, while the third line displays Y in one possible *Dot* format.

Although Y may be defined by using any of the several different formats, any algorithm or formula applied to Y will have the same outcome. Conversely, a discrete random variable with countably infinite support can only be input using a *Dot* support format. Since it is impossible to physically list every element of a countably infinite set, the *NoDot* support format for this type of random variable cannot be used.

The `Convert` procedure acts on discrete random variables in both the *Dot* and *NoDot* cases. It converts a discrete random variable X with a

- *Dot* support format to the standard APPL *Dot* support format. If Ω is the support of X, then the standard *Dot* second sublist support format is

$$[\min\{\Omega\} \; .. \; \max\{\Omega\}, \, k, \, x \to g(x)]$$

 where k is the support increment and $g(x)$ is the support transformation function.
- *NoDot* support format to the standard APPL *NoDot* support format with corresponding PDF. If $f(x)$ is the PDF and the support is $x = x_1, x_2, \ldots, x_n$, then the PDF first sublist and the standard *NoDot* second sublist support format are

$$[f(x_1), f(x_2), \ldots, f(x_n)], [x_1, x_2, \ldots, x_n]$$

Two examples of the `Convert` procedure for the *NoDot* and *Dot* formats are included for further clarification.

Example 7.11. Use the APPL `Convert` procedure to convert a Zipf random variable [106, page 182] with parameter $\alpha = 1$ to the standard *Dot* format.

The APPL statements below define X as the desired Zipf random variable and convert it to its standard *Dot* format.

```
> X := [[x -> 6 / (Pi * x) ^ 2], [1 .. infinity],
        ["Discrete", "PDF"]];
> Convert(X);
```

The converted Zipf random variable in the standard APPL *Dot* format is

$$\left[\left[x \to \frac{6}{(\pi x)^2} \right], [1 .. \infty, 1, x \to x], [\text{``Discrete''}, \text{``PDF''}] \right]$$

Example 7.12. Use the APPL `Convert` procedure to convert a Benford random variable (see Example 7.3) to the standard *NoDot* format.

The APPL statements below define X as a Benford random variable and convert it to its standard *NoDot* format.

```
> X := [[x -> log[10](1 + 1 / x)],
        [1, 2, 3, 4, 5, 6, 7, 8, 9],
        ["Discrete", "PDF"]];
> Convert(X);
```

The converted Benford random variable in the standard APPL *NoDot* format is

$$\left[\left[\frac{\ln(2)}{\ln(10)}, \frac{\ln(\frac{3}{2})}{\ln(10)}, \frac{\ln(\frac{4}{3})}{\ln(10)}, \frac{\ln(\frac{5}{4})}{\ln(10)}, \frac{\ln(\frac{6}{5})}{\ln(10)}, \frac{\ln(\frac{7}{6})}{\ln(10)}, \frac{\ln(\frac{8}{7})}{\ln(10)}, \frac{\ln(\frac{9}{8})}{\ln(10)}, \frac{\ln(\frac{10}{9})}{\ln(10)}\right],\right.$$

$$\left.[1, 2, 3, 4, 5, 6, 7, 8, 9], [\text{"Discrete"}, \text{"PDF"}]\right]$$

As in the case of continuous random variables, there are several commonly-used random variables that are pre-defined and automated for input into APPL: *BenfordRV, BernoulliRV, BingoCoverRV, BinomialRV, BirthdayRV, GeometricRV, HypergeometricRV, NegativeBinomialRV, PoissonRV, UniformDiscreteRV*, and *ZipfRV*. This list of distributions, along with their parameters, is displayed every time APPL is entered. A binomial random variable X with $n = 3$ and $p = 1/2$, for example, can be defined by the APPL statement

```
> X := BinomialRV(3, 1 / 2);
```

which is much more compact than the standard list-of-sublists approach:

```
> X := [[1 / 8, 3 / 8, 3 / 8, 1 / 8], [0, 1, 2, 3],
        ["Discrete", "PDF"]];
```

7.2 Simple Algorithms

Now that the data structure for representing discrete random variables in APPL has been established, we present some elementary algorithms that can be applied to these random variables. Specifically, the algorithms described in this section conduct the following elementary operations:

- convert any of the six functional forms of a random variable (PDF, CDF, SF, HF, CHF, or IDF) into another functional form,
- plot any of the six functions for any fully-specified distribution, e.g., the CDF,
- determine common summary characteristics of a random variable, such as the mean and variance,
- verify the validity of a particular PDF functional form.

Changing the Functional Form of a Random Variable. The functional form of a distribution can be converted to and from its six well-known forms, the PDF, CDF, SF, HF, CHF, and IDF. As in the case of continuous random variables, the procedures return functional forms when they are called with a single argument and scalars when an optional second argument is included. The names of the conversion procedures are PDF, CDF, SF, HF, CHF, and IDF, and the procedures can be applied to both discrete and continuous random variables.

The matrix in Table 7.2 shows how the PDF, CDF, SF, HF, and CHF distribution forms (given in the columns) can be found if one of the forms (given by the rows) is known. For example, if the CDF of a distribution is given, then its CHF can be determined for $x_i \in \Omega$ by

Table 7.2. A 5×5 transition matrix for determining $f(x_i)$, $F(x_i)$, $S(x_i)$, $h(x_i)$, or $H(x_i)$ from any of the others for discrete distributions, where $F(x_0) \equiv 0$, $h(x_0) \equiv 0$, and when $|\Omega|$ is finite, $S(x_{|\Omega|+1}) \equiv 0$ and $e^{-H(x_{|\Omega|+1})} \equiv 0$

	$f(x_i)$	$F(x_i)$	$S(x_i)$	$h(x_i)$	$H(x_i)$
$f(x)$	•	$\sum\limits_{j\,\mid\,x_j \leq x_i} f(x_j)$	$\sum\limits_{j\,\mid\,x_j \geq x_i} f(x_j)$	$\dfrac{f(x_i)}{\sum_{j\,\mid\,x_j \geq x_i} f(x_j)}$	$-\log\left(\sum\limits_{j\,\mid\,x_j \geq x_i} f(x_j)\right)$
$F(x)$	$F(x_i) - F(x_{i-1})$	•	$1 - F(x_{i-1})$	$\dfrac{F(x_i) - F(x_{i-1})}{1 - F(x_{i-1})}$	$-\log(1 - F(x_{i-1}))$
$S(x)$	$S(x_i) - S(x_{i+1})$	$1 - S(x_{i+1})$	•	$\dfrac{S(x_i) - S(x_{i+1})}{S(x_i)}$	$-\log S(x_i)$
$h(x)$	$h(x_i) \prod\limits_{j\,\mid\,x_j < x_i} [1 - h(x_j)]$	$1 - \prod\limits_{j\,\mid\,x_j \leq x_i} [1 - h(x_j)]$	$\prod\limits_{j\,\mid\,x_j < x_i} [1 - h(x_j)]$	•	$-\sum\limits_{j\,\mid\,x_j < x_i} \log(1 - h(x_j))$
$H(x)$	$e^{-H(x_i)} - e^{-H(x_{i+1})}$	$1 - e^{-H(x_{i+1})}$	$e^{-H(x_i)}$	$\dfrac{e^{-H(x_i)} - e^{-H(x_{i+1})}}{e^{-H(x_i)}}$	•

$$H(x_i) = -\log\big(S(x_i)\big) = -\log\big(1 - F(x_{i-1})\big),$$

where log is the natural logarithm (log base e) and $F(x_0) \equiv 0$. Some facts used to compute the entries in the matrix for $x_i \in \Omega$ are given in Leemis [72, pages 56–57, 73]:

- $h(x_i) = f(x_i)/S(x_i)$,
- $H(x_i) = -\log S(x_i)$,
- $S(x_i) = \displaystyle\prod_{j|x_j < x_i} \big(1 - h(x_j)\big)$, and
- $H(x_j) = -\displaystyle\sum_{j|x_j < x_i} \log\big(1 - h(x_j)\big)$.

An example of how to find one distribution form when another is known using the 5×5 matrix in Table 7.2 is given next.

Example 7.13. Given the hazard function

$$h(x) = 1/4 \qquad x = 1, 2, \ldots,$$

find $f(x_i)$ for $x_i = 1, 2, 3, \ldots$.

Using the $\big(h(x), f(x_i)\big)$ matrix element in Table 7.2,

$$f(x_i) = h(x_i) \prod_{j|x_j < x_i} [1 - h(x_j)] = \frac{1}{4} \prod_{j=1}^{x_i - 1} \left[1 - \frac{1}{4}\right] = \frac{1}{4}\left(\frac{3}{4}\right)^{x_i - 1}$$

for $x_i = 1, 2, 3, \ldots$, which is a geometric distribution with $p = 1/4$. The geometric distribution is the only discrete distribution having both a constant hazard function and the memoryless property. This calculation is made in APPL with the statements

```
> X := [[x -> 1 / 4], [1 .. infinity], ["Discrete", "HF"]];
> PDF(X);
```

The algorithms for the PDF, CDF, SF, HF, and CHF procedures are similar and utilize the formulas in Table 7.2. When changing the "current form" of a random variable to a "new form," the algorithm uses the Table 7.2 matrix element ("current form", "new form").

We now delve into some detail on how the PDF procedure works.

The PDF of the discrete random variable X is defined as $f(x) = \Pr(X = x)$ for all $x \in \Omega$. The PDF(X, [x]) procedure returns either

1. the PDF of the random variable X in its APPL list-of-sublists format if only one argument is provided in the procedure call; i.e., PDF(X), or
2. the probability value $\Pr(X = x)$ if both arguments are provided in the procedure call; i.e., PDF(X, x).

Example 7.14. In order to illustrate many of the conversion operations, we use a discrete random variable X with PDF

$$f(x) = \frac{x}{6} \qquad x = 1, 2, 3.$$

The formulas for its CDF, SF, HF, CHF, and IDF are:

$$F(x) = \frac{x^2 + x}{12} \qquad x = 1, 2, 3$$

$$S(x) = \frac{-x^2 + x + 12}{12} \qquad x = 1, 2, 3$$

$$h(x) = \frac{2x}{-x^2 + x + 12} \qquad x = 1, 2, 3$$

$$H(x) = -\log\left(\frac{-x^2 + x + 12}{12}\right) \qquad x = 1, 2, 3$$

$$F^{-1}(u) = \left\lceil -\frac{1}{2} + \frac{1}{2}\sqrt{1 + 48u} \right\rceil \qquad u = 1/6, 1/2, 1.$$

If we assign the variable X to the CDF of X using the *Dot* format, then the statement PDF(X) returns the PDF of X in its APPL list-of-sublists format:

```
> X := [[x -> (x ^ 2 + x) / 12], [1 .. 3],
        ["Discrete", "CDF"]];
> PDF(X);
```

yielding the PDF

$$\left[\left[x \to \frac{x}{6}\right], [1 .. 3], [\text{"Discrete"}, \text{"PDF"}]\right]$$

Alternatively, the statements

```
> X := [[x -> (x ^ 2 + x) / 12], [1 .. 3],
        ["Discrete", "CDF"]];
> PDF(X, 2);
```

compute $\Pr(X = 2)$, which is 1/3.

The algorithm in the PDF procedure first checks a discrete random variable X to determine if it has a *Dot* or *NoDot* support format. Although two distinct branches in the algorithm have been constructed to process random variables of each format separately, the basic functional relationships displayed in Table 7.2 are used in each branch. Whether X has a *Dot* or *NoDot* support format, one of the following bullets will be executed depending on X's distribution form (and support format in the case of IDF). We assume $\Omega = \{x_1, x_2, \dots\}$, where $x_1 < x_2 < \cdots$.

- If X has a PDF form, do nothing to X.
- If X has a CDF form, determine the PDF of X with the $(F(x), f(x_i))$ matrix element in Table 7.2:

$$f(x_i) = F(x_i) - F(x_{i-1}) \qquad x_i \in \Omega,$$

where $i = 1, 2, \ldots$ and $F(x_0) \equiv 0$.
- If X has a SF form, determine the PDF of X with the $(S(x), f(x_i))$ matrix element in Table 7.2:

$$f(x_i) = S(x_i) - S(x_{i+1}) \qquad x_i \in \Omega,$$

where $i = 1, 2, \ldots$ and $S(x_{|\Omega|+1}) \equiv 0$ when $|\Omega|$ is finite.
- If X has a HF form, determine the PDF of X with the $(h(x), f(x_i))$ matrix element in Table 7.2:

$$f(x_i) = h(x_i) \prod_{j | x_j < x_i} [1 - h(x_j)] \qquad x_i \in \Omega,$$

where $i = 1, 2, \ldots$ and $h(x_0) \equiv 0$.
- If X has a CHF form, determine the PDF of X with the $(H(x), f(x_i))$ matrix element in Table 7.2:

$$f(x_i) = e^{-H(x_i)} - e^{-H(x_{i+1})} \qquad x_i \in \Omega,$$

where $i = 1, 2, \ldots$ and $e^{-H(x_{|\Omega|+1})} \equiv 0$ when $|\Omega|$ is finite.
- If X has an IDF form with *Dot* support, then

 1. If there is a ceiling term as part of the inverse function, extract the expression under the ceiling in order to solve $F^{-1}(x) = y$. Referring back to X with PDF $f(x) = x/6$ for $x = 1, 2, 3$, the IDF of X (from Example 7.14) is

 $$F^{-1}(u) = \left\lceil -\frac{1}{2} + \frac{1}{2}\sqrt{1 + 48u} \right\rceil,$$

 for $u = 1/6, 1/2, 1$. Rewrite $F^{-1}(u)$ as

 $$F^{-1}(u) = -\frac{1}{2} + \frac{1}{2}\sqrt{1 + 48u}.$$

 This removal of the ceiling function can be done using the Maple `eval` procedure.

 2. Solve the equation

 $$F^{-1}(x) = y$$

 for x, yielding $F(y)$. For our particular X, solving $-\frac{1}{2} + \frac{1}{2}\sqrt{1 + 48x} = y$ for x yields $F(x) = \frac{x^2 + x}{12}$, replacing y with x.

3. Find the *appropriate* inverse. If there is more than one solution to the equation $F^{-1}(x) = y$, determine the correct inverse by testing which solution $F(x)$ correctly calculates $F\big(F^{-1}(s)\big) = s$ for s equal to one of the following. The choices for s are considered in the bulleted order.

 - if $\min\{\Omega\} = -\infty$ and $\max\{\Omega\} = \infty$, then $s = 1$.
 - if $\max\{\Omega\} \neq \infty$, then $s = \max\{\Omega\}$.
 - if $\min\{\Omega\} \neq -\infty$, then $s = \min\{\Omega\}$.
 - otherwise, $s = \dfrac{\min\{\Omega\} + \max\{\Omega\}}{2}$.

 See the beginning of Sect. 4.2 for the continuous analog of these conditions.

4. Convert the CDF of X to its PDF form by the using the $\big(F(x), f(x_i)\big)$ matrix element in Table 7.2.

5. Determine the range of the PDF of X. The minimum support value is $F^{-1}(x_1)$ (where x_1 is the minimum support value of the IDF here) and the maximum support value is $F^{-1}(1)$. Since the minimum support value of the IDF for the example random variable X is $x_1 = 1/6$, then $F^{-1}(1/6) = 1$. Also, $F^{-1}(1) = 3$.

- If X has an IDF form and a *NoDot* support, then the second sublist of the APPL list-of-sublists contains the CDF values of X and the first sublist contains the CDF support values. To compute the PDF of X, the first and second sublists of the IDF are swapped to form the CDF, and then the PDF is computed using the $\big(F(x), f(x_i)\big)$ matrix element in Table 7.2.

The IDF of our example random variable X is

$$F^{-1}(x) = \begin{cases} 1 & x = \frac{1}{6} \\ 2 & x = \frac{1}{2} \\ 3 & x = 1. \end{cases}$$

The APPL list-of-sublists is

$$\left[[1, 2, 3], \left[\frac{1}{6}, \frac{1}{2}, 1 \right], [\text{``Discrete''}, \text{``IDF''}] \right]$$

The CDF is found by swapping the first two sublists:

$$\left[\left[\frac{1}{6}, \frac{1}{2}, 1 \right], [1, 2, 3], [\text{``Discrete''}, \text{``CDF''}] \right]$$

Finally, the PDF is

$$f(1) = F(1) = \tfrac{1}{6},$$
$$f(2) = F(2) - F(1) = \tfrac{1}{3},$$
$$f(3) = F(3) - F(2) = \tfrac{1}{2}.$$

After the PDF has been computed, the algorithm checks whether one or two arguments were entered. If only one argument, X, is provided, i.e., PDF(X), then the PDF of X is returned in its list-of-sublists format. Otherwise, if two arguments, X and x, are provided, then the probability value $\Pr(X = x)$ is computed and returned. The algorithm distinguishes between the two support formats when determining this probability value:

- If X has the *Dot* support format, then

 1. If $x_i \in \Omega$, then compute and return the probability value $f(x_i)$ using the PDF formula already determined in the earlier part of the procedure.

 2. If the value $x_i \notin \Omega$, then return $\Pr(X = x_i) = 0$.

- If X has the *NoDot* support format, then

 1. Loop through the support values of the PDF searching for the target value x_i in the second sublist. If x_i is found in position j, then its corresponding probability value is in position j of the first sublist. Return the jth element in the first sublist.

 2. If the target value x_i is not found in the second sublist, then return $\Pr(X = x_i) = 0$.

Example 7.15 (Adapted from Hogg et al. [52, page 44]). Cast a die two independent times and let X equal the absolute value of the difference of the two resulting values. The CDF of X is

$$
F(x) = \begin{cases} \dfrac{1}{6} & x = 0 \\[2ex] \dfrac{11x - x^2 + 6}{36} & x = 1, 2, \ldots, 5. \end{cases}
$$

Determine the PDF of X.

Define the APPL *NoDot* format of $F(x)$ as the random variable X. Then use the PDF procedure to determine X's numeric PDF.

```
> X := [[1 / 6, 4 / 9, 2 / 3, 5 / 6, 17 / 18, 1],
        [0, 1, 2, 3, 4, 5], ["Discrete", "CDF"]];
> PDF(X);
```

The APPL *NoDot* format of $f(x)$ is

$$
\left[\left[\frac{1}{6}, \frac{5}{18}, \frac{2}{9}, \frac{1}{6}, \frac{1}{9}, \frac{1}{18} \right], [0, 1, 2, 3, 4, 5], [\text{"Discrete"}, \text{"PDF"}] \right]
$$

which can be converted by hand to the formulaic PDF:

$$f(x) = \begin{cases} \dfrac{1}{6} & x = 0 \\[2ex] \dfrac{6-x}{18} & x = 1, 2, \ldots, 5. \end{cases}$$

Plotting Distributions. Especially for newcomers to probability theory, plots of the various functional forms of a random variable can provide insight into the distribution of that random variable. The `PlotDist` procedure requires only one argument, a random variable X, but two additional optional arguments can be provided to indicate the plotting range, `low` and `high`. The procedure can plot a random variable in any of its six functional forms. The algorithm employed by the `PlotDist` procedure utilizes the list-of-sublists data structure to glean the necessary information for a plot. For discrete random variables, graphs for CDFs, SFs, CHFs, and IDFs, for instance, are plots of step functions with jumps at the support values specified in the second sublist. Open and closed circles at the ends of the steps indicate exclusion and inclusion, respectively, of support values.

Example 7.16 (Karian and Tanis [58, page 40]). Let

$$f(x) = (5-x)/10 \qquad x = 1, 2, 3, 4$$

be a PDF of a discrete random variable X. Use `PlotDist` to depict the CDF of X.

Instead of first converting X's PDF form into a CDF form by hand, one can directly define the PDF and compose the CDF conversion with `PlotDist`. This leaves the conversion work to APPL. The plot of the desired CDF can be executed with the statements

```
> X := [[x -> (5 - x) / 10], [1 .. 4], ["Discrete", "PDF"]];
> PlotDist(CDF(X));
```

The corresponding plot is displayed in Fig. 7.1.

Calculating Expected Values. If X is a discrete random variable with support $\Omega = \{x_1, x_2, \ldots\}$ and PDF $f(x)$, then the mean (or expected value) of X is defined by

$$\mu = E[X] = \sum_{\text{all } x_i \in \Omega} x_i \cdot f(x_i).$$

More generally, if $g(X)$ is any function of X, such as $g(X) = X^2$, then

$$E[g(X)] = \sum_{\text{all } x_i \in \Omega} g(x_i) \cdot f(x_i).$$

The `ExpectedValue` procedure requires one argument, a random variable X, and can be given an additional argument, the function `g(X)`. When only one

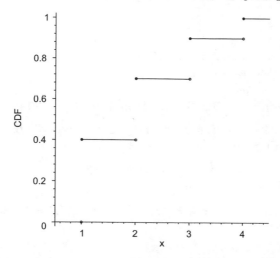

Fig. 7.1. CDF for $f(x) = (5 - x)/10$ for $x = 1, 2, 3, 4$

argument is provided for the `ExpectedValue` procedure, i.e., `ExpectedValue`
`(X)`, the mean of the random variable `X` is returned. When $g(x)$ is also
given, the `ExpectedValue` procedure calculates the mean of $g(X)$. By de-
fault, the procedure assumes $g(X) = X$ when no second argument is given.
The `Mean`, `Variance`, `Kurtosis`, and `Skewness` procedures all make use of the
`ExpectedValue` procedure. This is efficient because variance, the coefficient
of skewness, γ_1, and the coefficient of kurtosis, γ_2, for a random variable X
can be calculated as

$$\text{Var}[X] = E\left[(X - \mu)^2\right] = E\left[X^2\right] - \mu^2,$$

$$\gamma_1 = E\left[\left(\frac{X - \mu}{\sigma}\right)^3\right] = \frac{E\left[X^3\right] - 3\mu E\left[X^2\right] + 2\mu^3}{(\text{Var}[X])^{3/2}}, \text{ and}$$

$$\gamma_2 = E\left[\left(\frac{X - \mu}{\sigma}\right)^4\right] = \frac{E\left[X^4\right] - 4\mu E\left[X^3\right] + 6\mu^2 E\left[X^2\right] - 3\mu^4}{(\text{Var}[X])^2}.$$

Example 7.17 (Ross [106, page 178]). Find the expected value and
the variance of the number of times one must throw a die until the
outcome one has occurred four times.

Since the random variable of interest is a negative binomial with pa-
rameters $r = 4$ and $p = 1/6$, the statements

```
> r := 4;
> p := 1 / 6;
> X := NegativeBinomialRV(r, p);
> ExpectedValue(X);
> Variance(X);
```

yield the correct results: $E[X] = 24$ and $\text{Var}[X] = 120$.

Example 7.18 (Ross [106, page 198]). Let X be a binomial random variable with parameters n and p. Show that

$$E\left[\frac{1}{X+1}\right] = \frac{1 - (1-p)^{n+1}}{(n+1)p}.$$

Since we want to compute $E\left[1/(X+1)\right]$, let $g(X) = 1/(X+1)$. The APPL statements

```
> unassign('p');
> X := BinomialRV(n, p);
> ExpectedValue(X, x -> 1 / (x + 1));
```

produce the desired expected value.

Verifying the Validity of a PDF. Let X be a discrete random variable defined on the support $\Omega = \{x_1, x_2, \dots\}$. A discrete probability PDF must satisfy the following two conditions:

$$\sum_{\text{all } x_i \in \Omega} \Pr(X = x_i) = 1 \qquad \text{and} \qquad \Pr(X = x_i) > 0 \text{ for each } x_i \in \Omega.$$

The APPL VerifyPDF procedure has one argument, a random variable X, and the procedure's purpose is to verify that X satisfies these two conditions. When the call to the procedure VerifyPDF(X) is made, either a message is returned stating that X has a valid PDF or that it does not. The following example illustrates the use of this procedure.

Example 7.19 (Karian and Tanis [58, page 62]). Verify that the probabilities of a geometric random variable with parameter p sum to 1.

The random variable X can be defined as a geometric random variable with parameter p using the pre-defined APPL random variable GeometricRV(p). When the GeometricRV procedure receives a symbolic argument (parameter), it uses the Maple assume command to correctly assume that the symbolic parameter is inclusively between 0 and 1. Thus, the statements

```
> X := GeometricRV(p);
> VerifyPDF(X);
```

generate a message to the screen indicating that X has a valid PDF.

This concludes our introduction of the data structure and some simple algorithms for discrete random variables. The next two chapters introduce nontrivial algorithms for two useful and representative operations on discrete random variables: finding the probability distribution of sums of mutually independent random variables and finding the probability distribution of order statistics.

8

Sums of Independent Discrete Random Variables

This chapter presents an algorithm for computing the PDF of the sum of two independent discrete random variables, along with an implementation of the algorithm in APPL. Some examples illustrate the utility of this algorithm.

8.1 Preliminary Examples

An important operation in probability theory and statistical inference is to calculate the distribution of the sum of two independent random variables X and Y. Applications of sums of random variables appear in many areas of mathematics, probability theory, physics, and engineering. Most texts devote the majority of their attention to sums of continuous random variables, rather than discrete random variables. We focus here on the case in which X and Y are discrete random variables with integer-valued supports that are bounded below.

There are several approaches for computing the PDF of $Z = X + Y$ in the discrete case. The event $\{X + Y = z\}$, $z \in \mathbf{Z}$, can be written as the union of the disjoint events $\{X = \zeta, Y = z - \zeta\}$, $\{X = \zeta + 1, Y = z - (\zeta + 1)\}$, ..., $\{X = z - \zeta, Y = \zeta\}$, where ζ is the minimum of the union of the support values of X and Y. The PDF $f_Z(z)$ of the convolution of the PDFs of the independent random variables X and Y (with integer supports) can be computed as

© Springer International Publishing AG 2017
J.H. Drew et al., *Computational Probability*, International Series
in Operations Research & Management Science 246,
DOI 10.1007/978-3-319-43323-3_8

$$\Pr(Z = z) = \Pr(X + Y = z)$$

$$= \sum_{k=\zeta}^{z-\zeta} \Pr(X = k, Y = z - k)$$

$$= \sum_{k=\zeta}^{z-\zeta} \Pr(X = k)\Pr(Y = z - k).$$

The following example illustrates the use of this discrete convolution formula.

Example 8.1 ([106, page 287]). If X and Y are independent Poisson random variables with respective parameters λ_1 and λ_2, compute the PDF of $Z = X + Y$.

The PDF of a Poisson random variable X with parameter λ is

$$f_X(x) = \frac{\lambda^x e^{-\lambda}}{x!} \qquad x = 0, 1, 2, \ldots .$$

Since $\zeta = \min\{0, 1, 2, \ldots\} = 0$,

$$\Pr(Z = z) = \Pr(X + Y = z)$$

$$= \sum_{k=0}^{z} \Pr(X = k, Y = z - k)$$

$$= \sum_{k=0}^{z} \Pr(X = k) \cdot \Pr(Y = z - k)$$

$$= \sum_{k=0}^{z} \frac{e^{-\lambda_1}\lambda_1^k}{k!} \cdot \frac{e^{-\lambda_2}\lambda_2^{z-k}}{(z-k)!}$$

$$= e^{-(\lambda_1+\lambda_2)} \sum_{k=0}^{z} \frac{\lambda_1^k \lambda_2^{z-k}}{k!\,(z-k)!}$$

$$= \frac{e^{-(\lambda_1+\lambda_2)}}{z!} \sum_{k=0}^{z} \frac{z!}{k!\,(z-k)!} \lambda_1^k \lambda_2^{z-k}$$

$$= \frac{(\lambda_1 + \lambda_2)^z e^{-(\lambda_1+\lambda_2)}}{z!} \qquad z = 0, 1, 2, \ldots .$$

So $Z = X + Y$ has a Poisson distribution with parameter $\lambda_1 + \lambda_2$. As will be seen subsequently, the APPL procedure `Convolution` automates this process.

If one wants to sum more than two iid random variables, then the distribution function for Z can be determined by induction [42, page 286]. Let $Z_n = X_1 + X_2 + \cdots + X_n$ be the sum of n independent random variables with common PDF $f_X(x)$ defined on the integers. Then the PDF of Z_1 is $f_X(x)$. We can write

$$Z_i = Z_{i-1} + X_i,$$

for $i = 2, 3, \ldots, n$. Thus, since we know the distribution of X_i is $f_X(x)$, for $i = 1, 2, \ldots, n$, we can find the distribution function of Z_n by induction. When summing more than two iid random variables, the APPL procedure ConvolutionIID uses this inductive process to compute the PDF of the sum.

For example, let X_1, X_2, and X_3 have PDFs $f_X(x) = 1/6$ for $x = 1, 2, \ldots, 6$. If $Z_2 = X_1 + X_2$, then the PDF $f_{Z_2}(z)$ is

$$f_{Z_2}(z) = \begin{cases} \dfrac{z-1}{36} & z = 2, 3, \ldots, 7 \\ \dfrac{13-z}{36} & z = 8, 9, \ldots, 12. \end{cases}$$

The PDF of $Z_3 = X_1 + X_2 + X_3$ is the convolution of the PDFs of Z_2 and X_3. For example,

$$\Pr(Z_3 = 4) = \Pr(Z_2 = 3) \cdot \Pr(X_3 = 1) + \Pr(Z_2 = 2) \cdot \Pr(X_3 = 2)$$
$$= \frac{2}{36} \cdot \frac{1}{6} + \frac{1}{36} \cdot \frac{1}{6}$$
$$= \frac{1}{72}.$$

Due to the mathematical intractability in implementing the discrete convolution formula, $\sum_{k=\zeta}^{z-\zeta} \Pr(X = k, Y = z - k)$, for certain non-identically distributed random variables [e.g., $X \sim \text{Poisson}(\lambda)$, $Y \sim \text{geometric}(p)$] and the inefficiency in making computations with this formula for random variables with arbitrary supports (e.g., X with support $\{-216, -57, 23, 81\}$ and Y with support $\{-1002, -15, 2, 62, 211\}$), only certain convolutions can or should be computed using this formula. For random variables with arbitrary supports, the discrete convolution formula *can* be used, but it is often inefficient because one or both of the random variables have support values ranging over a large domain of non-adjacent integer values. The following example displays the inefficiency that can be encountered by using this formula, even for random variables with only a small number of support values.

Example 8.2. Suppose X and Y are independent discrete random variables with PDFs:

$$f_X(x) = \begin{cases} 0.15 & x = -3 \\ 0.25 & x = -1 \\ 0.1 & x = 2 \\ 0.3 & x = 6 \\ 0.2 & x = 8 \end{cases}$$

and

$$f_Y(y) = \begin{cases} 0.2 & y = -2 \\ 0.1 & y = 1 \\ 0.3 & y = 5 \\ 0.4 & y = 8. \end{cases}$$

Compute the PDF of $Z = X + Y$.

The support values for Z are

$$\{-5, -3, -2, 0, 2, 3, 4, 5, 6, 7, 9, 10, 11, 13, 14, 16\}.$$

We use the discrete convolution formula $\Pr(Z = z) = \sum_{k=\zeta}^{z-\zeta} \Pr(X = k, Y = z - k)$, where $\zeta = -3$, to compute $\Pr(Z = 4)$:

$$\Pr(Z = 4) = \sum_{k=-3}^{7} \Pr(X = k, Y = 4 - k)$$

$$\begin{aligned}
&= \Pr(X = -3) \cdot \Pr(Y = 7) + \Pr(X = -2) \cdot \Pr(Y = 6) + \\
&\quad \Pr(X = -1) \cdot \Pr(Y = 5) + \Pr(X = 0) \cdot \Pr(Y = 4) + \\
&\quad \Pr(X = 1) \cdot \Pr(Y = 3) + \Pr(X = 2) \cdot \Pr(Y = 2) + \\
&\quad \Pr(X = 3) \cdot \Pr(Y = 1) + \Pr(X = 4) \cdot \Pr(Y = 0) + \\
&\quad \Pr(X = 5) \cdot \Pr(Y = -1) + \Pr(X = 6) \cdot \Pr(Y = -2) + \\
&\quad \Pr(X = 7) \cdot \Pr(Y = -3) \\
&= 0.15 \cdot 0 + 0 \cdot 0 + 0.25 \cdot 0.3 + 0 \cdot 0 + 0 \cdot 0 + 0.1 \cdot 0 + \\
&\quad 0 \cdot 0.1 + 0 \cdot 0 + 0 \cdot 0 + 0.3 \cdot 0.2 + 0 \cdot 0 \\
&= 0 + 0 + 0.075 + 0 + 0 + 0 + 0 + 0 + 0 + 0.06 + 0 \\
&= 0.135.
\end{aligned}$$

The probabilities for the other support values of Z are computed in a similar fashion. Because of the tedious calculations needed to compute the PDF of Z by the discrete convolution formula, we compute it fully in the next example using moment generating functions (MGFs).

Unlike the discrete convolution formula, the algorithm to be presented in this chapter avoids all zero term computations by using the moment generating function technique, which is illustrated below.

Example 8.3. Suppose X and Y are the discrete random variables defined in Example 8.2 and $Z = X + Y$. Find the PDF of Z using the moment generating function technique (see, for example, Hogg et al. [52]).

Since X and Y are independent, the MGF of Z is

$$\begin{aligned}
M_Z(t) &= E\left[e^{tZ}\right] \\
&= E\left[e^{t(X+Y)}\right] \\
&= E\left[e^{tX} e^{tY}\right] \\
&= E\left[e^{tX}\right] E\left[e^{tY}\right] \\
&= M_X(t) M_Y(t).
\end{aligned}$$

The MGFs of X and Y, respectively, are

$$M_X(t) = E\left(e^{tX}\right) = 0.15e^{-3t} + 0.25e^{-t} + 0.1e^{2t} + 0.3e^{6t} + 0.2e^{8t}$$

and

$$M_Y(t) = E\left(e^{tY}\right) = 0.2e^{-2t} + 0.1e^{t} + 0.3e^{5t} + 0.4e^{8t}$$

for $-\infty < t < \infty$. Thus, the MGF of Z is

$$
\begin{aligned}
M_Z(t) = {}& 0.03e^{-5t} + 0.05e^{-3t} + 0.015e^{-2t} + 0.045 + 0.045e^{2t} + \\
& 0.01e^{3t} + 0.135e^{4t} + 0.06e^{5t} + 0.04e^{6t} + 0.16e^{7t} + 0.02e^{9t} + \\
& 0.04e^{10t} + 0.09e^{11t} + 0.06e^{13t} + 0.12e^{14t} + 0.08e^{16t}
\end{aligned}
$$

for $-\infty < t < \infty$. Then, the PDF of Z is found by matching the coefficients (mass values) with the associated coefficients in the exponents (support values):

z	-5	-3	-2	0	2	3	4	5	6	7	9	10	11	13	14	16
$f_Z(z)$	0.03	0.05	0.015	0.045	0.045	0.01	0.135	0.06	0.04	0.16	0.02	0.04	0.09	0.06	0.12	0.08

In complicated examples, especially those involving continuous random variables, using the moment generating function technique to obtain convolution functions can be more efficient than direct summation or integration. Along the same lines as the moment generating function technique, the probability generating function technique can be used for determining the PDF of the sum of discrete random variables with non-negative integer-valued supports. Unfortunately, implementing either of these techniques in a computer algebra system like Maple has drawbacks when the supports of the random variables X and/or Y are not integer-valued. These implementation issues are discussed in Sect. 8.4.

Besides the integration/summation and generating function methods already described, the characteristic functions of independent random variables can be used to compute convolutions (see Parzen [100, page 395]). In order to use this method, though, one must know the inversion formula of the PDF of a random variable in terms of its characteristic function. The complexity of the inversion makes this method unappealing for complicated or arbitrary distributions.

The purpose of this chapter is to present an algorithm for determining the distributions of sums of discrete random variables, especially those with finite arbitrary supports. Like the algorithm presented in Woodward [121], our algorithm handles well-known distributions, such as the binomial and Poisson. But, our algorithm was primarily written for arbitrary random variables whose supports have moderate cardinality. Computer algebra systems make it feasible to determine the distribution of a sum of these types of random variables.

Section 8.2 describes the algorithm for determining the PDF of a sum of two discrete random variables in conceptual terms. The algorithm that was constructed to compute the convolution appears in Sect. 8.3; implementation

issues that arose when the algorithm was coded in APPL are given in Sect. 8.4. Section 8.5 provides a collection of examples that can be solved with the convolution algorithm.

8.2 Conceptual Algorithm Development

One way to compute the convolution of the PDFs of two independent discrete random variables is by what we call the "brute force method." Let X have support x_1, x_2, \ldots, x_n and Y have support y_1, y_2, \ldots, y_m. This method does just what the name implies—it computes all possible sums between the support of X and the support of Y by brute force, e.g., $x_1 + y_1, x_1 + y_2, \ldots, x_1 + y_m, x_2 + y_1, x_2 + y_2, \ldots, x_n + y_{m-1}, x_n + y_m$. The sums are placed in a one-dimensional array, called s, of length $n \cdot m$. The corresponding probabilities for each of these sums, $f_X(x_1) \cdot f_Y(y_1), f_X(x_1) \cdot f_Y(y_2), \ldots, f_X(x_n) \cdot f_Y(y_m)$, are stored in an one-dimensional array called $Probs$, also of length $n \cdot m$. The probability in position $Probs_i$ corresponds to the sum in position s_i, $i = 1, 2, \ldots, n \cdot m$.

For example, let X and Y be the random variables introduced in Example 8.2. The arrays s and $Probs$ for the random variables X and Y are

$$s = [-5, -2, 2, 5, -3, 0, 4, 7, 0, 3, 7, 10, 4, 7, 11, 14, 6, 9, 13, 16],$$

$$Probs = [0.03, 0.015, 0.045, 0.06, 0.05, 0.025, 0.075, 0.1, 0.02, 0.01,$$
$$0.03, 0.04, 0.06, 0.03, 0.09, 0.12, 0.04, 0.02, 0.06, 0.08].$$

We assume that s is initially unsorted and may contain repeated values, such as 0, 4, and 7 in this particular example. The array s is then sorted and the elements in the array $Probs$ are rearranged in a corresponding way. After sorting, the arrays s and $Probs$ are

$$s = [-5, -3, -2, 0, 0, 2, 3, 4, 4, 5, 6, 7, 7, 7, 9, 10, 11, 13, 14, 16],$$

$$Probs = [0.03, 0.05, 0.015, 0.025, 0.02, 0.045, 0.01, 0.075, 0.06, 0.06,$$
$$0.04, 0.1, 0.03, 0.03, 0.02, 0.04, 0.09, 0.06, 0.12, 0.08].$$

Finally, the redundancies in s are removed and the appropriate probabilities corresponding to those redundancies are combined in $Probs$. The final arrays are

$$s = [-5, -3, -2, 0, 2, 3, 4, 5, 6, 7, 9, 10, 11, 13, 14, 16],$$

$$Probs = [0.03, 0.05, 0.015, 0.045, 0.045, 0.01, 0.135, 0.06, 0.04, 0.16,$$
$$0.02, 0.04, 0.09, 0.06, 0.12, 0.08].$$

One algorithm we employ to sort the array s is the *insertion sort*, as described in Weiss [116]. When $n \cdot m$ is small, the simplicity of the insertion

sort makes it an appropriate choice. Unfortunately, for random variables X and Y with larger support sizes n and m, such as $n = m = 10$, the insertion sort becomes inefficient. Since the insertion sort is an $O(N^2)$ algorithm, where $N = n \cdot m$ in our setting, it is not an appropriate method for sorting lists containing more than one hundred or so elements. For this reason, another sorting algorithm, *heapsort*, also described in Weiss [116], is used to sort larger arrays. Heapsort uses a heap, which is a binary tree with special properties, to sort the array s. Heapsort is an $O(N \cdot \log(N))$ algorithm (see, for example, Carrano et al. [15]).

Heapsort builds the array s as a maximum heap data structure. It then swaps the maximum element (the root) of the heap with the element in the last array position $s_{n \cdot m}$. The heap is rebuilt with the remaining unsorted elements in array positions s_1 through $s_{n \cdot m - 1}$. The maximum element of the new heap is then swapped with the element in the second to last position of the array s, which is position $s_{n \cdot m - 1}$. Now the last two positions in s are sorted in ascending order. The heap structure is again restored with the remaining unsorted elements in array positions s_1 through $s_{n \cdot m - 2}$. This swap and rebuild process repeats itself until all elements are removed from the unsorted region of the heap and placed in ascending order from the front to the back of the heap. Heapsort proved more efficient than insertion sort for large values of N.

Shellsort, an improved insertion sort, is the algorithm employed by the mathematical software package Maple to sort polynomials. Since Shellsort's "performance is quite acceptable in practice, even for N [the number of elements] in the tens of thousands" [116, page 260], we take advantage of Maple's sorting algorithm for polynomials (when possible) by using the moment generating function technique to compute the sum of discrete random variables. The MGFs for X and Y, denoted by $M_X(t)$ and $M_Y(t)$ respectively, are first computed. Next, the product of the MGFs, $M_Z(t)$, is computed. We manipulate the terms of the MGF with Maple's `expand` procedure so that they are written in a fashion that Maple interprets as polynomials terms. For example, if the MGF is $M_Z(t) = \frac{1}{3}e^{3t} + \frac{1}{6}e^{2t} + \frac{1}{2}e^{5t}$, then `expand(`$M_Z(t)$`)` returns $M_Z(t)$ as $\frac{1}{3}\left(e^t\right)^3 + \frac{1}{6}\left(e^t\right)^2 + \frac{1}{2}\left(e^t\right)^5$. The terms of the resulting expanded MGF are then sorted in descending order by the constant appearing in the exponent of each e^t term. Sorting the example expression $M_Z(t)$ returns $\frac{1}{2}\left(e^t\right)^5 + \frac{1}{3}\left(e^t\right)^3 + \frac{1}{6}\left(e^t\right)^2$. The probability and support values are extracted from the terms of the expression $M_Z(t)$, and the PDF of the sum is formed. The PDF for the example expression $M_Z(t)$ is

$$
f_Z(z) = \begin{cases} \frac{1}{6} & z = 2 \\ \frac{1}{3} & z = 3 \\ \frac{1}{2} & z = 5. \end{cases}
$$

Although in theory this is an ideal method, Maple recognizes that expressions with non-integer valued constants in the exponents, such as $\frac{1}{3}e^{3.1t} + \frac{1}{6}e^{2.5t} + \frac{1}{2}e^{5.4t}$, are not truly polynomials and will incorrectly sort such expressions.

Since the MGF $M_Z(t)$ may not always have integer constants in its exponents, the moment generating function technique for computing the PDF of a sum is only reasonable to use for integer supports. Using probability generating functions to compute the PDF of a sum of random variables results in the same complications. Further implementation issues faced by moment and probability generating functions are discussed in Sect. 8.4.

As suggested by Nicol [92], the sum array s can be constructed in such a way that the next largest sum element is placed in s as it is being built. Instead of constructing the array s first and then sorting it, our algorithm constructs s by sequentially appending the next ordered element. We refer to this method as the "moving heap method," and it involves building, deleting, and inserting sums into a minimum heap data structure. A minimum heap contains its smallest element in the root, rather than its largest as in a maximum heap data structure.

The idea behind this sorting algorithm is the construction of a two-dimensional "conceptual" array A, which has $m + 1$ rows and $n + 1$ columns. The array A is not instantiated to save on memory, but serves as a bookkeeping device to help explain the nature of the algorithm. As illustrated in Example 8.4, A is displayed in an unusual manner in order to resemble the axes in the Cartesian coordinate system. Without loss of generality, we assume that the supports of X and Y are arranged in increasing order; i.e., $x_1 < x_2 < \cdots < x_n$ and $y_1 < y_2 < \cdots < y_m$. The array cell (i, j) contains the sum $A_{i,j} = y_i + x_j$ for $i = 1, 2, \ldots, m$ and $j = 1, 2, \ldots, n$. The cells in row $m+1$ of A hold a 0 or 1 for each column $j = 1, 2, \ldots, n$ to indicate whether the cell in column j is "active," which means its entry is in the heap H, consisting of the candidates for the next element to be appended to s. Thus, $A_{m+1,j} = 0$ or 1 for $j = 1, 2, \ldots, n$. Likewise, the cells in column $n + 1$ of A also hold a 0 or 1 for each row $i = 1, 2, \ldots, m$ to indicate whether the cell in row i is "active," so $A_{i,n+1} = 0$ or 1 for $i = 1, 2, \ldots, m$. The cell $(m + 1, n + 1)$ is not used in the algorithm. Example 8.4 illustrates what is meant by an "active" cell. Occasionally, it will be convenient to have a notation for the matrix A with row $m + 1$ and column $n + 1$ deleted. This $m \times n$ matrix will be denoted by \hat{A}.

Since $x_j < x_{j+1}$ for $j = 1, 2, \ldots, n-1$ and $y_i < y_{i+1}$ for $i = 1, 2, \ldots, m-1$, the entry in cell (i, j) is always guaranteed to be less than both the entries in cells $(i+1, j)$ and $(i, j+1)$; i.e. $A_{i,j} < A_{i+1,j}$ and $A_{i,j} < A_{i,j+1}$. This result allows the algorithm to repeatedly move the smallest number of candidate entries for the next largest sum from the array A to the current heap H. Thus, this algorithm moves from the southwest cell to the northeast cell of the array \hat{A} repeatedly placing the next largest sum into s after choosing it as the minimum element in the current heap H.

Theorem 8.1. Let $x_1 < x_2 < \cdots < x_n$ and $y_1 < y_2 < \cdots < y_m$ be finite real numbers. Let the $m \times n$ array \hat{A} be arranged as shown in Fig. 8.1 and have entries $\hat{A}_{i,j} = y_i + x_j$, $i = 1, 2, \ldots, m$ and $j = 1, 2, \ldots, n$. Let the set C contain all (i, j) pairs such that $x_i + y_j \leq c$, where $c > y_1 + x_1$ is a fixed real number. Let P be the path from the northwest corner of \hat{A} to the southeast

y_m	$y_m + x_1$	$y_m + x_2$	\cdots	$y_m + x_n$
	\vdots	\vdots	\ddots	\vdots
y_2	$y_2 + x_1$	$y_2 + x_2$	\cdots	$y_2 + x_n$
y_1	$y_1 + x_1$	$y_1 + x_2$	\cdots	$y_1 + x_n$
	x_1	x_2		x_n

Fig. 8.1. Array \hat{A} where $x_1 < x_2 < \cdots < x_n$ and $y_1 < y_2 < \cdots < y_m$

corner of \hat{A} that separates C from its complement C'. The smallest element in C' must occur immediately northeast of a point at which the path P changes direction from southward to eastward.

Proof. Suppose the smallest element in C' occurs in the cell (r, s). Since $A_{r-1,s} < A_{r,s}$ and $A_{r,s-1} < A_{r,s}$, cells $(r-1, s)$ and $(r, s-1)$ are both in C. Thus, the path P must pass between cell (r, s) and cells $(r, s-1)$ and $(r-1, s)$, changing direction from southward to eastward as it does. □

Theorem 8.1 not only identifies the candidates for the next element to append to s, but it also shows how the list of candidates should be modified when its smallest element is removed. If $A_{p,q}$ is appended to s, then the path P corresponding to $c = A_{p,q}$ passes along the upper edge and right edge of cell (p, q), implying that P may change direction just southwest of $A_{p+1,q}$ and $A_{p,q+1}$. As an illustration, Fig. 8.2 displays the array \hat{A} associated with Example 8.2 with $c = 3$.

Since this process and its intricacies are best explained by an example, we'll reintroduce the discrete random variable X and Y which were first encountered in Example 8.2.

Example 8.4. Let X and Y have PDFs:

$$f_X(x) = \begin{cases} 0.15 & x = -3 \\ 0.25 & x = -1 \\ 0.1 & x = 2 \\ 0.3 & x = 6 \\ 0.2 & x = 8 \end{cases} \qquad f_Y(y) = \begin{cases} 0.2 & y = -2 \\ 0.1 & y = 1 \\ 0.3 & y = 5 \\ 0.4 & y = 8. \end{cases}$$

Fig. 8.2. Array \hat{A} corresponding to Example 8.2. The *thickened path* P from the northwest corner to the southeast corner of \hat{A} delimits the set $C = \{(i, j) \mid y_i + x_j \leq 3\}$ from its complement C'. The *circled entries* lie just to the northeast of points where the path turns from southward to eastward. These three entries are candidates for the smallest entry in the cells of C' for $c = 3$

Use the "moving heap method" to determine the PDF of $Z = X + Y$.

Construct the 5×6 array A. Set $A_{i,n+1} = A_{i,6} = 0$ for $i = 1, 2, 3, 4$ and $A_{m+1,j} = A_{5,j} = 0$ for $j = 1, 2, 3, 4, 5$. The smallest value in A is positioned in cell $(1, 1)$ and is $A_{1,1} = y_1 + x_1 = -5$. The algorithm designates the cell $(1, 1)$ as an "active" cell in A by setting $A_{m+1,1} = A_{5,1} = 1$ and $A_{1,n+1} = A_{1,6} = 1$. The zeros in the other cells of row five and column six remain. Figure 8.3 displays this initial array. The entries of A increase in value as one moves up, to the right, or a combination of both (as in the Cartesian coordinate system).

Fig. 8.3. Array A with active cell $(1, 1)$, which contains the entry $A_{1,1} = -5$

As in the brute force method, let the one-dimensional array s of length $n \cdot m$ hold the sums of the supports of the random variables X and Y. The corresponding probabilities for each of these sums will again be stored in the one-dimensional array called $Probs$, also of length $n \cdot m$. Clearly, the first (smallest) sum to be placed in the first position of array s is $A_{1,1}$. Accordingly, $f_X(x_1) \cdot f_Y(y_1) = 0.03$ is placed in the $Probs$ array in its first position, $Probs_1$. After setting $s_1 = A_{1,1} = -5$ and $Probs_1 = \Pr(Z = A_{1,1}) = f_X(x_1) \cdot f_Y(y_1) = 0.03$, the cell $(1, 1)$ becomes inactive. In order to reflect the absence of an active cell in the first row and first column, reset $A_{m+1,1} = A_{5,1} = 0$ and $A_{1,n+1} = A_{1,6} = 0$. The next two cells to become "active" (i.e., cells that may contain the next largest sum) in the array A are $(1, 2)$ and $(2, 1)$, with $A_{1,2} = y_1 + x_2 = -3$ and $A_{2,1} = y_2 + x_1 = -2$. Since cell $(1, 2)$ in A is now active, reset $A_{1,6} = 1$ and set $A_{5,2} = 1$. Similarly, since cell $(2, 1)$ is active, set $A_{2,6} = 1$ and reset $A_{5,1} = 1$. The purpose of these ones and zeros along the boundary of the A array is to assure that there is no more than one active cell in each row and column. Figure 8.4 contains the current state of array A.

Fig. 8.4. Array A after $A_{1,1}$ has been processed and added to the one-dimensional sum array s. The cells $(1, 2)$ and $(2, 1)$ are active, as indicated by the ones in cells $(1, 6)$, $(5, 2)$, $(2, 6)$, and $(5, 1)$

The values $A_{1,2}$ and $A_{2,1}$ now form the minimum heap H, resulting in $H = \{A_{1,2}, A_{2,1}\} = \{-3, -2\}$. Informally, a *heap* is a complete binary tree with a special ordering property of its nodes. A complete binary tree is a tree that is completely filled with the possible exception of the bottom level, which is filled from left to right. Figure 8.5 contains illustrations of structures which are and are not complete binary trees. Each node of the tree has one parent, except the *root* of the tree, which

Fig. 8.5. Six binary trees. The *top* three trees are complete binary trees and the *bottom* three are not

has no parent. In a *minimum* heap, the smallest element of the heap is contained in its root. In the upper-right tree in Fig. 8.5, a is the root of the tree. Nodes b and c are a's *children*, where b is the *left child* and c is the *right child*. (According to the definition of a complete binary tree, when a node above the bottom level of the tree has only one child, it must be a left child.) Node b is the parent to nodes d and e. The *height* of a tree is the number of nodes from the root to a node at the bottom level of the tree. For example, the heights of the top trees in Fig. 8.5 are three, three, and four, respectively.

Thus, a minimum heap is a complete binary tree with the special ordering property that each parent node contains a value less than or equal to the values in its children's nodes. Because of this ordering property, the smallest value in a minimum heap will always be at the root.

The binary heap H formed with the values $A_{1,2}$ and $A_{2,1}$ is in Fig. 8.6. The next sum to be entered into s in position s_2 is the root of the heap, i.e., the smallest element in H. Since $A_{1,2} = -3$ is the root, it is removed from the heap H and placed in s_2, while its corresponding probability is placed in $Probs_2$. Because the entry $A_{1,2}$ is removed from the array A, reset $A_{1,6} = 0$ and $A_{5,2} = 0$ to indicate that row one and column two no longer contain an active cell. After these changes, array A is displayed in Fig. 8.7.

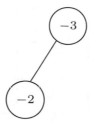

Fig. 8.6. Heap H containing entries $A_{1,2} = -3$ and $A_{2,1} = -2$

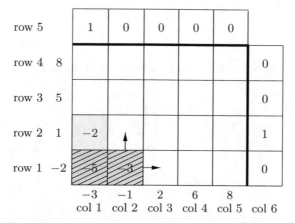

Fig. 8.7. Array A after $A_{1,2} = -3$ is processed and appended to s. Cell $(2, 1)$ is the only active cell. Candidates to become active are cells $(1, 3)$ and $(2, 2)$. Cell $(2, 2)$ cannot become active since row two already contains an active cell

After setting cell $(1, 2)$ to inactive, the two cells that are candidates to become active (if the corresponding row and column do not already contain an active cell) are cells $(1, 3)$ and $(2, 2)$. Since row two contains the active cell $(2, 1)$, and by design $A_{2,1} < A_{2,2}$, entry $A_{2,2}$ is not activated. However, cell $(1, 3)$ does become active, and its entry is $A_{1,3} = y_1 + x_3 = 0$. Hence, $A_{1,6} = 1$ and $A_{5,3} = 1$. After these changes, array A is displayed in Fig. 8.8.

The entry $A_{1,3}$ is inserted into the heap H, resulting in $H = \{A_{2,1}, A_{1,3}\} = \{-2, 0\}$. After the addition of $A_{1,3}$, the heap H is displayed in Fig. 8.9. The minimum element, $A_{2,1}$, is removed from the root of the heap and placed in the sum array s in position s_3. Its corresponding probability is placed in $Probs_3$.

The two cells that may enter the array A after the removal of the $A_{2,1}$ entry are in cells $(2, 2)$ and $(3, 1)$, as indicated by the arrows in Fig. 8.10. Both cells $(2, 2)$ and $(3, 1)$ become active, and their values are $A_{2,2} = y_2 + x_2 = 0$ and $A_{3,1} = y_3 + x_1 = 2$. Hence, $A_{2,6} = 1$, $A_{5,2} = 1$, $A_{3,6} = 1$, and $A_{5,1} = 1$, as displayed in Fig. 8.11. Entries

		-3	-1	2	6	8	
row 5		1	0	1	0	0	
row 4	8						0
row 3	5						0
row 2	1	-2					1
row 1	-2	-5	-3	0			1
		-3	-1	2	6	8	
		col 1	col 2	col 3	col 4	col 5	col 6

Fig. 8.8. Array A with active cells $(1, 3)$ and $(2, 1)$

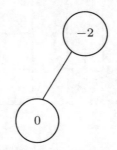

Fig. 8.9. Heap H containing entries $A_{2,2} = -2$ and $A_{1,3} = 0$

row 5		0	0	1	0	0	
row 4	8						0
row 3	5						0
row 2	1	-2					0
row 1	-2	-5	-3	0			1
		-3	-1	2	6	8	
		col 1	col 2	col 3	col 4	col 5	col 6

Fig. 8.10. Array A after $A_{2,2} = -2$ is removed. Cell $(3, 1)$ is the only active cell. Candidates to become active are cells $(2, 2)$ and $(3, 1)$

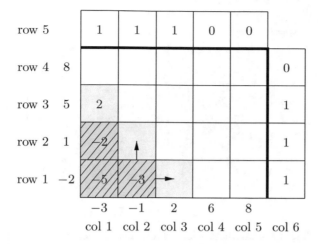

Fig. 8.11. Array A with active cells $(1, 3)$, $(2, 2)$, and $(3, 3)$

$A_{2,2}$ and $A_{3,1}$ are inserted into the heap H, and H is again rebuilt as $H = \{A_{1,3}, A_{2,2}, A_{3,1}\} = \{0, 0, 2\}$. Its structure is displayed in Fig. 8.12.

Moving ahead to the seventeenth pass through the construction of the array A, its appearance is displayed in Fig. 8.13. $A_{3,4} = 11$ is placed in s_{17}, and values $A_{3,5} = 13$ and $A_{4,4} = 14$ are activated in the array A and inserted into the heap H. Since $A_{3,5}$ is the root of the heap, it is deleted and placed in s_{18}. No new element is allowed to enter H, so the root element of the heap is now $A_{4,4} = 14$, and it is removed and placed in s_{19}. The last entry to be activated is $A_{4,5} = 16$, and it is placed in position s_{20} of the sum array s.

Thus, after 20 iterations of this process, s and $Probs$ arrays are

$$s = [-5, -3, -2, 0, 0, 2, 3, 4, 4, 5, 6, 7, 7, 7, 9, 10, 11, 13, 14, 16]$$

$$Probs = [0.03, 0.05, 0.015, 0.025, 0.02, 0.045, 0.01, 0.075, 0.06, 0.06,$$
$$0.04, 0.1, 0.03, 0.03, 0.02, 0.04, 0.09, 0.06, 0.12, 0.08]$$

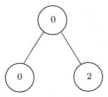

Fig. 8.12. Heap H with entries $A_{1,3} = 0$, $A_{2,2} = 0$ and $A_{1,3} = 2$

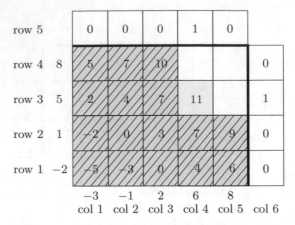

Fig. 8.13. Array A with its seventeenth active cell $(3, 4)$

which are the same arrays encountered by using the moment generating function technique. The redundancies are removed from s and the appropriate probabilities are combined in *Probs* to complete the algorithm. (This could have been embedded into the iterative steps of the algorithm to save memory.) Thus, the PDF of $Z = X + Y$ is the same as determined in Example 8.3.

8.3 Algorithm

The algorithm for the procedure `Convolution(X, Y)` returns the convolution of the PDFs of the random variables X and Y. A brief description of the algorithm for discrete random variables follows.

If X and Y are discrete, their supports, finite or infinite, dictate which of the methods described in Sect. 8.2 is used to compute the convolution. The convolution of the PDFs of X and Y with finite support is computed either using the `BruteForceMethod` or `MovingHeapMethod` procedures, whose algorithms appear in the appendices of Evans and Leemis [29]. The PDF of the convolution of Z is stored and returned in the usual list-of-sublists format. The list of elements $f(z_1), f(z_2), \ldots, f(z_{n \cdot m})$ are the probability values of Z, while $z_1, z_2, \ldots, z_{n \cdot m}$ are its support values. The one-dimensional array s is created to contain the unique sums extracted from the array A. Similarly, the one-dimensional array *Probs* is created to hold the probability values corresponding to the sums in s.

The zeros (if any) at the end of the *Probs* array do not represent probability values; they correspond to the zeros in the *Probs* array that are not support values. These extra zeros indicate that there are redundant values in the support of Z. The non-zero probability values are removed from *Probs* and placed

in the array $FinalProbs$. The support values that correspond to the removed probability values are removed and placed in the array $FinalSupport$.

If the supports of the random variables X and Y are infinite, either the discrete convolution formula is used to compute the convolution or the APPL procedure MGF is used to determine the MGF of the sum of X and Y.

If either X or Y has infinite support, while the other has finite support, the product of their MGFs is returned. If both X and Y have infinite support and the discrete convolution formula formed with their PDFs results in an intractable sum, then the product of their MGFs is returned. Otherwise, the discrete convolution formula is used to determine the convolution of their PDFs.

Unless the MGF of Z or the PDF of the sum of X and Y (with $n = 1$ and $m = 1$) has already been returned, the PDF $f_Z(z)$ is returned.

Procedure Convolution: Computes the PDF of the sum two independent random variables (indentation indicates nesting).

Input: The random variables X and Y in the usual APPL list-of-sublists format. Assume that the random variables X and Y are input in their PDF forms. The support of X is Ω_X and the support of Y is Ω_Y.

Output: The PDF of $Z = X + Y$.

Algorithm:
 if X and Y are continuous
 $f_Z(z) \leftarrow \int_{-\infty}^{\infty} f_X(z - y) f_Y(y) \, dy$
 else if X and Y are discrete
 if X and Y have finite support
 $n \leftarrow |\Omega_X|$
 $m \leftarrow |\Omega_Y|$
 if $n \cdot m \leq 100$
 $Z \leftarrow$ BruteForceMethod(X, Y)
 else
 $Z \leftarrow$ MovingHeapMethod(X, Y)
 if $(n = 1$ and $m = 1)$ i.e. $f_Z(z) = 1$ for $z = c \in \mathbb{R}$
 return$(f_Z(z))$
 dimension $s[n \cdot m]$ create the sums array s
 dimension $Probs[n \cdot m]$ create the probability array $Probs$
 for $i \leftarrow 1$ to $n \cdot m$
 $s_i \leftarrow 0$
 $Probs_i \leftarrow 0$
 $s_1 \leftarrow z_1$
 $s_2 \leftarrow z_2$
 $Probs_1 \leftarrow f(z_1)$
 $k \leftarrow 2$
 $j \leftarrow 2$

$\quad\quad$ while $(k < n \cdot m)$ do

$\quad\quad\quad$ $Probs_j \leftarrow Probs_j + f(z_k)$

$\quad\quad\quad$ if $z_k \neq z_{k+1}$ then $\quad\quad\quad\quad$ eliminate redundant support values

$\quad\quad\quad\quad$ $j \leftarrow j + 1$

$\quad\quad\quad\quad$ $s_j \leftarrow z_{k+1}$

$\quad\quad\quad$ $k \leftarrow k + 1$

$\quad\quad$ $Probs_j \leftarrow Probs_j + f(z_k)$

$\quad\quad$ $NumZeros \leftarrow 0$

$\quad\quad$ for $i \leftarrow n \cdot m$ to 1 by -1 while $Probs_i = 0$

$\quad\quad\quad$ $NumZeros \leftarrow NumZeros + 1$

$\quad\quad$ dimension $FinalProbs[1, n \cdot m - NumZeros]$

$\quad\quad$ dimension $FinalSupport[1, n \cdot m - NumZeros]$

$\quad\quad$ for $i \leftarrow 1$ to $(n \cdot m - NumZeros)$

$\quad\quad\quad$ $FinalProbs_i \leftarrow Probs_i$

$\quad\quad\quad$ $FinalSupport_i \leftarrow s_i$

$\quad\quad$ $f_Z(z) \leftarrow [FinalSupport, FinalProbs]$

\quad else if (X or Y has infinite support or X and Y have infinite support

$\quad\quad\quad\quad\quad\quad\quad\quad\quad\quad$ with intractable discrete convolution sum)

$\quad\quad$ $mgfx \leftarrow \text{MGF}(X)$

$\quad\quad$ $mgfy \leftarrow \text{MGF}(Y)$

$\quad\quad$ $mgfprod \leftarrow mgfx \cdot mgfy$

$\quad\quad$ return($mgfprod$)

\quad else $\quad\quad\quad\quad\quad\quad\quad\quad\quad\quad\quad\quad\quad$ discrete convolution formula

$\quad\quad$ $f_Z(z) \leftarrow \sum_{k=0}^{z} \big(f_X(z)\big) \cdot \big(f_Y(z-k)\big)$

else

\quad print("ERROR: X and Y must both be continuous or discrete")

\quad return

return$\big(f_Z(z)\big)$

8.4 Implementation Issues

The algorithm for Convolution has been implemented and included in APPL's suite of procedures. One portion of the algorithm for the Convolution procedure involves computing every possible sum between the support of the random variables X and Y. At the same time, the probabilities for those sums are calculated. Earlier we have referred to this as the "brute force method." The list of sums are ordered and redundancies combined. The corresponding probability values are repositioned to match their corresponding sums. One important reason for sorting the sums is that all other APPL procedures assume that discrete distributions, written in their list-of-sublists form, have supports listed in increasing order without repeated values. To be consistent with the APPL language and textbooks, the sums are sorted. Also, placing

the values of the support into a list in sorted order means that tied $y_i + x_j$ values *can* be combined dynamically as the algorithm proceeds.

The first sorting method used to sort the list of sums was the insertion sort. It was chosen because of its straightforward code and efficiency in computing convolutions when the support sizes of the random variables are small. Unfortunately, as the supports of random variables grow larger [e.g., random variables with 50 random numbers chosen on the interval $(-1, 1)$ as a support], the time used to compute the PDF of the convolution becomes unreasonably long. A faster sorting method for larger list sizes is heapsort. Heapsort is employed in the `Convolution` procedure for sorting the list of sums created by the "brute force method."

Maple uses Shellsort to sort polynomials. In order to use the Shellsort procedure in Maple, the MGFs of X and Y need to be computed. The product of the MGFs of X and Y is an expression composed of exponential terms e^{kt}, where $k \in \mathbb{R}$, $t > 0$. Letting $u = e^t$, the MGF of the product can be rewritten as a polynomial-type expression. For example, if $u = e^t$, then $M_Z(t) = \frac{1}{3}e^{3t} + \frac{1}{6}e^{2t} + \frac{1}{2}e^{5t}$ for $-\infty < t < \infty$ can be rewritten $M_Z(t) = \frac{1}{3}u^3 + \frac{1}{6}u^2 + \frac{1}{2}u^5$ for $u > 0$. The Shellsort procedure sorts the polynomial-like expression, and the PDF of the convolution of X and Y is retrieved from this expression. Instead of MGFs, probability generating functions (PGFs) can be used in the process. MGFs were chosen over PGFs since PGFs can only be formed when the discrete distribution has non-negative integer support.

The method of computing convolutions via their MGFs was abandoned after realizing that Maple can only sort "true" polynomial expressions in which variables are raised to non-negative integer powers. Thus, using MGFs would be too restrictive for a Maple implementation. Maple is unable to sort an expression that has non-integers in the powers of the polynomial terms. For example, Maple cannot sort the expression $\frac{1}{3}\left(e^t\right)^{3/2} + \frac{1}{6}\left(e^t\right)^{1/2} + \frac{1}{2}\left(e^t\right)^{5.5}$. Since the `Convolution` procedure was intended to be used on arbitrary distributions, which could indeed have negative, non-integer supports, the MGF method was removed from the procedure `Convolution`. The extra time involved in checking for appropriate values also affected the algorithm's efficiency.

Unfortunately, for random variables with large finite support sizes, heapsort was also inefficient. Nicol [92] suggested constructing a heap dynamically and sorting the list of sums sequentially, instead of waiting until the list of sums is built and then sorting it. The heap will always contain $\min\{m, n\}$ or fewer entries. After constructing the algorithm for this method, which is called `MovingHeapMethod`, the `Convolution` procedure was tested on random variables with large supports by using the `BruteForceMethod` with insertion sort, the `BruteForceMethod` with heapsort, and the `MovingHeapMethod`. A brief comparison analysis of the three methods suggested that `MovingHeapMethod` yielded the best times for computing convolution of random variables with large supports. Test cases of random variables with increasing support sizes were performed to confirm this claim.

The test involved generating random numbers between -1 and 1 and making them the support values for the random variables X and Y. The random numbers were sorted and placed into the second sublist in the list-of-sublists format for the random variables that conforms to the system of the APPL language. The probabilities, which had no effect on the efficiency of the different algorithms, were assigned to be equal for all support values; i.e., if X's support consisted of 50 values, then each support value's probability was $1/50$. The times for determining the PDF of the sum of random variables of increasing support sizes appears in Evans and Leemis [29].

When either one or both random variables' supports are infinite, either the convolution of their PDFs is computed via the discrete convolution formula or the product of their MGFs is determined. If one of the random variables has infinite support, while the other has finite support, the product of their MGFs is returned. At this time, APPL does not contain a procedure to convert the MGF of a random variable to its PDF form.

Having two random variables with infinite support does not guarantee that the PDF of their sum can be determined by the discrete convolution formula. Only tractable summations, such as the convolution formula for two Poisson random variables as in Example 8.1, can be computed. This means that instead of determining the PDF for the convolution of the PDFs of some random variables, such as a Poisson with parameter $\lambda = 5$ and a geometric with parameter $p = 0.3$, the `Convolution` procedure only computes the product of their MGFs.

8.5 Examples

The following examples use the algorithm described in Sect. 8.3 to determine the PDF of the sum of independent random variables. Examples for a variety of random variables are provided to illustrate the utility of the algorithm. Returning first to Examples 8.1 and 8.2 introduced in Sect. 8.1, we can use the `Convolution` procedure to determine the PDF of $Z = X + Y$.

Example 8.5. If X and Y are independent Poisson random variables with respective parameters λ_1 and λ_2, compute the PDF of $Z = X + Y$.

In APPL, define `X` as a Poisson random variable with parameter `lambda1` and `Y` as a Poisson random variable with parameter `lambda2`. The Poisson random variable is also predefined in APPL. The PDF $Z = X + Y$ is found with the statements

```
> X := PoissonRV(lambda1);
> Y := PoissonRV(lambda2);
> Z := Convolution(X, Y);
```

which returns the PDF of Z as

$$\left[\left[z \rightarrow \frac{(z+1)\,e^{-\lambda_1 - \lambda_2}\,(\lambda_2 + \lambda_1)^z}{(z+1)!}\right], [0\mathinner{\ldotp\ldotp}\infty], [\text{``Discrete''}, \text{``PDF''}]\right]$$

Using the Maple `simplify` procedure, the resulting PDF after simplification is

$$f(z) = \frac{(\lambda_2 + \lambda_1)^z\,e^{-\lambda_1 - \lambda_2}}{\Gamma(z+1)} \qquad z = 0, 1, \ldots ,$$

which is easy to recognize in its standard form as

$$f(z) = \frac{(\lambda_1 + \lambda_2)^z\,e^{-\lambda_1 - \lambda_2}}{z!} \qquad z = 0, 1, \ldots .$$

We are fortunate in the previous example that Maple can compute the PDF by the discrete convolution formula by simplifying the sum $\sum_{k=0}^{x} \frac{\lambda_1{}^k e^{-\lambda_1} \lambda_2{}^{x-k} e^{-\lambda_2}}{k!\,(x-k)!}$. Unfortunately, Maple can only simplify certain expressions, so in some instances we cannot compute the PDF by the discrete convolution formula.

Example 8.6. X and Y are independent discrete random variables with PDFs defined as

$$f_X(x) = \begin{cases} 0.15 & x = -3 \\ 0.25 & x = -1 \\ 0.1 & x = 2 \\ 0.3 & x = 6 \\ 0.2 & x = 8 \end{cases} \qquad f_Y(y) = \begin{cases} 0.2 & y = -2 \\ 0.1 & y = 1 \\ 0.3 & y = 5 \\ 0.4 & y = 8. \end{cases}$$

Find the PDF of $Z = X + Y$.

Define the random variables X and Y in APPL's list-of-sublists format. Compute the PDF of $Z = X + Y$ with the statements

```
> X := [[0.15, 0.25, 0.1, 0.3, 0.2], [-3, -1, 2, 6, 8],
        ["Discrete", "PDF"]];
> Y := [[0.2, 0.1, 0.3, 0.4], [-2, 1, 5, 8],
        ["Discrete", "PDF"]];
> Z := Convolution(X, Y);
```

which return the PDF of Z as

$$[[0.03, 0.05, 0.015, 0.045, 0.045, 0.01, 0.135, 0.06, 0.04, 0.16, 0.02, 0.04,$$
$$0.09, 0.06, 0.12, 0.08],$$
$$[-5, -3, -2, 0, 2, 3, 4, 5, 6, 7, 9, 10, 11, 13, 14, 16], [\text{``Discrete''}, \text{``PDF''}]]$$

The probability generating function, defined for a discrete random variable X as $G_X(t) = E\left[t^X\right]$, has been referred to several times in this chapter as an alternative to the moment generating function for determining the probability distribution of the sum of two discrete random variables. Its use will be illustrated in the next three examples.

Example 8.7. Let X and Y be independent random variables; X assumes three possible values, 0, 1, 3, with probabilities 1/2, 3/8, and 1/8; Y assumes two possible values, 0 and 1, with probabilities 1/3 and 2/3. Find the PDF of the random variable $Z = X + Y$ (from Burr [14, page 136]).

By hand, we can compute the PDF of Z with probability generating functions. The PGFs $G_X(t)$ and $G_Y(t)$ of X and Y, respectively, are

$$G_X(t) = \frac{1}{8}t^3 + \frac{3}{8}t + \frac{1}{2} \qquad -\infty < t < \infty,$$

$$G_Y(t) = \frac{2}{3}t + \frac{1}{3} \qquad -\infty < t < \infty.$$

Thus, the PGF of $Z = X + Y$ is the product of $G_X(t)$ and $G_Y(t)$:

$$G_Z(t) = \frac{1}{12}t^4 + \frac{1}{24}t^3 + \frac{1}{4}t^2 + \frac{11}{24}t + \frac{1}{6} \qquad -\infty < t < \infty,$$

and hence the PDF of Z is

$$f_Z(z) = \begin{cases} \frac{1}{6} & z = 0 \\ \frac{11}{24} & z = 1 \\ \frac{1}{4} & z = 2 \\ \frac{1}{24} & z = 3 \\ \frac{1}{12} & z = 4. \end{cases}$$

In APPL, defining X and Y as lists-of-sublists and then applying the Convolution procedure achieves the same result.

```
> X := [[1 / 2, 3 / 8, 1 / 8], [0, 1, 3],
        ["Discrete", "PDF"]];
> Y := [[1 / 3, 2 / 3], [0, 1], ["Discrete", "PDF"]];
> Z := Convolution(X, Y);
```

Other measures of a distribution, such as the mean and variance, can be found with the use of additional APPL procedures, as illustrated in the next example.

Example 8.8. Let X_1 and X_2 be observations of a random sample of size $n = 2$ drawn from a distribution with PDF $f(x) = \frac{x}{6}$, $x = 1, 2, 3$. Find the PDF of $Y = X_1 + X_2$, and determine the mean and variance of the sum [51, page 297].

The PGF of X_1 and X_2 is

$$G_{X_1}(t) = G_{X_2}(t) = \frac{1}{2}t^3 + \frac{1}{3}t^2 + \frac{1}{6}t \qquad -\infty < t < \infty.$$

Thus, $G_Y(t)$ is the product of $G_{X_1}(t)$ and $G_{X_2}(t)$:

$$G_Y(t) = \frac{1}{4}t^6 + \frac{1}{3}t^5 + \frac{5}{18}t^4 + \frac{1}{9}t^3 + \frac{1}{36}t^2 \qquad -\infty < t < \infty.$$

and $f_Y(y)$ is

$$f_Y(y) = \begin{cases} \frac{1}{36} & y = 2 \\ \frac{1}{9} & y = 3 \\ \frac{5}{18} & y = 4 \\ \frac{1}{3} & y = 5 \\ \frac{1}{4} & y = 6. \end{cases}$$

The mean and the variance of Y, respectively, are

$$E[Y] = G_Y'(t)\bigg|_{t=1} = \left[\frac{3}{2}t^5 + \frac{5}{3}t^4 + \frac{10}{9}t^3 + \frac{1}{3}t^2 + \frac{1}{18}t\right]_{t=1} = \frac{14}{3}$$

and

$$\text{Var}[Y] = G_Y''(1) + G_Y'(1) - \left[G_Y'(1)\right]^2 = \frac{10}{9}.$$

In APPL, the mean and variance of Y are computed with the statements

```
> X := [[x -> x / 6], [1 .. 3], ["Discrete", "PDF"]];
> Y := Convolution(X, X);
> Mean(Y);
> Variance(Y);
```

which return the mean and variance computed by hand.

Example 8.9. Find the probability of obtaining a total of 14 in a toss of four fair dice (from Hogg et al. [50, page 230]).

Let X be the PDF $f_X(x) = \frac{1}{6}$, $x = 1, 2, \ldots, 6$. The PGF G of X is

$$G_X(t) = \frac{1}{6}t^6 + \frac{1}{6}t^5 + \frac{1}{6}t^4 + \frac{1}{6}t^3 + \frac{1}{6}t^2 + \frac{1}{6}t \qquad -\infty < t < \infty.$$

The PDF of $Z = X_1 + X_2 + X_3 + X_4$ can be found by computing $\left[G_X(t)\right]^4$, which is

$$\begin{aligned}
\left[G_X(t)\right]^4 = &\frac{1}{1296}t^{24} + \frac{1}{324}t^{23} + \frac{5}{648}t^{22} + \frac{5}{324}t^{21} + \frac{35}{1296}t^{20} + \\
&\frac{7}{162}t^{19} + \frac{5}{81}t^{18} + \frac{13}{162}t^{17} + \frac{125}{1296}t^{16} + \frac{35}{324}t^{15} + \\
&\frac{73}{648}t^{14} + \frac{35}{324}t^{13} + \frac{125}{1296}t^{12} + \frac{13}{162}t^{11} + \frac{5}{81}t^{10} + \\
&\frac{7}{162}t^9 + \frac{35}{1296}t^8 + \frac{5}{324}t^7 + \frac{5}{648}t^6 + \frac{1}{324}t^5 + \frac{1}{1296}t^4
\end{aligned}$$

for $-\infty < t < \infty$. Thus, $\Pr(Z = 14) = 73/648$. In APPL, define X as a uniform discrete random variable (predefined in APPL) with parameter $1/6$. The procedure ConvolutionIID(X, n) computes the PDF of the convolution of n iid random variables X. This procedure contains a Maple for loop which calls Convolution n times.

```
> X := UniformDiscreteRV(1, 6);
> Z := ConvolutionIID(X, 4);
> PDF(Z, 14);
```

The APPL procedure PDF computes the probability that Z is 14 correctly as $73/648$.

Examples 8.10–8.12 are from the article "Getting Normal Probability Approximations without Using Normal Tables" by Thompson [112]. His article discusses an alternate approach to approximating probabilities involving sums of discrete random variables using the PDF for the normal distribution. He lets S denote the sum of n independent discrete random variables, and assumes that S can take on consecutive integer values. Letting $\mu = E[S]$ and $\sigma^2 = \text{Var}[S]$, he argues that for sufficiently large values of n, S is approximately normally distributed. Using the standard continuity correction, Thompson uses the approximation

$$\Pr(S = s) \cong \Pr(s - 0.5 < N(\mu, \sigma^2) < s + 0.5).$$

Calculating a midpoint approximation using a single subinterval, the *normal PDF approximation* is obtained, which is

$$\Pr(S = s) \cong \frac{1}{\sqrt{2\pi}\sigma} e^{-(s-\mu)^2/2\sigma^2}.$$

Instead of settling for approximations of the probabilities, we will show how APPL procedures, including Convolution, can retrieve exact solutions, while still giving the same insight into the problem.

Example 8.10. Suppose that X_1, X_2, \ldots, X_{20} are independent and that $X_i \sim \text{Bernoulli}\left(p_i = \frac{29+2i}{100}\right)$, $i = 1, 2, \ldots, 20$. Let $S = \sum_{i=1}^{20} X_i$. (Here S denotes the total number of successes obtained in a series of independent trials where the probability of success varies from trial to trial.) Give an exact probability table for S for $s = 2, 3, \ldots, 10$ (from Thompson [112, page 53]).

Using Thompson's notation, $\mu = 10$ and $\sigma^2 = 2367/500$, and so $\Pr(S = s) \cong \frac{1}{\sqrt{2\pi}\sqrt{2367/500}} e^{-(s-10)^2/(2367/250)}$, $s = 0, 1, \ldots, 20$.

Using APPL, we obtain the exact distribution of the random variable S by calling Convolution inside a Maple for loop:

```
> p := (29 + 2 * 1) / 100;
> S := BernoulliRV(p);
> for i from 2 to 20 do
>    p := (29 + 2 * i) / 100:
>    X := BernoulliRV(p):
>    S := Convolution(S, X):
> od:
> S;
```

Table 8.1 contains the exact probabilities found with APPL and the normal PDF approximations for $s = 2, 3, \ldots, 10$.

Table 8.1. Exact probabilities and normal PDF approximations of $\Pr(S = s)$ for $s = 2, 3, \ldots, 10$

s	True $\Pr(S = s)$ from APPL using `Convolution`	Approximation of true $\Pr(S = s)$	Normal PDF approximation
2	$\frac{20465876514498922257889307137290 11}{1600000000000000000000000000000000000000}$	0.0001	0.0002
3	$\frac{67058104438186111727196296204396 7}{800000000000000000000000000000000000000}$	0.0008	0.0010
4	$\frac{12306309890051216090420607156481161}{3200000000000000000000000000000000000000}$	0.0038	0.0041
5	$\frac{13130118961411820609429234497062639}{1000000000000000000000000000000000000000}$	0.0131	0.0131
6	$\frac{13845545992556016094922419904605161}{4000000000000000000000000000000000000000}$	0.0346	0.0338
7	$\frac{14429186684261724023997491367619439}{2000000000000000000000000000000000000000}$	0.0721	0.0709
8	$\frac{19319652859308915302509324590493029 3}{16000000000000000000000000000000000000000}$	0.1207	0.1202
9	$\frac{6554941445025712560001435444760796 9}{400}$	0.1639	0.1650
10	$\frac{7253130084768895124176352940113025 41}{400}$	0.1813	0.1834

Example 8.11. There are 20 girls and 30 boys in Group 1, 25 girls and 25 boys in Group 2, and 10 girls and 10 boys in Group 3. If 10 children are randomly chosen from each group and S denotes the total number of girls chosen, give an exact probability table for S for $s = 7, 8, \ldots, 21$ (from Thompson [112, pages 53–54]).

Let X_1, X_2, and X_3 be the three independent hypergeometric random variables, and let $S = X_1 + X_2 + X_3$. The mean and variance of S are $\mu = E[S] = E[X_1] + E[X_2] + E[X_3] = 14$ and $\sigma^2 = \text{Var}[S] = \text{Var}[X_1] + \text{Var}[X_2] + \text{Var}[X_3] = 101/19$ (since X_1, X_2, and X_3 are

independent). Table 8.2 shows the normal PDF approximation values
$\Pr(S = s) \cong \frac{1}{\sqrt{2\pi}\sqrt{101/19}} e^{-(s-14)^2/(202/19)}$ for $s = 7, 8, \ldots, 21$.

Using the APPL Convolution procedure, we can readily produce the
PDF of S with the statements

```
> X1 := HypergeometricRV(50, 20, 10);
> X2 := HypergeometricRV(50, 25, 10);
> X3 := HypergeometricRV(20, 10, 10);
> Y := Convolution(X1, X2, X3);
```

Table 8.2. The exact probabilities and normal PDF approximations for $\Pr(S = s)$
for $s = 7, 8, \ldots, 21$

s	True $\Pr(S = s)$ from APPL using Convolution	Approximation of true $\Pr(S = s)$	Normal PDF approximation
7	$\frac{4641594894759547665}{3082276280132202064912}$	0.0015	0.0017
8	$\frac{97479371530863990}{17512933409842057187}$	0.0056	0.0059
9	$\frac{12613791756912076515}{770569070033050516228}$	0.0164	0.0165
10	$\frac{748495252604110945911}{1926422675082626290570}$	0.0389	0.0384
11	$\frac{57967137494173367365}{770569070033050516228}$	0.0752	0.0742
12	$\frac{2096975232909133615}{17512933409842057187}$	0.1197	0.1188
13	$\frac{22076335771392253895}{140103467278736457496}$	0.1576	0.1575
14	$\frac{317244095646532855}{1843466674720216546}$	0.1721	0.1730
15	$\frac{9955623438355053449}{63683394217607480680}$	0.1563	0.1575
16	$\frac{217921905682010165}{1843466674720216546}$	0.1182	0.1188
17	$\frac{1894259194489549345}{25473357687042992272}$	0.0744	0.0742
18	$\frac{71588441634588035}{1843466674720216546}$	0.0388	0.0384
19	$\frac{10756216836381565}{641205799902684016}$	0.0168	0.0165
20	$\frac{1208983087163529637}{202781334219223820060}$	0.0060	0.0059
21	$\frac{280730797358534065}{16222506737537905648}$	0.0017	0.0017

The exact values for $s = 7, 8, \ldots, 21$ are shown in Table 8.2.

In the next example, APPL is able to find the distribution of the sum of a large number (150) of random variables. While not impossible, computing the actual distribution by hand is tremendously tedious and time-consuming.

Example 8.12. Let the discrete random variable $S = X_1 + X_2 + \cdots + X_{150}$, where the X_i's are independent, $\Pr(X_i = -1) = \Pr(X_i = 0) = \Pr(X_i = 1) = 1/3$, $i = 1, 2, \ldots, 150$. Find a normal approximation to $\Pr(S = 5)$ and its exact value (from Thompson [112, page 54]).

Since the mean and variance of S are $\mu = 0$ and $\sigma^2 = 100$, using the normal PDF approximation we obtain approximately

$$\Pr(S = 5) \cong \frac{1}{20} \frac{\sqrt{2}e^{-1/8}}{\sqrt{\pi}} \cong 0.03521.$$

The APPL code

```
> X := [[1 / 3, 1 / 3, 1 / 3], [-1, 0, 1],
        ["Discrete", "PDF"]];
> S := ConvolutionIID(X, 150);
> PDF(S, 5);
```

yields the exact PDF for S. The statement PDF(S, 5) returns

$$\Pr(S = 5) = \frac{160709987007649212790999852367465829596098558279031212787052332840770}{45677590745077404064777874376752672121786802517249749853726469790033929},$$

which is approximately 0.03518. Since the support values of the parent population are -1, 0, and 1, $\Pr(S = 5)$ can also be computed using a combinatorics approach:

$$\Pr(S = 5) = \sum \sum \sum_{\substack{\{(p,q,r)|p+q+r=150, \\ 0 \le p \le 150, \\ 0 \le q \le 150, \\ 0 \le r \le 150, \\ r-p=5\}}} \binom{150}{p, q, r} \left(\frac{1}{3}\right)^p \left(\frac{1}{3}\right)^q \left(\frac{1}{3}\right)^r$$

or equivalently

$$\Pr(S = 5) = \sum_{p=0}^{72} \binom{150}{p, 145 - 2p, 5 + p} \left(\frac{1}{3}\right)^{150},$$

yielding the same result as the APPL code. The true power of APPL is not demonstrated in this particular example because of the special nature of the parent populations. The APPL approach allows for unequal mass values and more general support values. Also, more than three support values can be used in the APPL approach. Although a combinatorics method may still be used on these types of examples, the resulting expressions will not be as easy to derive or to evaluate as the expression obtained in the sum above.

The `Convolution` algorithm that was constructed (with heaps) to compute the PDF of the sum was extended to determine the PDF of the product also. Because of negative and positive support values for random variables X and Y, the product algorithm must split the array \hat{A} into four quadrants before forming heaps. The quadrants are split based on X's and Y's negative and non-negative support values using the support from Example 8.2. Figure 8.14 illustrates the evolution of the algorithm in the most general case when both X and Y have negative and positive support values. The product algorithm works from the corners of quadrants two and four toward the center where the quadrants meet, and then works outward from the center through quadrants one and three. For the product algorithm, the heap can receive values from two competing quadrants at a time. The `Product` procedure has been implemented for discrete random variables and is a part of APPL.

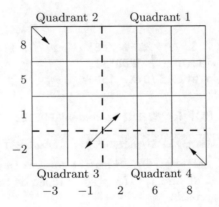

Fig. 8.14. Array A split into four quadrants for the `Product` procedure

This concludes our discussion concerning the algorithm to calculate the distribution of the sum of two discrete random variables. The next chapter considers another non-trivial algorithm associated with discrete random variables—one which determines the distribution of order statistics drawn from discrete populations.

9

Order Statistics for Random Sampling from Discrete Populations

This chapter presents an algorithm for computing the PDF of order statistics drawn from discrete parent populations, along with an implementation of the algorithm in APPL. Several examples illustrate the utility of this algorithm.

9.1 Notation and Taxonomy

Over a thousand references cited in the survey text on order statistics by David and Nagaraja [20] reflect the interest in the theory and application of order statistics. Many popular mathematical statistics texts, such as Hogg et al. [52], Casella and Berger [16], and Rice [103], address only distributions of order statistics for iid random variables drawn from continuous populations due to the mathematical tractability of their theoretical formulas.

Results for order statistics drawn from discrete parent populations are sparse and usually specialized to fit one particular discrete population (e.g., Young [123], Srivastava [110], Ciardo et al. [17]). Arnold et al. [5] devote a chapter to discrete order statistics. We present an algorithm for determining distributions of order statistics drawn from a general discrete parent population. The algorithm presented in this chapter handles discrete parent populations with finite or infinite support, and sampling with or without replacement. Application areas, bootstrapping in particular, are postponed until Chap. 10.

Let X_1, X_2, \ldots, X_n be iid discrete random variables defined on support Ω, each with CDF $F_X(x)$ and PDF $f_X(x)$. Let $X_{(1)} \leq X_{(2)} \leq \cdots \leq X_{(n)}$ denote these random variables rearranged in weakly increasing order. Thus, $X_{(r)}$ is the rth smallest number in the sample, $r = 1, 2, \ldots, n$, which is known as the rth "order statistic." Because order statistics are random variables, it is possible to compute probabilities associated with values in their support. Since X_1, X_2, \ldots, X_n is a random sample from a discrete population, the PDF of the rth order statistic cannot always be expressed as a single formula, as in

© Springer International Publishing AG 2017
J.H. Drew et al., *Computational Probability*, International Series in Operations Research & Management Science 246,
DOI 10.1007/978-3-319-43323-3_9

the continuous case (see Sect. 3.2). When working with discrete random variables, the computation of the PDF of the rth order statistic will fall into one of several categories, depending on the sampling convention (with or without replacement), the random variable's support (finite or infinite), and the random variable's distribution (equally likely or nonequally likely probabilities). A taxonomy of these categories appears in Fig. 9.1.

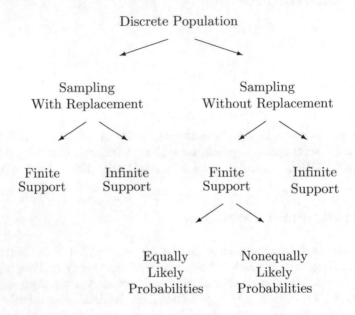

Fig. 9.1. Categorization of discrete order statistics by sampling convention, support, and probability distribution

The APPL procedure `OrderStat(X, n, r, ["wo"])` has three required arguments: a random variable `X`, the number of items `n` randomly drawn from the population with PDF $f_X(x)$, and the index `r` of the desired order statistic. An optional fourth argument `"wo"` can be specified to indicate that the items are drawn from the population *without* replacement. `OrderStat` can be used for continuous and discrete populations; the algorithm associated with discrete populations is detailed in this chapter. The implementation steps of the `OrderStat` algorithm are explained in detail in the following two sections ("Without Replacement" and "With Replacement") and further details and applications are given in Evans et al. [30]. In both cases, the output of the algorithm is $f_{X_{(r)}}(x)$, the PDF of the rth order statistic, where n items have been sampled either with or without replacement from a population with PDF $f_X(x)$. Each of the five branches displayed in the tree in Fig. 9.1 is discussed, followed by an example illustrating the particular sampling convention, support, and population PDF combinations. The particular branch from Fig. 9.1 being illustrated in each example is given in italics at the beginning of the example.

9.2 Sampling Without Replacement

Providing the string "wo" as the fourth argument to the OrderStat procedure indicates that items are drawn from the discrete population without replacement. Random variables with finite or infinite support are handled by two separate techniques.

If the population distribution has finite support, the population size is denoted by N. To specify this support in a compact form, it is temporarily assumed to be (without loss of generality) the ordered set $\{1, 2, \ldots, N\}$. For example, if

$$f_X(x) = \begin{cases} 0.5 & x = 7 \\ 0.2 & x = 11 \\ 0.3 & x = 15, \end{cases}$$

then $N = 3$, and the support of X is temporarily assumed to be $\{1, 2, 3\}$, instead of $\{7, 11, 15\}$. The original support is reinstated after the order-statistic probabilities are computed.

If the population with PDF $f_X(x)$ has equally likely probability values, e.g., rolling a fair die, then by combinatorial analysis the PDF of the rth order statistic when n items are sampled is [5, page 54]

$$f_{X_{(r)}}(x) = \frac{\binom{x-1}{r-1}\binom{N-x}{n-r}}{\binom{N}{n}} \qquad x = r, r+1, \ldots, r+N-n.$$

Example 9.1 (Sampling Without Replacement; Finite Support; Equally Likely Probabilities). Draw 15 cards at random and without replacement from a deck of 25 cards numbered $1, 2, \ldots, 25$. Find the probability that the card numbered 10 is the median of the cards selected [50, page 231].

Let $f_X(x) = \frac{1}{25}$ for $x = 1, 2, \ldots, 25$. The population size is $N = 25$, the size of the sample is $n = 15$, and the sample median $X_{(8)}$ corresponds to $r = 8$. The probability that the sample median equals 10 is given by the above formula as

$$\begin{aligned} \Pr(X_{(8)} = 10) &= \frac{\binom{9}{7}\binom{15}{7}}{\binom{25}{15}} \\ &= \frac{1053}{14858} \\ &\cong 0.0709. \end{aligned}$$

In addition to just computing one probability value as obtained by hand above, the OrderStat algorithm returns the PDF of the 8th order statistic (when $n = 15$ items are sampled from the population without replacement), which is

$$f_{X_{(8)}}(x) = \frac{1}{3268760} \binom{x-1}{7}\binom{25-x}{7} \qquad x = 8, 9, \ldots, 18.$$

The APPL code to compute the PDF of the sample median and the probability that the median equals 10 is

```
> X := UniformDiscreteRV(1, 25);
> Y := OrderStat(X, 15, 8, "wo");
> PDF(Y, 10);
```

If X has finite support N and nonequally likely probabilities, there are three cases to consider when computing the PDF of the rth order statistic:

1. One item is sampled: $n = 1$. The PDF of the rth order statistic is the same as the population PDF; i.e., $f_{X_{(r)}}(x) = f_X(x)$ for $r = 1, 2, \ldots, N$.

2. The entire population is sampled: $n = N$. The PDF of the rth order statistic is $f_{X_{(r)}}(x) = 1$ for $x = r$.

3. More than one item, but not the entire population, is sampled: $n = 2, 3, \ldots, N - 1$. In this nontrivial case, an $n \times N$ array, *ProbStorage*, is defined that eventually contains the probabilities $f_{X_{(r)}}(x)$ for $x = 1, 2, \ldots, N$ and $r = 1, 2, \ldots, n$. The value of $f_{X_{(r)}}(x)$ is found in row r and column x of the array. For example, if $X \sim \text{binomial}(5, \frac{1}{4})$ and $n = 3$ items are sampled (without replacement) from this population, then the *ProbStorage* array is 3×6 and has the form displayed in Table 9.1. The algorithm's implementation requires use of two APPL combinatorial procedures, `NextCombination` and `NextPermutation`, which were specifically created for use in the `OrderStat` algorithm. Following is a brief description of the algorithm for the without-replacement/finite support/nonequally-likely-probabilities case for $n = 2, 3, \ldots, N - 1$.

 (a) The first lexicographical combination of n items sampled from the sequence of integers 1 to N is formed; it is $\{1, 2, \ldots, n\}$. For $X \sim \text{binomial}(5, \frac{1}{4})$, the first lexicographical combination of $n = 3$ items from $\{1, 2, \ldots, 6\}$ is $\{1, 2, 3\}$.

Table 9.1. Initial form of the *ProbStorage* array for $X \sim \text{binomial}(5, \frac{1}{4})$, where $n = 3$ items are sampled without replacement. Bullets denote positive probability values; zeros denote impossible situations

r \ x	1	2	3	4	5	6
1	•	•	•	•	0	0
2	0	•	•	•	•	0
3	0	0	•	•	•	•

(b) Given a combination consisting of n distinct integers, the algorithm generates all possible permutations of that combination and their corresponding probabilities. For the example $X \sim \text{binomial}(5, \frac{1}{4})$, the first combination is $\{1, 2, 3\}$ and its permutations are $[1, 2, 3]$, $[1, 3, 2]$, $[2, 1, 3]$, $[2, 3, 1]$, $[3, 1, 2]$, $[3, 2, 1]$. The probabilities obtained for each of these permutations, respectively, are $\frac{13286025}{150351872}$, $\frac{13286025}{204334592}$, $\frac{13286025}{119164928}$, $\frac{13286025}{110607872}$, $\frac{13286025}{197270528}$, and $\frac{13286025}{134730752}$. As an example, the probability of obtaining the permutation $[3, 1, 2]$ is computed as

$$f_X(3) \cdot \frac{f_X(1)}{1 - f_X(3)} \cdot \frac{f_X(2)}{1 - \big(f_X(3) + f_X(1)\big)}$$

or

$$\frac{135}{512} \cdot \frac{\frac{243}{1024}}{1 - \frac{135}{512}} \cdot \frac{\frac{405}{1024}}{1 - \big(\frac{135}{512} + \frac{243}{1024}\big)} = \frac{13286025}{197270528}.$$

(c) After the probability of each permutation generated in step (b) is computed, each permutation is "rewritten" lexicographically to determine the corresponding order-statistic probabilities. The permutation $[3, 1, 2]$, for example, is rewritten as $[1, 2, 3]$. The algorithm then adds the probability of obtaining that particular permutation to the appropriate three cells in the *ProbStorage* array, which is denoted by $\sum(r, x) + q$ in Table 9.2. The (r, x) cell accumulates the probabilities of the various permutations for which the rth order statistic has the value x. For example, the final value in the $(1, 1)$ cell represents the probability that the first order statistic assumes the value one. In the binomial$(5, \frac{1}{4})$ example, the probability of obtaining the permutation $[3, 1, 2]$, $q = \frac{13286025}{197270528}$, is added to the *ProbStorage* cells $(1, 1)$, $(2, 2)$, and $(3, 3)$, as illustrated in Table 9.2.

(d) After all $n!$ permutations of a given combination are exhausted, the procedure NextCombination determines the next lexicographical combination. Given this new combination, the algorithm repeats steps (b) and (c). This process iterates through the $\binom{N}{n}$ combinations of size n chosen from a population of size N.

Table 9.2. Computation of the *ProbStorage* array for $X \sim \text{binomial}\big(5, \frac{1}{4}\big)$, where $n = 3$ items are sampled without replacement. Given the permutation $[3, 1, 2]$, its probability $q = \frac{13286025}{197270528}$ is added to the current probability sums $\sum(r, x)$ in cells $(1, 1)$, $(2, 2)$, and $(3, 3)$. Zeros denote impossible situations

r \ x	1	2	3	4	5	6
1	$\sum(1,1) + q$	$\sum(1,2)$	$\sum(1,3)$	$\sum(1,4)$	0	0
2	0	$\sum(2,2) + q$	$\sum(2,3)$	$\sum(2,4)$	$\sum(2,5)$	0
3	0	0	$\sum(3,3) + q$	$\sum(3,4)$	$\sum(3,5)$	$\sum(3,6)$

Example 9.2 (Sampling Without Replacement; Finite Support; Nonequally Likely Probabilities). Let X be a random variable with PDF

$$f_X(x) = \begin{cases} p_1 & x = 1 \\ p_2 & x = 10 \\ p_3 & x = 100 \\ p_4 & x = 1000, \end{cases}$$

where $p_1 + p_2 + p_3 + p_4 = 1$ and $p_i > 0$, $i = 1, 2, 3, 4$. Find the distribution of the median when $n = 3$ items are sampled without replacement from the population.

Since sampling is without replacement, the median can only assume the values $x = 10$ and $x = 100$. The APPL code to compute the distribution of the median is

```
> X := [[p1, p2, p3, p4], [1, 10, 100, 1000],
        ["Discrete", "PDF"]];
> OrderStat(X, 3, 2, "wo");
```

which returns the PDF

$$f_{X_{(2)}}(x) = \begin{cases} \dfrac{p_4 \, p_2 \, p_1}{(1 - p_4)(1 - p_2 - p_4)} + \dfrac{p_4 \, p_2 \, p_1}{(1 - p_4)(1 - p_1 - p_4)} + \\ \dfrac{p_4 \, p_2 \, p_1}{(1 - p_2)(1 - p_2 - p_4)} + \dfrac{p_4 \, p_2 \, p_1}{(1 - p_2)(1 - p_1 - p_2)} + \\ \dfrac{p_4 \, p_2 \, p_1}{(1 - p_1)(1 - p_1 - p_4)} + \dfrac{p_4 \, p_2 \, p_1}{(1 - p_1)(1 - p_1 - p_2)} + \\ \dfrac{p_3 \, p_2 \, p_1}{(1 - p_3)(1 - p_2 - p_3)} + \dfrac{p_3 \, p_2 \, p_1}{(1 - p_3)(1 - p_1 - p_3)} + \\ \dfrac{p_3 \, p_2 \, p_1}{(1 - p_2)(1 - p_2 - p_3)} + \dfrac{p_3 \, p_2 \, p_1}{(1 - p_2)(1 - p_1 - p_2)} + \\ \dfrac{p_3 \, p_2 \, p_1}{(1 - p_1)(1 - p_1 - p_3)} + \dfrac{p_3 \, p_2 \, p_1}{(1 - p_1)(1 - p_1 - p_2)} & x = 10 \\[4pt] \dfrac{p_4 \, p_3 \, p_2}{(1 - p_4)(1 - p_4 - p_3)} + \dfrac{p_4 \, p_3 \, p_2}{(1 - p_4)(1 - p_2 - p_4)} + \\ \dfrac{p_4 \, p_3 \, p_2}{(1 - p_3)(1 - p_4 - p_3)} + \dfrac{p_4 \, p_3 \, p_2}{(1 - p_3)(1 - p_2 - p_3)} + \\ \dfrac{p_4 \, p_3 \, p_2}{(1 - p_2)(1 - p_2 - p_4)} + \dfrac{p_4 \, p_3 \, p_2}{(1 - p_2)(1 - p_2 - p_3)} + \\ \dfrac{p_4 \, p_3 \, p_1}{(1 - p_4)(1 - p_4 - p_3)} + \dfrac{p_4 \, p_3 \, p_1}{(1 - p_4)(1 - p_1 - p_4)} + \\ \dfrac{p_4 \, p_3 \, p_1}{(1 - p_3)(1 - p_4 - p_3)} + \dfrac{p_4 \, p_3 \, p_1}{(1 - p_3)(1 - p_1 - p_3)} + \\ \dfrac{p_4 \, p_3 \, p_1}{(1 - p_1)(1 - p_1 - p_4)} + \dfrac{p_4 \, p_3 \, p_1}{(1 - p_1)(1 - p_1 - p_3)} & x = 100. \end{cases}$$

If the population distribution from which the sample is drawn has a countably infinite support, e.g., $\{1, 2, \ldots\}$, then the pattern established for finding

the PDF of the rth order statistic in the finite-support case will not work. At this time, beyond the trivial case when $n = 1$ item is sampled, where $f_{X_{(r)}}(x) = f_X(x)$ for $r = 1, 2, \ldots, N$, the OrderStat algorithm computes the PDF of the minimum order statistic when at most $n = 2$ items are sampled without replacement from a discrete population with infinite support. The $n = 2$ case is considered in the following paragraph. Work on the OrderStat algorithm to incorporate the pattern for values of $n \geq 3$ is an open research question.

When $n = 2$ items are sampled without replacement, the probability that the minimum order statistic has value x, for $x = 1, 2, \ldots$, is given by

$$\Pr(X_{(1)} = x) = \Pr(X_1 = x)\Pr(X_2 \geq x + 1 \mid X_1 = x) +$$

$$\sum_{y=x+1}^{\infty} \Pr(X_1 = y)\Pr(X_2 = x \mid X_1 = y).$$

Thus, the PDF of $X_{(1)}$ when $n = 2$ items are sampled is

$$f_{X_{(1)}}(x) = f_X(x)\left[\frac{S_X(x+1)}{1 - f_X(x)} + \sum_{y=x+1}^{\infty} \frac{f_X(y)}{1 - f_X(y)}\right] \qquad x = 1, 2, \ldots,$$

where $S_X(x)$ is the survivor function defined by $S_X(x) = \Pr(X \geq x)$.

Example 9.3 (Sampling Without Replacement; Infinite Support; Non equally Likely Probabilities). Let $X \sim \text{geometric}(\frac{1}{2})$. Find the probability that the minimum order statistic is five when $n = 2$ items are sampled without replacement from the population.

The PDF of the minimum order statistic as calculated using the OrderStat algorithm is

$$f_{X_{(1)}}(x) = \frac{\left(\frac{1}{4}\right)^x}{1 - \left(\frac{1}{2}\right)^x} + \frac{1}{2}\left(\frac{1}{2}\right)^{x-1} \sum_{y=x+1}^{\infty} \frac{1}{2}\frac{\left(\frac{1}{2}\right)^{y-1}}{1 - \frac{1}{2}\left(\frac{1}{2}\right)^{y-1}}$$

for $x = 1, 2, \ldots$. By substitution, it can easily be determined that $\Pr(X = 5) \cong 0.0020$. The PDF of the second order statistic and the probability that it assumes the value of five can be computed in APPL with the statements

```
> X := GeometricRV(1 / 2);
> Y := OrderStat(X, 2, 1, "wo");
> evalf(PDF(Y, 5));
```

9.3 Sampling with Replacement

If the optional fourth argument "wo" is *not* provided to the OrderStat procedure, the items are assumed to be sampled from the discrete population with replacement. When sampling with replacement, the PDF $f_X(x)$, CDF

$F_X(x)$, and SF $S_X(x)$ are needed to determine the distribution of the rth order statistic.

If the random variable X has finite support Ω, then without loss of generality we can assume that $\Omega = \{1, 2, \ldots, N\}$. The PDF of $X_{(r)}$ when n items are sampled with replacement from this finite population is given by

$$
f_{X_{(r)}}(x) = \begin{cases} \sum_{w=0}^{n-r} \binom{n}{w} [f_X(1)]^{n-w} [S_X(2)]^w & x = 1 \\ \sum_{u=0}^{r-1} \sum_{w=0}^{n-r} \binom{n}{u, n-u-w, w} [F_X(x-1)]^u [f_X(x)]^{n-u-w} [S_X(x+1)]^w & \\ & x = 2, 3, \ldots, N-1 \\ \sum_{u=0}^{r-1} \binom{n}{u} [F_X(N-1)]^u [f_X(N)]^{n-u} & x = N. \end{cases}
$$

This formula is not valid in the special case when the discrete population consists of only one item, that is, when $N = 1$. The PDF of the rth order statistic in this case is simply $f_{X_{(r)}}(1) = 1$.

The formula for the PDF of $X_{(r)}$ is a direct result of the following observation. For the rth order statistic to equal x, there must be u values less than or equal to $x - 1$ and w values greater than or equal to $x + 1$, where $u = 0, 1, \ldots, r - 1$ and $w = 0, 1, \ldots, n - r$. The other $n - u - w$ values must be equal to x. The CDF procedure is used to determine the probability of obtaining a value less than or equal to $x - 1$, the SF procedure is used to determine the probability of obtaining a value greater than or equal to $x + 1$, and the PDF procedure is used to determine the probability of obtaining the value x. The multinomial coefficient expresses the number of combinations resulting from a specific choice of u and w.

Using an example from Arnold et al. [5, page 43], let X be a discrete uniform random variable with PDF $f_X(x) = \frac{1}{4}$, $x = 1, 2, 3, 4$. Then the CDF and SF of X, respectively, are

$$
F_X(x) = \begin{cases} 0 & x < 1 \\ \dfrac{\lfloor x \rfloor}{4} & 1 \le x < 4 \\ 1 & x \ge 4 \end{cases} \qquad S_X(x) = \begin{cases} 1 & x \le 1 \\ 1 - \dfrac{\lfloor x \rfloor}{4} & 1 < x \le 4 \\ 0 & x > 4. \end{cases}
$$

Suppose $n = 5$ values are sampled with replacement from this population. To calculate $f_{X_{(2)}}(3)$, i.e., the probability that the second order statistic is $x = 3$, evaluate the sum

$$
f_{X_{(2)}}(3) = \sum_{u=0}^{1} \sum_{w=0}^{3} \binom{5}{u, 5-u-w, w} [F_X(2)]^u [f_X(3)]^{5-u-w} [S_X(4)]^w.
$$

The first term in the summation is the probability of drawing all threes. The second term, in which $u = 0$ and $w = 1$, is the probability of drawing four threes and a value greater than or equal to four (which can only be the value four in this example). This collection of four threes and one four can be drawn five different ways. The subsequent terms in the sum have similar meanings.

Example 9.4 (Sampling with Replacement; Finite Support). A fair die is cast eight independent times. Find the PDF of the smallest of the eight numbers obtained, $X_{(1)}$ [50, page 230].

To compute the numeric PDF by hand,

$$f_{X_{(1)}}(x) = \sum_{w=0}^{7} \binom{8}{w} \left(\frac{1}{6}\right)^{8-w} \left(1 - \frac{x}{6}\right)^{w}$$

is calculated for $x = 1, 2, \ldots, 6$. For example, the probability that the first order statistic is $x = 4$ is

$$\begin{aligned}
f_{X_{(1)}}(4) &= \sum_{w=0}^{7} \binom{8}{w} \left(\frac{1}{6}\right)^{8-w} \left(\frac{1}{3}\right)^{w} \\
&= \frac{1}{1679616} + \frac{1}{104976} + \frac{7}{104976} + \frac{7}{26244} + \frac{35}{52488} + \\
&\qquad \frac{7}{6561} + \frac{7}{6561} + \frac{4}{6561} \\
&= \frac{6305}{1679616} \\
&\cong 0.0038.
\end{aligned}$$

Similar calculations for $x = 1, 2, \ldots, 6$ yield the PDF of the first order statistic as

$$f_{X_{(1)}}(x) = \begin{cases} \dfrac{1288991}{1679616} & x = 1 \\[2mm] \dfrac{36121}{186624} & x = 2 \\[2mm] \dfrac{58975}{1679616} & x = 3 \\[2mm] \dfrac{6305}{1679616} & x = 4 \\[2mm] \dfrac{85}{559872} & x = 5 \\[2mm] \dfrac{1}{1679616} & x = 6. \end{cases}$$

The APPL code required to generate this PDF is:

```
> X := UniformDiscreteRV(1, 6);
> OrderStat(X, 8, 1);
```

The `OrderStat` algorithm yields the same results, but in addition returns the PDF of the minimum order statistic as a polynomial:

$$f_{X_{(1)}}(x) = -\frac{1}{209952}\,x^7 + \frac{91}{419904}\,x^6 - \frac{889}{209952}\,x^5 + \frac{38675}{839808}\,x^4$$

$$-\frac{63217}{209952}\,x^3 + \frac{496951}{419904}\,x^2 - \frac{543607}{209952}\,x + \frac{4085185}{1679616}$$

for $x = 1, 2, \ldots, 6$.

If the support of X is countably infinite, i.e., $\Omega = \{1, 2, \ldots\}$, then the calculation of the PDF of $X_{(r)}$ is similar to the finite-support case. The main difference is that the formula used in the finite-support case is now used for values of x that are unbounded:

$$f_{X_{(r)}}(x) = \begin{cases} \displaystyle\sum_{w=0}^{n-r} \binom{n}{w} \left[f_X(1)\right]^{n-w} \left[S_X(2)\right]^w & x = 1 \\ \displaystyle\sum_{u=0}^{r-1}\sum_{w=0}^{n-r} \binom{n}{u, n-u-w, w} \left[F_X(x-1)\right]^u \left[f_X(x)\right]^{n-u-w} \left[S_X(x+1)\right]^w \\ & x = 2, 3, \ldots \end{cases}$$

Because the formula assumes that the support of X is $\{1, 2, \ldots\}$, the algorithm only works for distributions with infinite right-hand tails.

Since it is physically impossible to execute the formula for infinitely many values of x, a general expression for $f_{X_{(r)}}(x)$ in terms of x is obtained by taking advantage of Maple's ability to sum and simplify complicated symbolic expressions. Maple computes (in symbolic terms) the double summation of a multinomial coefficient multiplied by the product of the CDF, PDF, and SF raised to various powers.

Example 9.5 (Sampling with Replacement; Infinite Support). Define a geometric random variable X with parameter p $(0 < p < 1)$ to be the trial number of the first success in repeated independent Bernoulli(p) trials. The PDF of X is $f_X(x) = p\,q^{x-1}$, for $x = 1, 2, \ldots$, where $q = 1 - p$. Margolin and Winokur [81] tabulated values for the mean and variance of the rth order statistic for a random sample drawn from a geometric population distribution for samples of size $n = 1, 5\,(5)\,20$ for $r = 1\,(1)\,5\,(5)\,20$, where $r = 1, 2, \ldots, n$ and $q = 0.25\,(0.25)\,0.75$. The values are calculated to two decimal places.

Margolin and Winokur's treatment of order statistics for the geometric population distribution is based on the use of recurrence relations. The formulas they use to compute the first and second moments of the rth order statistic for a random sample drawn from a geometric population distribution (with parameter $p = 1 - q$) when n items are sampled (with replacement) are

$$E\left[X_{(r)}\right] = n\binom{n-1}{r-1}\sum_{j=0}^{r-1}\frac{(-1)^j\binom{r-1}{j}}{(n-r+j+1)\left(1-q^{n-r+j+1}\right)}$$

and

$$E\left[X_{(r)}^2\right] = n\binom{n-1}{r-1}\sum_{j=0}^{r-1}\frac{(-1)^j\binom{r-1}{j}\left(1+q^{n-r+j+1}\right)}{(n-r+j+1)\left(1-q^{n-r+j+1}\right)^2}.$$

The OrderStat algorithm can easily produce the *exact value* of any of the rounded figures given in their tables. If $X \sim$ geometric$(\frac{1}{4})$, for example, then the exact values of the mean and variance of the third order statistic when $n = 5$ items are sampled can be found with the following APPL code

```
> X := GeometricRV(1 / 4);
> Y := OrderStat(X, 5, 3);
> Mean(Y);
> Variance(Y);
```

yielding

$$E\left[X_{(3)}\right] = \frac{3257984}{1011395} \cong 3.22$$

and

$$\text{Var}\left[X_{(3)}\right] = \frac{13665723739776}{5114599230125} \cong 2.67.$$

As shown in Example 9.5, the OrderStat procedure can be combined with the Mean and Variance procedures to compute the mean and variance of various order statistics. APPL might be used in this context to eliminate the need for some of the existing tables in the statistical literature. In addition, Maple can compute results using symbolic parameters. If $X \sim$ geometric(p), for example, then the variance of the first order statistic when $n = 6$ observations are sampled randomly from the geometric distribution with replacement can be computed with the APPL statements

```
> X := GeometricRV(p);
> Y := OrderStat(X, 6, 1);
> Variance(Y);
```

which yield

$$\text{Var}\left[Y_{(1)}\right] = \frac{p^6 - 6p^5 + 15p^4 - 20p^3 + 15p^2 - 6p + 1}{p^7 - 6p^6 + 15p^5 - 20p^4 + 15p^3 - 6p^2}.$$

9.4 Extension

One natural extension to the `OrderStat` algorithm is a `RangeStat` algorithm, which finds the PDF of the range of a sample of n items drawn from a discrete population, either with or without replacement.

Let X be a discrete random variable with PDF $f_X(x)$ and CDF $F_X(x)$, and let Y be a random variable representing the range of X. We can assume (without loss of generality) that the support of X is $\{1, 2, \ldots, N\}$, where N is a positive integer or infinity. If we are sampling with replacement (which implies that the n draws are independent), then the probability that Y assumes the value y, where $y = 0, 1, \ldots, N - 1$, is given by Burr [14]:

$$
\Pr(Y = y) = \begin{cases}
\displaystyle\sum_{x=1}^{N} [f_X(x)]^n & y = 0 \\
\displaystyle\sum_{x=1}^{N-y} \{[\Pr(x \leq X \leq x+y)]^n - [\Pr(x+1 \leq X \leq x+y)]^n \\
\quad -[\Pr(x \leq X \leq x+y-1)]^n + [\Pr(x+1 \leq X \leq x+y-1)]^n\} = \\
\displaystyle\sum_{x=1}^{N-y} \{[F_X(x+y) - F_X(x-1)]^n - [F_X(x+y) - F_X(x)]^n \\
\quad -[F_X(x+y-1) - F_X(x-1)]^n + [F_X(x+y-1) - F_X(x)]^n\} \\
\hspace{6cm} y = 1, 2, \ldots, N - 1.
\end{cases}
$$

Of the four terms being summed in the $y = 1, 2, \ldots, N-1$ case, the first term is the probability that all sampled elements lie between x and $x+y$ inclusive. The second term removes the probability that these elements do not include x, because this would result in a range that is less than y. The third term removes the probability that these elements do not include $x+y$, because this would also result in a range that is less than y. The fourth term adds back the probabilities of those elements that include neither x nor $x + y$, which were removed by both the second and third terms.

In the case of sampling without replacement, the `RangeStat` algorithm basically follows the same steps as the `OrderStat` algorithm, including use of the procedures `NextCombination` and `NextPermutation`. In the trivial case when the entire population is sampled, the PDF of the range Y is $f_Y(y) = 1$ for $y = N - 1$. If $n = 2, 3, \ldots, N$, where N is a positive integer less than or equal to n, then a single-dimensional array of length $N - n + 1$ is defined that will eventually contain the values of $f_Y(y)$ for $y = n - 1, n, \ldots, N - 1$. As in the `OrderStat` algorithm, the first lexicographical combination of n items sampled from the sequence of integers 1 to N is formed. Given such a combination, the algorithm generates all possible permutations of that combination. The probability for each permutation is calculated (as described in Sect. 9.1), and then the maximum and minimum values of that permutation

are determined. The permutation's range, which is the difference between its maximum and minimum values, is computed and then the appropriate array cell is incremented by that permutation's probability.

This concludes our survey of data structures and selected algorithms in APPL. The remaining portion of the monograph concerns the application of the language to problems in computational probability.

Part IV

Applications

When you can measure what you are speaking about and
express it in numbers, you know something about it;
but when you cannot measure it, when you cannot express
it in numbers, your knowledge is of the meager and
unsatisfactory kind.

—Lord Kelvin

Reliability and Survival Analysis

The remaining chapters contain dozens of computational probability applications using APPL. The applications range in complexity from brief examples to results and algorithms requiring long derivations. This chapter surveys some applications in reliability and the closely-related field of survival analysis.

Section 10.1 presents four brief examples concerning the probabilistic analysis of systems of components: (1) analyzing a k-out-of-n system; (2) determining the mean time to failure of a parallel system with symbolic parameters; (3) determining the conditional survivor function for a system with a piecewise survival distribution; and (4) determining the probability of mission success for a unit and two cold standby units. Section 10.2 contains an algorithm for determining a lower confidence bound on the system reliability from binary failure data collected on each of the independent components in a coherent system via bootstrapping. Section 10.3 uses APPL to avoid the use of resampling in performing a bootstrap analysis in a survival analysis setting. Section 10.4 uses APPL in bootstrapping applications in systems reliability modeling.

10.1 Systems Analysis

Reliability engineers often compare competing system designs by the use of probabilistic analysis. This section considers a series of four short examples that use APPL procedures to manipulate random variables associated with component lifetimes in order to compute measures of performance associated with the system, such as the mean time to failure or the system reliability. We begin with the analysis of a k-out-of-n system.

© Springer International Publishing AG 2017
J.H. Drew et al., *Computational Probability*, International Series
in Operations Research & Management Science 246,
DOI 10.1007/978-3-319-43323-3_10

Example 10.1. Consider a 4-out-of-6 system that operates when four or more of its six components are functioning. Let the six component lifetimes be iid Weibull random variables with SF

$$S(t) = e^{-(5t)^2} \qquad\qquad t \geq 0.$$

Use APPL to find
(a) the hazard function of the time to system failure,
(b) the mean time to system failure,
(c) the variance of the time to system failure,
(d) the probability that the system survives to time 0.1.

Let X denote the component lifetimes and T denote the system lifetime. The APPL code to calculate the four quantities of interest is

```
> X := WeibullRV(5, 2);
> T := OrderStat(X, 6, 3);
> HF(T);
> Mean(T);
> Variance(T);
> SF(T, 1 / 10);
```

The first statement defines the random variable X, which is associated with the lifetimes of the components, each with a Weibull time to failure with scale parameter 5 and shape parameter 2. Since this system has six components, the system failure occurs upon the third failure. Therefore, the system lifetime is the third order statistic of six independent Weibull random variables, denoted by T. The HF procedure returns the hazard function of the system time to failure, which is

$$h_T(t) = \frac{3000t\left(1 - e^{-25\,t^2}\right)^2}{10\,e^{-50\,t^2} - 24\,e^{-25\,t^2} + 15} \qquad\qquad t \geq 0.$$

The next statement computes the mean of the system time to failure, which is

$$E[T] = -\frac{12\sqrt{5\pi}}{25} + \frac{\sqrt{6\pi}}{6} + \frac{3\sqrt{\pi}}{4} \cong 0.1505.$$

The next statement computes the variance of the system time to failure, which is

$$\mathrm{Var}[T] = \frac{37}{1500} - \frac{11287\pi}{6000} + \frac{4\pi\sqrt{30}}{25} + \frac{18\pi\sqrt{5}}{25} - \frac{\pi\sqrt{6}}{4} \cong 0.002003.$$

Finally, the probability that the system survives to time 0.1 is determined by the SF procedure, which yields

$$S_T(0.1) = 10\,e^{-3/2} - 24\,e^{-5/4} + 15\,e^{-1} \cong 0.8734.$$

Following each of the final three APPL statements with the Maple statement `evalf(%);` computes the floating-point approximations associated with the exact values.

All of the parameters in Example 10.1 were numeric, which resulted, for example, in numeric values for the system mean time to failure and probability of system survival to time 0.1. A similar analysis can be applied to achieve a more general solution for a system with symbolic parameters, as illustrated in Example 10.2.

Example 10.2. Find the expected time to system failure of a parallel system, where each of the n components has an independent exponential(λ) lifetime.

As in the previous example, let X denote the component lifetimes and T denote the system lifetime. Since the reliability function of a parallel system of n identical components, each with reliability p, is

$$r(p) = 1 - (1-p)^n$$

and the survivor function of an exponential random variable with failure rate λ is

$$S_X(t) = e^{-\lambda t} \qquad t > 0,$$

the system survivor function is

$$S_T(t) = 1 - \left(1 - e^{-\lambda t}\right)^n \qquad t > 0.$$

The mean time to system failure is given by the expression

$$E[T] = \int_0^\infty \left(1 - \left(1 - e^{-\lambda t}\right)^n\right) dt.$$

This quantity can be calculated, for instance when $n = 8$, with the APPL statements

```
> X := ExponentialRV(lambda);
> T := MaximumIID(X, 8);
> Mean(T);
```

which return a mean time to system failure of $E[T] = 761/(280\lambda)$. The MaximumIID(X, 8) procedure call is equivalent to a call to OrderStat(X, 8, 8). This solution did not require that the user know the functional form of the reliability function of a parallel system. If, on the other hand, the modeler did know the functional form of the reliability function, the solution can be checked using entirely Maple statements:

```
> assume(lambda > 0);
> s := 1 - (1 - exp(-lambda * t)) ^ 8;
> int(s, t = 0 .. infinity);
```

which also return a mean time to system failure of $E[T] = 761/(280\lambda)$. The advantage afforded by APPL is that the mathematical details associated with the derivation can be bypassed.

Examples 10.1 and 10.2 have considered components and systems with lifetimes characterized by having positive support values. Situations occasionally arise where the component lifetimes have varying support, perhaps due to a warranty or burn-in program. The next example considers a system which has varying supports, making the bookkeeping tedious. APPL relieves the modeler of the tedium of keeping track of the varying supports.

Example 10.3. Consider a three-component system with component 1 in parallel with the series arrangement of components 2 and 3, as shown in the block diagram in Fig. 10.1. Let T_1, T_2, and T_3 be the independent lifetimes of the three components, where

$$T_1 \sim U(2, 10) \qquad T_2 \sim U(4, 14) \qquad T_3 \sim U(6, 18).$$

Find the 90th percentile of the distribution of the failure time of such a system that has survived to time 8.

Fig. 10.1. Three-component system

Although this particular problem is conceptually straightforward, keeping track of the various supports for the components is tedious. Let $T = \max\{T_1, \min\{T_2, T_3\}\}$ denote the system failure time and let T_c denote the system failure time conditioned on survival to time $t = 8$. The APPL code to solve this problem is given below.

```
> T1 := UniformRV(2, 10);
> T2 := UniformRV(4, 14);
> T3 := UniformRV(6, 18);
> T  := Maximum(T1, Minimum(T2, T3));
> TC := Truncate(T, 8, infinity);
> IDF(TC, 0.90);
```

This code returns the 90th percentile of the conditional distribution of the system lifetime as

$$\frac{32 - \sqrt{46}}{2} \cong 12.6088.$$

This code illustrates three new APPL procedures that have not yet been introduced. The first three statements define the distribution of T_1, T_2, and T_3. The next statement calculates the PDF of T using the

Minimum and Maximum procedures, which each take on two indepen-
dent, but not necessarily identically distributed, random variables as
arguments. The next statement uses the Truncate procedure to calcu-
late the conditional distribution given survival to time 8. Finally, the
IDF procedure is used to calculate the 90th percentile of T_c. The entire
distribution of T_c has been computed in case one wants to perform
further operations.

The APPL code given above is the ideal way to solve this particular
problem. If one were uneasy about the correctness of the solution, it
could be verified by Monte Carlo simulation. This Monte Carlo simula-
tion is possible in APPL and Maple, although it is not computationally
efficient in either language since Maple is an interpreted language. The
Maple code below generates 10,000 system lifetimes and estimates the
quantity of interest. Surprisingly, Maple does not contain a random
number generator for producing pseudorandom numbers. Instead, we
use Maple's **rand** function, which generates a random 12-digit integer.

```
> _seed := 17:
> variates := []:
> for i from 1 to 10000 do
>    t1 := IDF(UniformRV(2, 10), rand() / 10 ^ 12):
>    t2 := IDF(UniformRV(4, 14), rand() / 10 ^ 12):
>    t3 := IDF(UniformRV(6, 18), rand() / 10 ^ 12):
>    t  := max(t1, min(t2, t3)):
>    if (t > 8) then
>      variates := [op(variates), t]:
>    fi:
> od:
> variates := sort(variates):
> n := nops(variates):
> print(evalf(variates[floor(0.90 * n)]));
```

The first statement initializes the random number generator seed. The
second statement initializes **variates** to an empty list. The **for** loop
cycles through the 10,000 replications of the simulation. The random
variates $T_1 \sim U(2, 10)$, $T_2 \sim U(4, 14)$, and $T_3 \sim U(6, 18)$ are generated
in the first three lines of code within the **for** loop using the IDF
procedure. This is followed by the calculation of the (unconditional)
system failure time $T = \max\{T_1, \min\{T_2, T_3\}\}$. If the value of T
exceeds 8, then it is concatenated onto the right of the list **variates**.
Outside of the **for** loop, the elements in **variates** are sorted using the
Maple **sort** command. The **nops** function is used to find **n**, the number
of elements in **variates**. Finally, the estimated 90th percentile of the
conditional system lifetime is printed. When this code was executed
five times (with **_seed** fixed prior to the first execution) the following
five *estimates* for the conditional failure time were generated:

12.57 12.62 12.63 12.57 12.58.

The output of these five runs of Monte Carlo simulation hovers around the theoretical value of 12.61, which supports the analytic solution provided by APPL. A more careful statistical analysis would involve calculating a confidence interval from the five estimates.

The three previous examples all considered systems with "warm" standby units, i.e., all components are on-line until failure. Some systems have cold standby units, as illustrated in the next example.

Example 10.4. Let T_1, T_2, T_3 be independent gamma random variables that represent the lifetime of a unit (T_1) and the lifetimes of two backup cold standby units $(T_2$ and $T_3)$ taken on a mission of duration 10. If $T_1 \sim \text{gamma}(1/2, 3)$, $T_2 \sim \text{gamma}(1/4, 5)$, and $T_3 \sim \text{gamma}(1/6, 7)$, find the probability that the mission is successful.

Let the system lifetime $T = T_1 + T_2 + T_3$. The APPL code required to solve this problem is

```
> T1 := GammaRV(1 / 2, 3);
> T2 := GammaRV(1 / 4, 5);
> T3 := GammaRV(1 / 6, 7);
> T  := Convolution(T1, T2, T3);
> SF(T, 10);
```

If there were more components with distributions following the same pattern and a Maple `for` loop is preferable. The following equivalent APPL code could be used, replacing the upper limit of the `for` by the number of components.

```
> T := GammaRV(1 / 2, 3);
> for i from 2 to 3 do
>   T := Convolution(T, GammaRV(1 / (2 * i), 2 * i + 1)):
> od:
> SF(T, 10);
```

In either case, the probability of a successful mission, i.e., $\Pr(T \geq 10)$ is returned as

$$-\frac{57415240}{3} e^{-5/2} + \frac{2129286487}{256} e^{-5/3} + \frac{739}{256} e^{-5} \cong 0.9995.$$

If this probability of mission success is not high enough, more cold standby units could be added to the system.

10.2 Lower Confidence Bound on System Reliability

The examples to this point in the monograph have been drawn mainly from applied probability. This section takes up a problem in reliability that has a statistical nature.

We consider the problem of determining a $(1 - \alpha)100\%$ lower confidence bound p_L on the system reliability for a coherent system of k components with independent lifetimes. Components of type i are subjected to a life test, $i = 1, 2, \ldots, k$. The failure data is (y_i, n_i), where y_i is the number of components of type i that pass the test and n_i is the number of components of type i being tested, $i = 1, 2, \ldots, k$. We assume throughout that the components fail independently, e.g., no common-cause failures. Bootstrapping is used to estimate a lower confidence bound on the system reliability.

A three-component ($k = 3$) series system is used as an example throughout this section, although the techniques described here apply to any coherent system of k independent components. A block diagram for the series system is given in Fig. 10.2.

Fig. 10.2. Three-component series system

Binary data is collected on each component in the system. The number of components tested and the number of passes for each type of component for the example problem are given in Table 10.1.

Table 10.1. Failure data for a three-component series system example

Component number	$i = 1$	$i = 2$	$i = 3$
Number passing (y_i)	21	27	82
Number on test (n_i)	23	28	84

The point estimate for the system reliability is

$$\frac{21}{23} \cdot \frac{27}{28} \cdot \frac{82}{84} = \frac{1107}{1288} \cong 0.8595.$$

There is no known exact confidence interval for this system reliability. We will consider only a *lower confidence bound* on the system reliability here, because this is typically the quantity of interest to a reliability engineer when assessing risk. The *bootstrapping* technique proposed by Efron and Tibshirani [26] can be used to calculate a lower 95% confidence interval bound. Other authors, such as Martin [85] and Padgett and Tomlinson [96], have proposed the use

of bootstrapping for determining confidence limits. The approach used here differs conceptually from the standard application of bootstrapping, in which the standard error of a *single* unknown distribution is oftentimes estimated by resampling iid data. In our setting, there are k *different* distributions—one for each component—and we resample component reliabilities and combine using the reliability function to yield the system reliability.

Bootstrapping resamples some fixed number B of the systems, calculates the system reliability, then calculates the αBth ordered system reliability. More specifically, for the three-component series system of interest, the bootstrapping algorithm follows the following steps:

- For the first component, the data set (21 ones and 2 zeros) is sampled with replacement 23 times.
- These values are summed and divided by 23 yielding a reliability estimate for the first component.
- The previous two steps are repeated for components 2 and 3.
- The product of the reliability estimates for the three components are multiplied (because the components are arranged in series and their failures are independent) to give a system reliability estimate.

The four steps listed above are repeated B times. The B system reliability estimates are then sorted. Finally, the αBth ordered system reliability is output, which is used as a lower bound on the system reliability.

The algorithm for estimating the $(1 - \alpha)100\%$ lower confidence interval bound is given in Table 10.2, where \tilde{p}_i is a bootstrap estimate for the reliability of component i and z_j is a bootstrap estimate of the system reliability. Indentation is used to indicate begin–end blocks in the pseudocode. Generating a binomial random variate is appropriate because the resampling from the data set is performed with replacement. The returned value $z_{\alpha B}$ is the order statistic associated with the z_j's generated in the outside loop. See Law [70] for handling the case when αB is not an integer.

This algorithm has been implemented in R as a function named `seriessystemboot`, which is given next.

Table 10.2. Bootstrap algorithm for calculating a $(1 - \alpha)100\%$ lower confidence bound for the reliability of a k-component series system

for $j \leftarrow 1$ to B	resampling loop
\quad for $i \leftarrow 1$ to k	loop through components
$\quad\quad \tilde{p}_i \leftarrow \text{binomial}(n_i, y_i/n_i)/n_i$	component i reliability
$\quad z_j \leftarrow \prod_{i=1}^{k} \tilde{p}_i$	calculate system reliability
sort \mathbf{z}	sort the system reliability values
return $z_{\alpha B}$	return the estimate for the lower bound

```
> seriessystemboot <- function(n, y, alpha) {
>     k <- length(n)
>     b <- 10000
>     z <- rep(1, b)
>
>     point <- prod(y) / prod(n)
>
>     for (j in 1:b) {
>       for (i in 1:k) {
>         z[j] <- z[j] * rbinom(1, n[i], y[i] / n[i]) / n[i]
>       }
>     }
>     z <- sort(z)
>     pl <- z[floor(alpha * b)]
>     c(point, pl)
> }
```

The first two arguments, n and y, are vectors of length k, and the third argument, alpha, is a real number between 0 and 1. Thus, the R function call

```
> seriessystemboot(c(23, 28, 84), c(21, 27, 82), 0.05)
```

prints a point estimate and a 95% lower confidence interval bound on the system reliability for the three-component series system considered in this section. After a call to set.seed(3) to initialize the random number seed, five calls to seriessystemboot yielded the following estimates for the 95% lower bound on the system reliability p_L:

 0.7426057 0.7482993 0.7456744 0.7486690 0.7453416.

The dispersion associated with these five estimates is due to the finite choice of B, i.e., $B = 10,000$. Based on these results, the third digit after the decimal point can't be reported with certainty, so the 95% lower bound on the system reliability is best reported for these simulations as $p_L = 0.75$. A modeler would be uncertain how many digits to report on the estimate in a general setting. If the effect of the choice of B could be eliminated, there would be no need for concern about the precision of the estimates. So a major drawback of this approach is our inability to determine the size of the resampling error as a function of the size of B.

Resampling error can be eliminated using APPL. The APPL statements given below utilize the Product and Transform procedures.

```
> X1 := BinomialRV(23, 21 / 23);
> X1 := Transform(X1, [[x -> x / 23], [0, infinity]]);
> X2 := BinomialRV(28, 27 / 28);
> X2 := Transform(X2, [[x -> x / 28], [0, infinity]]);
> X3 := BinomialRV(84, 82 / 84);
> X3 := Transform(X3, [[x -> x / 84], [0, infinity]]);
> T := Product(X1, X2, X3);
```

This alternative approach to determining a lower 95% bootstrap confidence interval bound for the system reliability is equivalent to using an infinite value for B. This is equivalent to deriving the entire PDF of the bootstrap estimate of the system reliability. Since \tilde{p}_i can assume any one of $n_i + 1$ values, there are a possible $24 \cdot 29 \cdot 85 = 59{,}160$ potential mass values for the random variable T determined by the two calls to the Product procedure. Of these, only 6633 are distinct because the Product procedure combines repeated values. Since the random variable T from the APPL code plays an analogous role to the vector \mathbf{z} from the bootstrap algorithm in Table 10.2, the lower 95% bootstrap confidence interval bound is the 0.05 fractile of the distribution of T, which is $p_L = 6723/9016 \cong 0.746$. This result using APPL is consistent with the standard resampling approach for finding the lower confidence interval limit based on the five results presented earlier (one equals 6723/9016 exactly, two are above 6723/9016, and two are below 6723/9016).

Bootstrapping worked well in this example, but will encounter problems when systems with perfect test results on one or more components are encountered. The problem that arises when component i has perfect test results (i.e., $y_i = n_i$) is that the sample size n_i becomes irrelevant. A test in which 2 components out of 2 pass the test is equivalent to one in which 100 components out of 100 pass the test from the perspective of the bootstrapping algorithm. Efforts to overcome the problem associated with perfect component test results by placing a beta conjugate prior distribution on component reliabilities are outlined in Leemis [73].

10.3 Survival Analysis

Survival analysis is a branch of statistics that concerns "time-to-event" data. For biological organisms, the interest might be in the time to death or the remission time of a cancer. In reliability, the interest might be in the time to failure or the time to repair for a system. In actuarial science, the interest might be in the time to death of the insured or the time to first claim on the insurance of a structure. In sociology, the interest might be in the duration of a strike or the time to commit another crime after release from prison in a recidivism study. In economics, survival analysis is known as duration analysis or duration modeling. The key element in a survival analysis study is a nonnegative random variable (typically the time to some event) with significant dispersion and oftentimes with an associated vector of covariates. Two popular textbooks in survival analysis are Lawless [71] and Kalbfleisch and Prentice [57].

In this section, we investigate bootstrapping in survival analysis. More specifically, we consider the simple question of comparing some aspect of two populations. Using Efron and Tibshirani's [26] notation, this section again considers eliminating the generation of B bootstrap samples when performing

a bootstrap analysis by calculating the exact distribution of the statistic of interest via APPL. There are several reasons for considering this approach:

- Although bootstrap methods are relatively easy to apply, determining the number of bootstrap repetitions, B, to employ is a common problem facing practitioners [4]. A practitioner needs to be concerned about problem-specific requirements for B, e.g., "B in the range of 50 to 200 usually makes \hat{se}_{boot} a good standard error estimator, even for estimators like the median" [26, page 14] or "B should be \geq 500 or 1000 in order to make the variability of [the estimated 95th percentile] acceptably low" for estimating 95th percentiles [26, page 275]. In fact, "one can obtain a 'different answer' from the same data merely by using different simulation draws if B is too small, but computational costs can be great if B is chosen to be extremely large" [3, page 23]. Because of this, Andrews and Buchinsky [3] introduced a three-step method to determine B to attain a specified level of accuracy.
- Exact values are always preferred to approximations. There is no point in adding resampling error to the already-existing sampling error unnecessarily.
- A bootstrapping novice can easily confuse the sample size n and number of bootstrap samples B. Eliminating the resampling of the data set B times simplifies the conceptualization of the bootstrap process.
- In many situations, computer time is saved using the exact approach.

By way of example, this section shows how the OrderStat and RangeStat procedures, along with other APPL procedures, can be used to perform exact bootstrap analysis. The use of these algorithms eliminates the resampling variability that is present in the usual bootstrap procedure. Exact results in the bootstrapping literature are limited: "The only exact solutions that are available for the bootstrap variance of an L-statistic (e.g., mean, trimmed mean, median, quick estimators of location and scale, upper and lower quartiles) are for the specific cases of the sample mean and sample median (when the sample size is odd)" [53, page 89]. The application areas that will be presented here are the estimation of standard errors for the median, mean, and range, and interval estimation for the range.

Estimation of Standard Errors. The standard error of the sample mean, s/\sqrt{n}, is useful when comparing means, but standard errors for comparing other quantities (e.g., fractiles) are often intractable. The following examples consider the estimation of standard errors associated with the rat-survival data given in Table 10.3 from Efron and Tibshirani [26, page 11]. Seven rats are given a treatment and their survival times, given in days, are shown in the first row of the table. Nine other rats constitute a control group, and their survival times are shown in the second row of the table.

Table 10.3. Rat-survival data

Group	Data	n	Median	Mean	Range
Treatment	16, 23, 38, 94, 99, 141, 197	7	94	86.86	181
Control	10, 27, 30, 40, 46, 51, 52, 104, 146	9	46	56.22	136

In the first example, we compare the medians of the two populations using bootstrapping.

Example 10.5. Consider first the estimation of the standard error of the difference between the medians of the two samples. The standard bootstrap approach to estimating the standard error of the median for the treatment group is to generate B bootstrap samples, each of which consists of seven samples drawn with replacement from 16, 23, 38, 94, 99, 141, and 197. The sample standard deviation of the medians of these B bootstrap samples is an estimate of the standard error of the median. Using the R commands

```
> tr = c(16, 23, 38, 94, 99, 141, 197)
> B = 50
> medians = numeric(B)
> for (i in 1:B) {
>    bootstrap = sample(tr, 7, replace = TRUE)
>    medians[i] = median(bootstrap)
> }
> sd(medians)
```

yielded an estimated standard error of 41.18 for the treatment data with $B = 50$ bootstrap replicates. Table 10.4 shows the estimated standard errors for several B values. There is considerable resampling error introduced for smaller values of B. The $B = +\infty$ column of Table 10.4 corresponds to the *ideal bootstrap estimate of the standard error* of $\hat{\theta}$, or $\mathrm{se}_{\hat{F}}(\hat{\theta}^*) = \lim_{B \to +\infty} \hat{\mathrm{se}}_B$, to use the terminology and notation in Efron and Tibshirani [26, page 46].

Table 10.4. Bootstrap estimates of the standard error of the median

	$B = 50$	$B = 100$	$B = 250$	$B = 500$	$B = 1000$	$B = +\infty$
Treatment	41.18	37.63	36.88	37.90	38.98	37.83
Control	20.30	12.68	9.538	13.10	13.82	13.08

We now consider the possibility of avoiding resampling altogether using APPL. An additional APPL procedure was constructed, namely `BootstrapRV(data)` (henceforth referred to as `BootstrapRV`), whose only argument is a list of n data values named `data`. The `BootstrapRV`

procedure creates a discrete random variable X that can assume the provided data values, each with probability $\frac{1}{n}$. If we put the treatment data values 16, 23, 38, 94, 99, 141, and 197 in a list and provide this list to the BootstrapRV procedure, it creates a discrete random variable X that can assume these data values each with probability $\frac{1}{7}$. The following APPL code calculates the bootstrap standard deviation of the median:

```
> X := BootstrapRV([16, 23, 38, 94, 99, 141, 197]);
> Y := OrderStat(X, 7, 4);
> sqrt(Variance(Y));
```

The random variable X has a discrete uniform distribution with support values equal to the data values from the treatment data set. The random variable Y is the fourth order statistic (the median) in seven draws with replacement from the treatment group of the rat-survival data set. Since the median in a resampling experiment must take on one of the seven data values, the support of Y contains seven elements. APPL returns the distribution of the random variable Y as

$$f_Y(y) = \begin{cases} \dfrac{8359}{823543} & y = 16 \\[2mm] \dfrac{80809}{823543} & y = 23 \\[2mm] \dfrac{196519}{823543} & y = 38 \\[2mm] \dfrac{252169}{823543} & y = 94 \\[2mm] \dfrac{196519}{823543} & y = 99 \\[2mm] \dfrac{80809}{823543} & y = 141 \\[2mm] \dfrac{8359}{823543} & y = 197. \end{cases}$$

The last APPL statement takes the square root of the variance of Y, yielding the standard error as $\frac{2}{823543}\sqrt{242712738519382} \cong 37.8347$. This is the upper-right-hand entry in Table 10.4. In a similar fashion, the *ideal bootstrap estimate of the standard error* of the median can be calculated in the control case giving the lower-right-hand entry in Table 10.4 as $\frac{1}{387420489}\sqrt{25662937134123797402} \cong 13.0759$. Finally, note that the seemingly large difference between the two sample medians $(94 - 46 = 48)$ is only $48/\sqrt{37.83^2 + 13.08^2} \cong 1.19$ standard-deviation units away from zero, indicating that the observed difference in the medians is not statistically significant. Had the standard bootstrap procedure been applied with $B = 50$ bootstrap replications, Table 10.4

indicates that the number of standard-deviation units would have been estimated to be $48/\sqrt{41.18^2 + 20.30^2} \cong 1.05$. Although the conclusion in this case is the same, the difference between using $B = 50$ and $B = +\infty$ could potentially result in different conclusions for the same data set.

Unlike the sample median, the standard error of the mean happens to be tractable. In spite of this fact, we continue with the previous analysis and attempt to compare the sample means to illustrate how other APPL procedures can be used for comparing means.

Example 10.6. In a similar fashion to the previous example involving medians, the following R code uses $B = 50$ bootstrap replications to estimate the standard error of the mean for the treatment group.

```
> tr = c(16, 23, 38, 94, 99, 141, 197)
> B = 50
> means = numeric(B)
> for (i in 1:B) {
>   bootstrap = sample(tr, 7, replace = TRUE)
>   means[i] = mean(bootstrap)
> }
> sd(means)
```

This code returned 23.89, which is the the upper-left-hand entry in Table 10.5. When this code is executed for other values of B and for the control group, we see the usual pattern of nonmonotonic convergence as B increases in Table 10.5, which is consistent with stochastic simulation.

We again consider the use of APPL to avoid resampling. The support of the distribution of the mean is much larger than the previous example because there are $7^7 = 823{,}543$ different samples with replacement that could be drawn, resulting in hundreds (762, to be exact) of possible sample means. The APPL code to calculate the exact value of the standard error of the mean for the treatment group is

```
> X := BootstrapRV([16, 23, 38, 94, 99, 141, 197]);
> W := ConvolutionIID(X, 7);
> g := [[x -> x / 7], [-infinity, infinity]];
> Y := Transform(W, g);
> sqrt(Variance(Y));
```

Table 10.5. Bootstrap estimates of the standard error of the mean

	$B = 50$	$B = 100$	$B = 250$	$B = 500$	$B = 1000$	$B = +\infty$
Treatment	23.89	24.29	23.16	24.36	23.75	23.36
Control	17.07	13.83	13.40	13.13	13.55	13.35

After the bootstrap random variable X is created as in the previous example, the APPL procedure `ConvolutionIID` sums $n = 7$ of these iid X's to create a random variable W. The APPL procedure `Transform` transforms the random variable W using the transformation $Y = W/n$ to calculate the PDF of the sample mean, which is

$$
f_Y(y) = \begin{cases}
1/7^7 = 1/823543 & y = 16 \\[2mm]
\binom{7}{1}/7^7 = 1/117649 & y = 17 \\[2mm]
\binom{7}{2}/7^7 = 3/117649 & y = 18 \\[2mm]
\binom{7}{3}/7^7 = 5/117649 & y = 19 \\[2mm]
\binom{7}{1}/7^7 = 1/117649 & y = 134/7 \\[2mm]
\binom{7}{4}/7^7 = 5/117649 & y = 20 \\[2mm]
\vdots & \vdots \\[2mm]
1/7^7 = 1/823543 & y = 197,
\end{cases}
$$

and the standard error as $\frac{2}{49}\sqrt{327649} \cong 23.3635$. This, of course, is equal to $\sqrt{\frac{n-1}{n}}\frac{s}{\sqrt{n}} = \sqrt{\frac{6}{7}}\frac{s}{\sqrt{7}}$, where s is the standard deviation of the treatment survival times. This fact is the fortunate consequence of the mathematical tractability of the standard error for means. Other, less fortunate situations can be handled in a similar manner.

Similar APPL code for the treatment case yielded an estimated standard error in the $B = +\infty$ case of $\frac{1}{27}\sqrt{129902} \cong 13.3489$. The difference between the treatment and control group means, which is $86.86 - 56.22 = 30.64$ days, is not statistically significant since it is only $30.64/\sqrt{23.36^2 + 13.35^2} \cong 1.14$ standard deviations away from zero.

The previous two examples have estimated the standard errors of measures of central tendency (the median and mean). Estimation of the standard error of a measure of dispersion, the sample range R, for the rat survival data will now be considered.

Example 10.7. As in the previous two examples, the R code given below creates a bootstrap estimate of the standard error of the range in the treatment case with $B = 50$.

```
> tr = c(16, 23, 38, 94, 99, 141, 197)
> B = 50
> ranges = numeric(B)
> for (i in 1:B) {
>    bootstrap = sample(tr, 7, replace = TRUE)
>    ranges[i] = diff(range(bootstrap))
> }
> sd(ranges)
```

Table 10.6 contains the standard error estimates for other values of B and for the control group.

To determine whether there is a statistically significant difference between the sample ranges for the treatment and control groups in the rat-survival data, we again use the APPL code below for estimating the standard error of the range for the treatment group.

```
> X := BootstrapRV([16, 23, 38, 94, 99, 141, 197]);
> Y := RangeStat(X, 7);
> sqrt(Variance(Y));
```

The bootstrap estimate of the standard error of the range R for the treatment case corresponding to $B = +\infty$ is $\frac{2}{16807}\sqrt{88781509983} \cong$ 35.4569. The distribution of the sample range Y from bootstrapping returned by APPL is

$$f_Y(y) = \begin{cases} 1/117649 & y = 0 \\ 18/117649 & y = 5 \\ 18/117649 & y = 7 \\ 18/117649 & y = 15 \\ 276/117649 & y = 22 \\ 18/117649 & y = 42 \\ 276/117649 & y = 47 \\ 36/117649 & y = 56 \\ 276/117649 & y = 61 \\ 276/117649 & y = 71 \\ 1734/117649 & y = 76 \\ 1734/117649 & y = 78 \\ 6792/117649 & y = 83 \\ 276/117649 & y = 98 \\ 3468/117649 & y = 103 \\ 6792/117649 & y = 118 \\ 20010/117649 & y = 125 \\ 6792/117649 & y = 159 \\ 20010/117649 & y = 174 \\ 48828/117649 & y = 181. \end{cases}$$

Similar APPL statements for the control case yielded an estimated standard error for the range as $\frac{2}{4782969}\sqrt{5666287777334555} \cong 31.4762$.

Table 10.6. Bootstrap estimates of the standard error of the range

	$B = 50$	$B = 100$	$B = 250$	$B = 500$	$B = 1000$	$B = +\infty$
Treatment	39.57	35.46	34.70	35.50	33.81	35.46
Control	36.70	29.84	30.95	30.07	30.14	31.48

The observed difference in the ranges between the treatment and control groups, $181 - 136 = 45$ days, is not statistically significant since it is only $45/\sqrt{35.46^2 + 31.48^2} \cong 0.95$ standard deviations away from zero.

Confidence Intervals. A final example is presented here to show how one of the shortcomings of the standard bootstrap approach can be overcome by using the parametric bootstrap method. APPL can again be used to eliminate resampling error.

The previous three examples estimated the standard errors of two measures of central tendency and a measure of dispersion. The next example constructs two confidence intervals for the sample range of the rat-treatment group.

Example 10.8. Let R be the range of the $n = 7$ observations in the treatment group from Table 10.3. The APPL statements

```
> X := BootstrapRV([16, 23, 38, 94, 99, 141, 197]);
> Y := RangeStat(X, 7);
> IDF(Y, 0.025);
> IDF(Y, 0.975);
```

yield a 95% confidence interval of $76 < R < 181$. This confidence interval has the unappealing property that the point estimator, $\hat{R} = 197 - 16 = 181$ days, is also the upper limit of the confidence interval.

Trosset [113] suggested an alternative method for computing a confidence interval for the range R, which involves parametric bootstrapping. First, an exponential distribution with rate λ is fit to the treatment group data using the APPL MLE (maximum likelihood estimator) procedure. The procedure identifies the parameter estimate for the distribution as $\hat{\lambda} = 7/608$. The (continuous) distribution of the sample range of $n = 7$ observations drawn from an exponential population with parameter $\hat{\lambda} = 7/608$ is then computed. The confidence interval is determined with the IDF procedure. The following APPL statements yield a 95% confidence interval for the range R for $n = 7$ observations as $68 < R < 475$.

```
> treatment := [16, 23, 38, 94, 99, 141, 197];
> X := ExponentialRV(lambda);
> lambdahat := op(MLE(X, treatment, [lambda]));
> Y := ExponentialRV(lambdahat);
```

```
> Z := RangeStat(Y, 7);
> IDF(Z, 0.025);
> IDF(Z, 0.975);
```

This section has considered the use of bootstrapping for simple comparisons between two populations. In conclusion, for moderate sample sizes and test statistics that are tractable for APPL, the exact approach to bootstrapping can eliminate resampling error and oftentimes reduce computation time.

10.4 Applying Bootstrap Methods to System Reliability

This section presents a fully enumerated bootstrap method to find the empirical system lifetime distribution for a coherent system modeled by a reliability block diagram. Given failure data for individual components of a coherent system, the bootstrap empirical system lifetime distribution derived here will be free of resampling error. We also derive distribution-free expressions for the bias associated with the bootstrap method for estimating the mean system lifetimes of parallel and series systems with statistically-identical components. We show that bootstrapping underestimates the mean system lifetime for parallel systems and overestimates the mean system lifetime for series systems, although both bootstrap estimates are asymptotically unbiased. The expressions for the bias are evaluated for several popular parametric lifetime distributions.

Background. Bootstrapping as a statistical method is ubiquitous. In most applications, one is left with two types of error: random sampling error associated with the data set and resampling error associated with the bootstrapping process. Resampling error is generally contained, but not eliminated, by increasing the number of bootstrap iterations, B. The "ideal bootstrap" sets $B = \infty$ [26], and is sometimes calculable by fully enumerating all possible outcomes of a bootstrap instead of resampling the data. In this section, we present a fully enumerated bootstrap method that has this advantage of eliminating resampling error. We will apply the method to models of a coherent system, represented by a reliability block diagram (RBD). This fully enumerated bootstrap produces the distribution of a discrete random variable that approximates the system's continuous lifetime distribution. We show that there is inherent bias that can be computed exactly under various parametric assumptions. System designers can use these methods to investigate system parameters to include in the design, for example, the effect of changing the number of components in series, parallel or combinations of both to achieve system reliability improvement.

This section is organized as follows. In the first subsection, we investigate the simplest forms of RBDs consisting of two components whose lifetimes are iid random variables and the components are placed in parallel (or series) to

illustrate the nature of the underlying mathematics in the enumerated bootstrap. In the next subsection we present the steps of a resampled bootstrap and present a simple example of how bias is induced by the minimum and maximum operations. In the next subsection, we present the general result for bias for parallel and series systems. We also present some of the difficulties encountered in generalizing to systems consisting of more than two components. The final subsection presents some applications of a fully enumerated bootstrap for an RBD example.

The literature concerning bootstrapping of RBDs is sparse. Leemis [73] uses enumerated bootstrapping of binary data that represents the availability of components in a RBD to construct a lower confidence bound for three components in series. Doss and Chiang [22] present a traditional bootstrap analysis of RBDs to include a comparison of resampling schemes. There is ample literature on enumerated bootstraps, but little of it applies to RBD models.

Parallel systems have two or more components operating simultaneously that perform identical tasks as backups to each other. This system configuration occurs quite often in design. Series systems of iid components are less frequently observed. Individual fins in a turbine engine could be considered to comprise a series system: when the first one fails the system fails. Clearly, the independence of the fins could be questioned. Consider the RBD in Fig. 10.3, a model of a computer server system (e-mail or brokerage servers perhaps). The series configuration could represent the major subsystems of power supply, computation hardware, and external memory. Each subsystem is made more resilient by having two components with iid lifetimes operating simultaneously in parallel. Thus, A_1 and A_2 represent the iid lifetimes of two power supplies, and the other components represent computer and memory subsystems. The system lifetime S is given by the algebraic relationship

$$S = \min\left\{\max\left\{A_1, A_2\right\}, \max\left\{B_1, B_2\right\}, \max\left\{C_1, C_2\right\}\right\}.$$

Given the hypothetical case that all lifetime CDFs of the components are known to be $F_A(t)$, $F_B(t)$, and $F_C(t)$ respectively, the exact CDF of S is

$$F_S(t) = 1 - \left[1 - F_A^2(t)\right] \cdot \left[1 - F_B^2(t)\right] \cdot \left[1 - F_C^2(t)\right] \qquad t > 0.$$

Now consider the more common case in which the CDFs of the components are not known, but life test data is available for each component. Let n_a

Fig. 10.3. Block diagram of a series arrangement of parallel subsystems

be the sample size of the data set for component A, with similar defini-
tions for n_b and n_c. These data sets, $\{a_1, a_2, \ldots, a_{n_a}\}$, $\{b_1, b_2, \ldots, b_{n_b}\}$, and
$\{c_1, c_2, \ldots, c_{n_c}\}$, provide the supports for three different discrete random vari-
ables $A^\star, B^\star,$ and C^\star, each having equal probability for these support values
(assuming that the data values are distinct). Thus, $A^\star, B^\star,$ and C^\star represent
the random variables associated with the bootstrap samples from these three
data sets. The discrete CDFs of $A^\star, B^\star,$ and C^\star, denoted by $F_{A^\star}(t)$, $F_{B^\star}(t)$,
and $F_{C^\star}(t)$, are the same as the empirical distribution functions (EDFs) of
their corresponding lifetime data sets. Thus, a fully enumerated bootstrap of
the system can be represented algebraically by

$$S^\star = \min\{\max\{A_1^\star, A_2^\star\},\ \max\{B_1^\star, B_2^\star\},\ \max\{C_1^\star, C_2^\star\}\}.$$

The discrete CDF [that approximates the unknown, but true continuous CDF
$F_S(t)$] is

$$F_{S^\star}(t) = 1 - \left(1 - [F_{A^\star}(t)]^2\right) \cdot \left(1 - [F_{B^\star}(t)]^2\right) \cdot \left(1 - [F_{C^\star}(t)]^2\right) \qquad t > 0.$$

It is this CDF that is the fully enumerated bootstrapped distribution of the
system. We will show that such bootstrapped systems have bias resulting from
taking minimums and maximums of discrete random variables.

We use the following notation in the section:

X A continuous random variable with PDF $f_X(t)$ and CDF $F_X(t)$ that
represents a component lifetime.

\mathbf{X} A random sample of n component lifetimes.

$X_{(i)}$ The ith ordered element of \mathbf{X}, $i = 1, 2, \ldots, n$.

S A continuous random variable representing the system lifetime of
a system consisting of two components arranged in parallel, each
with an independent lifetime distributed according to X, i.e., $S = \max\{X_1, X_2\}$.

T A continuous random variable representing the system lifetime of
a system consisting of two components arranged in series, each
with an independent lifetime distributed according to X, i.e., $T = \min\{X_1, X_2\}$.

X^\star A random variable distributed according to the EDF associated with
the random sample \mathbf{X}.

S^\star A random variable representing the system lifetime of a system con-
sisting of two components arranged in parallel, each with an indepen-
dent lifetime distributed according to X^\star, i.e., $S^\star = \max\{X_1^\star, X_2^\star\}$.
The distribution of S^\star is used to approximate the unknown distri-
bution of S. Also, it is known that S^\star converges almost surely to S
asymptotically by the Glivenko–Cantelli theorem (see Billingsley [11],
for example).

T^\star A random variable representing the system lifetime of a system con-
sisting of two components arranged in series, each with an independent
lifetime distributed according to X^\star, i.e., $T^\star = \min\{X_1^\star, X_2^\star\}$.

Two and m iid Components in Parallel or in Series. In this subsection, we present the mathematics underlying bootstrapping for simple systems. Consider a system with two identical components in parallel, in other words, a system with two constantly operating components with identically distributed failure times that act as real-time backups to one another. We assume that prior testing of n components has produced a failure-time data set $\mathbf{x} = \{x_1, x_2, \ldots, x_n\}$, from which we define X^\star with CDF $F_{X^\star}(t) = i/n$ for $t \in [x_{(i)}, x_{(i+1)})$, $i = 1, 2, \ldots, n$ and $x_{(1)}, x_{(2)}, \ldots, x_{(n)}$ are the ordered observations. Knowing only the failure-time data (with no assumption of an underlying parametric failure-time distribution), we can approximate the true system lifetime using the random variable $S^\star = \max\{X_1^\star, X_2^\star\}$. The CDF of S^\star is

$$
\begin{aligned}
F_{S^\star}(t) &= \Pr\left(S^\star \leq t\right) \\
&= \Pr(\max\{X_1^\star, X_2^\star\} \leq t) \\
&= \Pr(X_1^\star \leq t) \cdot \Pr(X_2^\star \leq t) \\
&= F_{X^\star}(t) \cdot F_{X^\star}(t) \\
&= \frac{i^2}{n^2} \qquad t \in [x_{(i)}, x_{(i+1)}),\ i = 0, 1, 2, \ldots, n-1.
\end{aligned}
$$

At each support point $x_{(i)}$ the value of the PDF can be found by subtracting the appropriate CDF values:

$$
\begin{aligned}
f_{S^\star}(t) &= F_{S^\star}(t) - F_{S^\star}(x_{(i-1)}) \\
&= \frac{i^2 - (i-1)^2}{n^2} \\
&= \frac{2i - 1}{n^2} \qquad t = x_{(i)},\ i = 1, 2, \ldots, n.
\end{aligned}
$$

Note that one must assign $x_{(0)} = 0$ and $F_{S^\star}(x_{(0)}) = 0$ for the indexing to be correct. We note several interesting properties of the PDF of S^\star:

- The PDF of S^\star has the same probability values $f_{S^\star}(t)$ regardless of the values x_i in the data set, as long as the data values are distinct (i.e., there are no repeated or tied values) for a fixed sample size n.
- The supports of S^\star and X^\star are identical.
- The PDF of S^\star is an increasing function over the values of its support with initial mass value $p_{S^\star}(x_{(1)}) = 1/n^2$, constant increments of $2/n^2$, and final mass value $p_{S^\star}(x_{(n)}) = (2n-1)/n^2$ for distinct data values.

Similar methods can be used to derive the empirical system distribution S_m^\star, which is the lifetime of a system of m components with iid lifetimes arranged in parallel, using X^\star to approximate the lifetime distribution of each. The CDF and PDF of S_m^\star are

$$
F_{S_m^\star}(t) = \frac{i^m}{n^m} \qquad t \in [x_{(i)}, x_{(i+1)}),\ i = 1, 2, \ldots, n
$$

and
$$f_{S_m^\star}(t) = \frac{i^m}{n^m} - \frac{(i-1)^m}{n^m} \qquad t = x_{(i)}, \ i = 1, 2, \ldots, n.$$

Using a binomial expansion and simplifying, the PDF can be expressed as:

$$f_{S_m^\star}(t) = n^{-m} \sum_{j=1}^{m} \binom{m}{j} (-1)^{j-1} i^{m-j} \qquad t = x_{(i)}, \ i = 1, 2, \ldots, n.$$

The minimum lifetime of iid components can be derived in a similar fashion. Let $T^\star = \min\{X_1^\star, X_2^\star\}$ be the analogous discrete random variable for a system of two iid components in series, given a component failure-time data set. The CDF and PDF of T^\star are

$$F_{T^\star}(t) = 1 - \left(1 - \frac{i}{n}\right)^2 \qquad t \in [x_{(i)}, x_{(i+1)}), \ i = 1, 2, \ldots, n$$

and
$$f_{T^\star}(t) = \frac{2n - 2i + 1}{n^2}, \qquad t = x_{(i)}, \ i = 1, 2, \ldots, n.$$

The PDF of T^\star is now a decreasing function over its support, beginning with $p_{T^\star}(x_{(1)}) = (2n-1)/n^2$ and ending with $p_{T^\star}(x_{(n)}) = 1/n^2$, with decrements of $2/n^2$ for distinct data values. An extension for a system of m components in series can be derived.

Bootstrapping and Bias for a Parallel System. We now illustrate how bias enters into bootstrapping. Consider the case of a two-component parallel system with iid unit exponential component lifetimes. The interest is in the mean system lifetime. The analyst is unaware of the true lifetime distribution, and only has the values of $n = 3$ component lifetimes. This small value of n is for enumeration purposes only; bootstrapping applications usually rely on larger sample sizes. What is the bias associated with the use of bootstrapping to estimate the mean system lifetime?

The unit exponential component distribution, represented by X in our notation, has expected value $E[X] = 1$ and CDF $F_X(t) = 1 - \exp(-t), \ t > 0$. In this case, the system lifetime S has CDF

$$F_S(t) = \Pr(S \leq t) = \left(1 - \exp(-t)\right)^2 \qquad t > 0.$$

So the true expected system lifetime is

$$E[S] = \int_0^\infty \left(1 - F_S(t)\right) dt = \frac{3}{2},$$

which uses the technique for calculating the mean of a nonnegative random variable from Meeker and Escobar [86, page 77]. The analyst, however, knows only (a) that the component lifetimes are independent, (b) that the components have identical lifetime distributions, and (c) the values of three component lifetimes: x_1, x_2, and x_3. Because there is no knowledge of the true

lifetime distribution, the analyst might consider a simple resample-style boot-strapping, rather than a parametric approach or our fully enumerated ap-proach. To further simplify the example we will set the bootstrap resample value artificially low: $B = 3$. In this setting, a resampling bootstrap proceeds as follows:

1. Three values are sampled with replacement from the data set, representing three lifetimes from component 1, y_1, y_2, and y_3.
2. Three additional values are sampled with replacement from the data set, representing three lifetimes from component 2, z_1, z_2, and z_3.
3. Pairwise maximums of the values from the two previous steps are taken, representing three system lifetimes, $s_1 = \max\{y_1, z_1\}$, $s_2 = \max\{y_2, z_2\}$, and $s_3 = \max\{y_3, z_3\}$.
4. The three system lifetimes, which approximate three true system lifetimes, are averaged in order to estimate the mean system lifetime.

To illustrate these steps with numerical values, consider the data set of the $n = 3$ component observations $0.8, 1.7, 0.4$. The steps given above might go as follows:

1. The component 1 lifetimes sampled are $0.8, 0.4, 0.8$.
2. The component 2 lifetimes sampled are $1.7, 1.7, 0.4$.
3. The pairwise maximums representing system lifetimes are $1.7, 1.7, 0.8$.
4. The estimate for the mean system lifetime is $(1.7 + 1.7 + 0.8)/3 = 1.4$.

To calculate the bias in this estimate, consider the more general case in which the value sampled for the lifetime of a component is equally likely to be one of the data values X_1, X_2, or X_3. Therefore, there are $3 \cdot 3 = 9$ different outcomes by the multiplication rule, as illustrated below.

Component 1	$X_{(1)}$	$X_{(1)}$	$X_{(1)}$	$X_{(2)}$	$X_{(2)}$	$X_{(2)}$	$X_{(3)}$	$X_{(3)}$	$X_{(3)}$
Component 2	$X_{(1)}$	$X_{(2)}$	$X_{(3)}$	$X_{(1)}$	$X_{(2)}$	$X_{(3)}$	$X_{(1)}$	$X_{(2)}$	$X_{(3)}$
Maximum	$X_{(1)}$	$X_{(2)}$	$X_{(3)}$	$X_{(2)}$	$X_{(2)}$	$X_{(3)}$	$X_{(3)}$	$X_{(3)}$	$X_{(3)}$

Since the nine outcomes are equally likely, the probabilities associated with the first, second, and third order statistic corresponding to the maximum of the two lifetimes in the bootstrap sample are $1/9$, $3/9$, and $5/9$. Therefore, the expected system lifetime using bootstrapping is

$$E\left[S^\star\right] = \frac{1}{9}E\left[X_{(1)}\right] + \frac{3}{9}E\left[X_{(2)}\right] + \frac{5}{9}E\left[X_{(3)}\right].$$

Using the standard order statistic results (see, for example Arnold et al. [5]) for random variables from a continuous population to calculate these expected values:

$$E\left[S^\star\right] = \frac{1}{9} \cdot \frac{1}{3} + \frac{3}{9} \cdot \frac{5}{6} + \frac{5}{9} \cdot \frac{11}{6} = \frac{4}{3}.$$

So the bootstrapping approach underestimates the true expected system life-time of $3/2$. Specifically the bias is $E\left[S^\star\right] - E\left[S\right] = 4/3 - 3/2 = -1/6$. The next

subsection uses the prior section's insights concerning bias to prove a general result that quantifies the bias in the bootstrap estimate for the mean system lifetime for parallel and series systems.

Bias Result for Two and m iid Components in Parallel or in Series. The example in the previous subsection has shown that bootstrapping can underestimate the mean system lifetime of a simple parallel system. The following theorems indicate that this result is true in general and that the bootstrap estimate is asymptotically unbiased. Theorem 10.1 establishes that bootstrapping underestimates the mean system lifetime for a two-component parallel system. In fact, since $E[S] > E[X]$, the bias will always be negative. The result can be interpreted as follows: the bias is $-1/n$ times the expected improvement of increasing from one component to two components in parallel.

Theorem 10.1. For a system consisting of two independent components with statistically-identical lifetimes and finite population means arranged in parallel, the bias resulting from the bootstrap estimation of the mean system lifetime is

$$E[S^\star] - E[S] = -\frac{E[S] - E[X]}{n}.$$

Proof. The conditional CDF of X^\star, given a random sample \mathbf{X}, is the EDF

$$F_{X^\star|\mathbf{X}}(t|\mathbf{x}) = \begin{cases} 0 & t < x_{(1)} \\ i/n & x_{(i)} \le t < x_{(i+1)}, \ i = 1, 2, \ldots, n-1 \\ 1 & t \ge x_{(n)}. \end{cases}$$

The conditional CDF of $S^\star = \max\{X_1^\star, X_2^\star\}$ is

$$\begin{aligned} F_{S^\star|\mathbf{X}}(t|\mathbf{x}) &= \Pr(\max\{X_1^\star, X_2^\star\} \le t | \mathbf{X} = \mathbf{x}) \\ &= \Pr(X_1^\star \le t | \mathbf{X} = \mathbf{x}) \cdot \Pr(X_2^\star \le t | \mathbf{X} = \mathbf{x}) \\ &= F_{X^\star|\mathbf{X}}^2(t|\mathbf{x}) \\ &= \begin{cases} 0 & t < x_{(1)} \\ i^2/n^2 & x_{(i)} \le t < x_{(i+1)}, \ i = 1, 2, \ldots, n-1 \\ 1 & t \ge x_{(n)}. \end{cases} \end{aligned}$$

The conditional PDF of S^\star on its support is

$$f_{S^\star|\mathbf{X}}(t|\mathbf{x}) = F_{S^\star|\mathbf{X}}(t|\mathbf{x}) - \lim_{\tau \to t^-} F_{S^\star|\mathbf{X}}(\tau|\mathbf{x}) = \frac{i^2 - (i-1)^2}{n^2} = \frac{2i-1}{n^2}$$

for $t = x_{(i)}$, $i = 1, 2, \ldots, n$. Applying the law of total probability and the result concerning the probability distribution of order statistics associated with a continuous population, the PDF of the bootstrap system lifetime is

$$f_{S^*}(t) = \sum_{i=1}^{n} p_{S^*|\mathbf{X}}(t|\mathbf{x})f_{X_{(i)}}(t)$$

$$= \sum_{i=1}^{n} \frac{2i-1}{n^2} \cdot \frac{n!}{(n-i)!(i-1)!} f_X(t)F_X^{i-1}(t)\left(1 - F_X(t)\right)^{n-i} \quad (10.1)$$

on its support. Using the binomial theorem to expand $\left(1 - F_X(t)\right)^{n-i}$ and then re-ordering the sums, we can re-write (10.1) as

$$f_{S^*}(t) = \sum_{i=1}^{n} \frac{2i-1}{n^2} \cdot \frac{n!}{(n-i)!(i-1)!} f_X(t)F_X^{i-1}(t) \cdot \sum_{j=0}^{n-i} \binom{n-i}{j}(-1)^{n-i-j}F_X^{n-i-j}(t)$$

$$= \sum_{j=0}^{n-1} \frac{f_X(t)n!}{j!n^2} \sum_{i=1}^{n-j} \frac{(2i-1)(-1)^{n-i-j}}{(n-i-j)!(i-1)!}F_X^{i-1}(t)F_X^{n-i-j}(t)$$

$$= \frac{f_X(t)}{n} \sum_{j=0}^{n-1} \frac{(n-1)!(n-j-1)!F_X^{n-j-1}(t)}{j!(n-j-1)!} \sum_{i=1}^{n-j} \frac{(2i-1)(-1)^{n-i-j}}{(n-i-j)!(i-1)!}$$

$$= \frac{f_X(t)}{n} \sum_{j=0}^{n-1} \binom{n-1}{j}F_X^{n-j-1}(t) \sum_{i=1}^{n-j} \binom{n-j-1}{n-i-j}(2i-1)(-1)^{n-i-j} \quad (10.2)$$

on its support. Making the substitution $\alpha = n - i - j$ in the right-most summation,

$$\sum_{i=1}^{n-j} \binom{n-j-1}{n-i-j}(2i-1)(-1)^{\alpha} = \sum_{\alpha=0}^{n-j-1} \binom{n-j-1}{\alpha}(2(n-j-\alpha)-1)(-1)^{\alpha}$$

$$= -2\sum_{\alpha=0}^{n-j-1} \binom{n-j-1}{\alpha}\alpha(-1)^{\alpha}+$$

$$(2n-2j-1)\sum_{\alpha=0}^{n-j-1} \binom{n-j-1}{\alpha}(-1)^{\alpha}. \quad (10.3)$$

Using the binomial expansion of $(1+t)^{n-j-1}$ and its derivative,

$$(1+t)^{n-j-1} = \sum_{\alpha=0}^{n-j-1} \binom{n-j-1}{\alpha}t^{\alpha},$$

$$(n-j-1)t(1+t)^{n-j-2} = \sum_{\alpha=0}^{n-j-1} \binom{n-j-1}{\alpha}\alpha t^{\alpha},$$

we can simplify the summations for certain values of j in (10.3):

$$\sum_{\alpha=0}^{n-j-1} \binom{n-j-1}{\alpha}(-1)^{\alpha} = \begin{cases} 1 & j = n-1 \\ (1-1)^{n-j-1} = 0 & j = 0, 1, \ldots, n-2, \end{cases}$$

and

$$\sum_{\alpha=0}^{n-j-1} \binom{n-j-1}{\alpha} \alpha(-1)^{\alpha} = (-1) \sum_{\alpha=0}^{n-j-1} \binom{n-j-1}{\alpha} \alpha(-1)^{\alpha-1}$$

$$= \begin{cases} -1 & j = n-2 \\ -(n-j-1)(1-1)^{n-j-2} = 0 & j = 0, 1, \ldots, n-1, n-3. \end{cases}$$

Substituting these results into (10.3), we have

$$\sum_{i=1}^{n-j} \binom{n-j-1}{n-i-j} (2i-1)(-1)^{\alpha} = \begin{cases} 1 & j = n-1 \\ 2 & j = n-2 \\ 0 & j = 0, 1, \ldots, n-3. \end{cases}$$

Therefore, the PDF of S^{\star} from (10.2) simplifies to

$$f_{S^{\star}}(t) = \frac{f_X(t)\big(2(n-1)F_X(t)+1\big)}{n} = \frac{2(n-1)f_X(t)F_X(t)+f_X(t)}{n}$$

on its support. Using the identities

$$E[X] = \int_0^{\infty} t f_X(t)\, dt,$$

$$F_S(t) = F_X^2(t),$$

$$E[S] = \int_0^{\infty} t F_S'(t)\, dt = \int_0^{\infty} 2t f_X(t) F_X(t)\, dt,$$

gives a general result for $E[S^{\star}]$:

$$E[S^{\star}] = \frac{(n-1)\int_0^{\infty} 2t f_X(t) F_X(t)\, dt + \int_0^{\infty} t f_X(t)\, dt}{n} = \frac{(n-1)E[S]+E[X]}{n}.$$

The bias result follows. □

Theorem 10.1 states that bootstrapping underestimates the mean system lifetime for a two-component parallel system consisting of components with independent lifetimes. Theorem 10.2 states that bootstrapping overestimates the mean system lifetime for a two-component series system consisting of components with independent lifetimes. Furthermore, the magnitudes of the two biases in the parallel and series cases are equal. A similar proof construction applies for Theorem 10.2, the series case, and the more general result for m items in parallel or series. Because these proofs are lengthy, they have been omitted.

Theorem 10.2. For a system consisting of two independent components with statistically-identical lifetimes and finite population means arranged in series, the bias resulting from the bootstrap estimation of the mean system lifetime is

$$E\left[T^{\star}\right] - E\left[T\right] = \frac{E\left[X\right] - E\left[T\right]}{n} = -\left(E\left[S^{\star}\right] - E\left[S\right]\right),$$

which, as indicated, is the negative of the result for the bootstrapping bias for a parallel system.

Both proofs begin by exploiting the underlying properties of X^{\star} and S^{\star} outlined earlier. The difficult part of the proofs is finding $E\left[S^{\star}\right]$, after that the result readily follows. Table 10.7 contains expressions for the bias for several popular survival distributions with positive support, positive scale parameter η, and positive shape parameter κ. Note that for the gamma distribution, $_2F_1(\cdot)$ refers to the regularized hypergeometric function, which is reviewed with a bibliography by Weisstein [117] and can be calculated in numerous software packages including Mathematica and Maple (see, Wolfram [119] and Maplesoft [80]). The exact form of $_2F_1(\cdot)$ is the following infinite sum:

$$_2F_1(a,b,c,z) = \sum_{n=0}^{\infty} \frac{(a)_n (b)_n}{(c)_n} \cdot \frac{z^n}{n!},$$

where $(\cdot)_n$ is the Pochhammer symbol, which is shorthand for

$$(x)_n = \frac{\Gamma(x+n)}{\Gamma(x)}.$$

This series converges when $|z| < 1$. Continuity arguments allow for calculating $_2F_1(\cdot)$ at $z = -1$. The gamma distribution bias formula when $\kappa = 1$ simplifies to the corresponding exponential distribution bias formula given in the table.

Table 10.7. Bootstrapping bias for parametric survival distributions

Distribution	$f(t)$	$E\left[S^{\star}\right] - E[S]$
Gamma(η, κ)	$\dfrac{x^{\kappa-1}\exp(-t/\eta)}{\eta^{\kappa}\Gamma(\kappa)}$	$-\dfrac{\eta\Gamma\left(1+2\kappa\right)\, _2F_1(\kappa, 1+2\kappa, 2+\kappa, -1)}{n\Gamma(\kappa)}$
Weibull(η, κ)	$\dfrac{\kappa}{\eta}\left(\dfrac{t}{\eta}\right)^{\kappa-1}\exp\left[-\left(\dfrac{t}{\eta}\right)^{\kappa}\right]$	$-\dfrac{\eta\left(1-2^{-1/\kappa}\right)\Gamma\left(1+1/\kappa\right)}{n}$
exponential(η)	$\eta^{-1}\exp(-t/\eta)$	$-\dfrac{\eta}{2n}$
log logistic(η, κ)	$\dfrac{\kappa(t/\eta)^{\kappa-1}}{\eta[1+(t/\eta)^{\kappa}]^2}$	$-\dfrac{\eta\pi}{n\kappa^2\sin(\pi/\kappa)},\quad \kappa > 1$

Some distributions do not have closed form expressions for bias, such as the log normal distribution. Even in these cases of no closed form, numerical methods yield bias calculations. Given that $F_X(t)$ is available, the bias can be calculated analytically or numerically with the following application of Theorem 10.1:

$$-\frac{1}{n}\big(E[S] - E[X]\big) = -\frac{1}{n}\left(\int_0^\infty \big(1 - F_X^2(t)\big)\, dt - \int_0^\infty \big(1 - F_X(t)\big)\, dt\right)$$

$$= -\frac{1}{n}\int_0^\infty F_X(t)\big(1 - F_X(t)\big)\, dt.$$

This result has a useful graphical interpretation. The integrand is the multiplication of the individual component's CDF times its SF, $1 - F_X(t)$. The integrand is plotted in Fig. 10.4, which also contains the CDF and SF of a typical lifetime distribution. The area under that integrand, labeled β, is proportional to the bootstrap bias, specifically, the bias is $-\beta/n$. More importantly, if one substitutes the parametric CDF and SF with the respective empirical CDF and SF, $F_n(t)$ and $1 - F_n(t)$, then an empirical estimate for bias can also be calculated. Thus, one must estimate $E[S]$ with $E[S^\star]$, producing the empirical estimate of bias as

$$E[S^\star] - E[S] \cong -\frac{1}{n}\big(E[S^\star] - E[X^\star]\big).$$

This estimate of bias has a smaller bias of order $1/n^2$, typically less than one tenth of one percent of a component's expected lifetime for reasonable sample sizes. As we shall see, the empirical bias estimates are within the variability of bias results for various assumptions of parametric distributions.

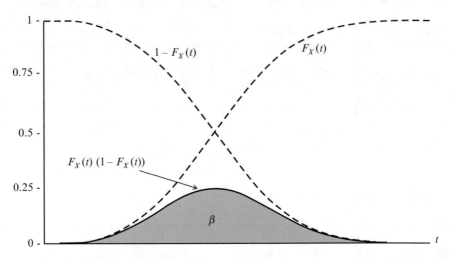

Fig. 10.4. The CDF and SF of a lifetime distribution with its associated bias, which is proportional to the area β

We can generalize the results of this section for systems consisting of m components in parallel (or series), but the same simplifications do not follow. If we let $S_m^\star = \max\{X_1^\star, X_2^\star, \ldots, X_m^\star\}$ be the bootstrap estimate of system lifetime for m components in parallel, then, following a similar approach to the one used in proving Theorem 10.1, we can find the exact bootstrap bias for a system consisting of m components in parallel:

$$E\left[S_m^\star\right] - E\left[S_m\right] = \int_0^\infty t f_{S_m^\star}(t)\, dt - \int_0^\infty t f_{S_m}(t)\, dt$$

$$= \int_0^\infty \sum_{i=1}^n \frac{i^m - (i-1)^m}{n^m} \cdot \frac{n!}{(n-i)!(i-1)!} t f_X(t) F_X^{i-1}(t)\bigl(1 - F_X(t)\bigr)^{n-i}\, dt$$

$$- \int_0^\infty t m f_X(t) F_X^{m-1}(t)\, dt.$$

For example, substituting $m = 3$ and following a similar approach to proving Theorem 10.1 yields

$$E\left[S_3^\star\right] - E\left[S_3\right] = \frac{(-3n+2)E\left[S_3\right] + 3(n-1)E\left[S_2\right] + E\left[X\right]}{n^2}. \tag{10.4}$$

We can produce analogous general expressions for bias related to bootstrap estimates for series system lifetimes and letting $m = 3$:

$$E\left[T_3^\star\right] - E\left[T_3\right] = \frac{(-3n+2)E\left[T_3\right] + 3(n-1)E\left[T_2\right] + E\left[X\right]}{n^2}. \tag{10.5}$$

Empirical estimates for $m = 3$ can be similarly constructed by substituting $E\left[S_m^\star\right]$ and $E\left[T_m^\star\right]$ for the appropriate $E\left[S_m\right]$ and $E\left[T_m\right]$ on the right-hand side of (10.4) and (10.5).

Interestingly, assuming different underlying parametric distributions results in fairly consistent bias estimates. Table 10.8 shows the effect of assuming different distributional families for each component in a two-component parallel system. In each case, the sample size was fixed at $n = 25$ and the distribution's parameters were chosen so that the data had identical population first and second moments μ and σ^2. In other words the different distributions are similar, at least in center and spread. Notice that the empirical estimates of bias are within the range of parametric biases, making the concern of the bias' bias less pressing. For the Weibull and gamma distributions, the third and fourth columns of the table represent data with an increasing failure rate (IFR) hazard function, indicative of items that degrade with time, and the last column corresponds to a decreasing failure rate (DFR), indicative of items where "used" is better than "new." In the IFR case the bias is on the order of about 0.1% of the mean failure time. In the DFR case, the biases vary significantly. Even in the worst case (gamma) the bias is about 0.3% of the mean value of the sample. The exponential distribution with means 62, 3, and 20 has associated bootstrap biases -1.24, -0.06, and -0.4, respectively. The exponential distribution lacks the advantage of a second parameter, causing

Table 10.8. Bootstrapping bias for parametric survival distributions for an m-component parallel system for three population parameter settings and sample size $n = 25$ with μ and σ^2 as shown

Moments and distribution	m	$\mu = 62$ $\sigma^2 = 10$	$\mu = 3$ $\sigma^2 = 0.1$	$\mu = 20$ $\sigma^2 = 2000$
Empirical	2	−0.0703	−0.0070	−0.6410
Gamma	2	−0.0713	−0.0071	−0.6381
log logistic	2	−0.0696	−0.0069	−0.3705
log normal	2	−0.0713	−0.0071	−0.5249
Weibull	2	−0.0693	−0.0070	−0.5992
Weibull	3	−0.0894	−0.0093	−1.4516

the bias to be of greater magnitude. Notice that in the Weibull cases in which $m = 3$, the bias is greater than for $m = 2$. In general, we have observed that bias increases with m, a conjecture that we have not proven.

A Practical Example. In this subsection we present an example of a fully enumerated bootstrap on the system in Fig. 10.3. The fully enumerated bootstrap (outlined in Efron and Tibshirani [26]) is generally untenable for typical operations involving sums of random variables. Minimum and maximum operations, however, are much more amenable to a fully enumerated bootstrap. Consider, for example, S^\star and T^\star. The size of their support is limited from above by the total number of distinct observations in the sample \mathbf{X}. For more complicated coherent systems, the support size of the system bootstrap random variable is similarly constrained because each element in the support of the system lifetime distribution is one of the component lifetimes. The following example illustrates the limited support of RBD systems. The open source APPL software allows for such fully enumerated bootstraps of RBDs as well as other systems.

Referring to Fig. 10.3, let the three types of components, A, B, and C, also have previous life test data sets of size $n_A = 45$, $n_B = 60$, and $n_C = 45$, respectively, for each component. The failure times are stored in the Maple lists `alist`, `blist`, and `clist`. Assume that a system designer wants to estimate the lifetime distribution of a system with the series arrangement of parallel subsystems shown in Fig. 10.3. The fully enumerated bootstrap lifetime distribution for this system, $S^\star = \min\{\max\{A_1^\star, A_2^\star\}, \max\{B_1^\star, B_2^\star\}, \max\{C_1^\star, C_2^\star\}\}$, can be found with the following APPL commands:

```
> A := BootstrapRV(alist);
> B := BootstrapRV(blist);
> C := BootstrapRV(clist);
> Sstar := Minimum(MaximumIID(A, 2), Minimum(MaximumIID(B, 2),
              MaximumIID(C, 2)));
```

Notice that the embedded `Minimum` commands, a necessity since that command only takes on two arguments. The `BootstrapRV` procedure transforms a list into a discrete uniform random variable in APPL's list-of-lists format. For a specific set of data, the fully enumerated bootstrapped system was produced and a portion of the 138-element PDF for S^\star is as follows:

$$
f_{S^\star}(t) = \begin{cases}
1/2025 & t = 2.268 \\
1/675 & t = 3.678 \\
\quad \vdots & \\
2009/7290000 & t = 10.051 \\
3599/810000 & t = 11.262 \\
\quad \vdots & \\
10057/20250000 & t = 173.620 \\
1157/750000 & t = 179.461.
\end{cases}
$$

This PDF and the associated CDF are graphed in Figs. 10.5 and 10.6. The graph does not include the vertical lines of a typical plot because they would eclipse some of the points in the plot. Also, each point in the PDF plot has a unique value on the horizontal axis, even though the pattern appears to have points on top of other points. Notice how $F_{S^\star}(t)$ approximates a smooth curve, that of some underlying, but unknown, continuous $F_S(t)$. Since the PDF of S^\star is exactly determined, all the typical characteristics of S^\star can be calculated, e.g., the mean, variance, skewness, percentiles. For example $E[S^\star] = 78.78$ and the 5th percentile of the system lifetime is 31.95. Also, note that the support size of S^\star is limited above by the total number of the individual component failure times, i.e., $n_{S^\star} = 138 \le n_A + n_B + n_C = 45 + 60 + 45 = 150$. The algorithm to find $F_{S^\star(t)}$ executes quickly, in approximately $O(n_A + n_B + n_C)$ time. In general, for all m, parallel and series systems will execute in $O(n)$ time, as minimum and maximum operations only use the observed failure times. No new values are created that could become support values in S^\star. This is unlike a cold backup system, with system lifetime $T = X_1 + X_2$, in which a fully enumerated bootstrap would execute very slowly, on the order of $O(n^2)$, since all n^2 possible support values would need to be enumerated. The authors have done extensive research in these more difficult algebraic structures, creating PDFs with thousands of support values for relatively simple models.

Estimating bias in S^\star is possible, but it requires simulation and distributional assumptions. Unlike the case of iid components in series or parallel, there are no easily formed analytic results for non-iid cases, in part due to the innumerable potential configurations of the more complex systems. Using an "estimate then simulate" approach, one can calculate a bias estimate for the system. Making parametric assumptions is helpful in this process, as we will outline next.

Consider the lifetime data available for the individual components A, B, and C. The "estimate" step of the bias estimation process is as follows:

Fig. 10.5. The PDF of the estimated lifetime of the complex system

Fig. 10.6. The CDF of the estimated lifetime of the complex system

1. Assume a Weibull model for each of the three component types and calculate the associated maximum likelihood estimates $\hat{\eta}_A$, $\hat{\kappa}_A$, $\hat{\eta}_B$, $\hat{\kappa}_B$, $\hat{\eta}_C$, and $\hat{\kappa}_C$ from each lifetime data set.
2. Form the estimated CDF for the system

$$\hat{F}_S(t) = 1 - \left[1 - \left(1 - \exp(t^{\hat{\kappa}_A}/\hat{\eta}_A)\right)^2\right] \cdot \left[1 - \left(1 - \exp(t^{\hat{\kappa}_B}/\hat{\eta}_B)\right)^2\right] \cdot \left[1 - \left(1 - \exp(t^{\hat{\kappa}_C}/\hat{\eta}_C)\right)^2\right]$$

for $t > 0$.
3. Find the expected system lifetime of the estimated system

$$\hat{E}[S] = \int_0^\infty \left(1 - \hat{F}_S(t)\right) dt.$$

The "simulate" step of the process is as follows:

1. For each i iteration of the simulation, draw n_A simulated component lifetimes from the Weibull($\hat{\eta}_A$, $\hat{\kappa}_B$) distribution, and likewise n_B and n_C lifetimes from the respective distributions for components B and C.
2. Calculate the empirical distribution for S_i^\star from these component lifetimes using the appropriate APPL `Minimum` and `Maximum` commands outlined earlier in this section.
3. Calculate the mean $E[S_i^\star]$ for the system.
4. Calculate this iteration's simulated bias, $b_i = E[S_i^\star] - \hat{E}[S]$.
5. Repeat this simulation many times, storing each b_i.
6. Find the mean and standard deviation of the bias estimates, \bar{b} and $s_{\bar{b}}$.

This simulation was conducted for 2500 iterations using the values of `alist`, `blist`, and `clist` for the original component lifetimes. In this case $\bar{b} = -0.3553$, $s_{\bar{b}} = 4.513$ and a 95% confidence interval for b was $(-0.5323, -0.1783)$.

To validate that the simulation presents estimates that are reasonable, we have conducted two different evaluations. First, we simulated this "estimate then simulate" process in its own set of simulations. In this effort we established true η and κ parameters for each component. We then simulated 20 sets of life-time data for each component. For each of the 20 sets of data we conducted the "estimate then simulate" process 500 times to produce 20 values of \bar{b}, $\hat{E}[S]$, and 20 confidence intervals. An ANOVA of the 20×500 bias values, versus the factor $\hat{E}[S]$ showed that the \hat{b} estimates were statistically similar. We also simulated the effect of reducing the sample sizes for the lifetime data of some of the components. As expected, the bias estimates grew both in expected magnitude and standard deviation when the size of component failure time data sets decreased. This algorithm could be altered to fit specific purposes. For example one could make the original procedure more robust by averaging over several parametric model assumptions, should that level of confidence be necessary.

The second evaluation of the method involved comparing known bias with simulated bias. We applied this "estimate then simulate" process to iid components in parallel (the subsystems in Fig. 10.3) and compared the simulated bias estimates to the true, analytic biases. In each case the 95% confidence intervals for simulated bias captured the analytic bias. This gives us rationale to opine that the two methods produce stochastically similar bias estimates for these parallel subsystems. Furthermore, the agreement of the analytic and simulated results argues in favor of applying the "estimate then simulate" method to the full system as a way of estimating bias.

Further, bias can be approximated by either calculating the empirical bias estimates or by making parametric assumptions on the components and using Table 10.7. We will outline the steps to calculate such bias. First, we will explain the empirical method in detail because it has the advantage of not making parametric assumptions. Then we will outline a similarly based method using the Weibull distribution as the assumed distribution of component lifetimes. In each case, one will estimate the bias b in units of time, then convert that bias to the proportion of the effect of bias to the expected value of the system $p = (\mu_X + b)/\mu_X$. To account for the combination of series and parallel structure to the system, the overall bias proportion can be expressed as a product of these proportions. A scalar transformation will then produce the unbiased system distribution.

The empirical estimate of bias is calculated as follows:

- Establish the discrete distributions of the subsystems that are in parallel by using APPL to calculate the PDF for the following random variables: $A2^\star = \max\{A_1^\star, A_2^\star\}$ which is calculated with the APPL code `A2star := MaximumIID(A, 2);`. Similarly define and calculate $B2^\star$ and $C2^\star$.
- Find the expected values $E[A2^\star]$, $E[B2^\star]$ and $E[C2^\star]$.
- Estimate the bias of each subsystem using the empirical method: for the $A2^\star$ subsystem bias $b_{A2} \cong -\frac{1}{n_a}(E[A2^\star] - E[A^\star]) = -1.2013$. Similar calculations produce $b_{B2} \cong -0.4561$ and $b_{C2} \cong -0.5491$. Note, $E[A^\star]$ is the sample mean of the data set for component A.
- Calculate the proportion of bias to the expected value of the subsystems. The proportion of the effect of bias to $E[A2^\star]$ would be $p_{A2} = (E[A2^\star] + b_{A2})/E[A2^\star] = 0.99287$. Similar calculations produce $p_{B2} = 0.99611$ and $p_{C2} = 0.9951$. In other words, $E[A2^\star]$ is under estimating the true $E[A2]$, i.e., $E[A2^\star] \cong 0.99287 E(A2)$.

The series configuration is harder to approximate, as the results stated earlier only apply to iid components in series, and we have different distributions for $A2^\star$, $B2^\star$, and $C2^\star$. However, one can see the effect of assuming three different scenarios, where the three series components were all iid composed of $A2^\star$ components, then all $B2^\star$ components, then all $C2^\star$ components. That way if one of these components is particularly more biased than the others, we have upper and lower bounds for system bias. Using the series bias formula for $m = 3$ outlined in the previous section, we calculate the following empirical based biases for the three different assumptions.

- For the case where three components in series are all of type $A2^\star$ (we abbreviate $A2{:}3$), the bias is $b_{A2{:}3} = 1.4697$ and the proportion of bias to the mean system lifetime is $p_{A2{:}3} = 1.01786$. Likewise, if all components in series are of type $B2^\star$, then $b_{B2{:}3} = 0.5896$ and $p_{B2{:}3} = 1.00740$. Likewise, $b_{C2{:}3} = 0.7328$ and $p_{C2{:}3} = 1.00927$.
- Combine the net effect of the parallel contribution of bias and the series contribution of bias by multiplying the appropriate proportions. For $A2 : 3$, we see that the parallel bias offsets the series bias, which can be calculated as $p_{A2} \cdot p_{A2{:}3} = 0.99287 \cdot 1.01786 = 1.0106$. Thus, for the assumption of three $A2$ items in series the net effect is that $E\left[S^\star\right] \cong 1.0106\,E\left[S\right]$, slightly overestimating the system average lifetime. The B and C proportions are calculated to be 1.00347 and 1.00433, respectively.
- Now one can estimate system bias in whatever way makes the most sense. To be conservative, one can use the first bias proportion estimate, 1.0106. The system distribution can be adjusted for bias by the simple transformation $S^\star_{\text{unbiased}} = S^\star/1.0106$, which can be done in APPL.
- Alternatively, one could argue the bias is perhaps the average of the three cases, and could then unbias S^\star with a different constant.

Note that we choose to unbias these systems by using proportions, not subtractions. The subtraction method is problematic, as the support of S^\star could shift into the negative numbers.

Alternatively, one could estimate bias using parametric assumptions. The process is similar to that described above, except one would have to estimate parameters of the assumed distribution along the way. For the parametric method, assume, for the sake of argument, an underlying Weibull distribution for each component. One can estimate $\hat{\eta}$ and $\hat{\kappa}$ for each component and use Table 10.1 to derive the bias for the parallel subsystems to be $b_{A2} = -1.25$, $b_{B2} = -0.45$, and $b_{C2} = -0.54$. Similar follow-on steps would mirror the empirical process, and one would again see that the S^\star distribution slightly overestimates the true $E\left[S\right]$. For reasonable sample sizes, the bias variations between the empirical method and a parametric method are small in comparison to the sampling error of the component lifetimes, thus, it is probably most practical to choose the non-parametric method based on the empirical estimates of bias.

In addition to modeling S^\star and estimating its bias, a further use of the enumerated bootstrap could be exploring the effect of system design on the system lifetime distribution. Consider again the system in Fig. 10.3. A system designer might want to improve that system by investigating how many components of type A in parallel it would take to increase both the system mean lifetime and the system 5th percentile by, say, 12%. The iterative calculations in APPL and Maple [80], using the code below, are used to calculate the answer.

```
> n := 2;
> target := 1.12;
```

```
> m := 78.78;
> p5 := 31.95;
> mux := m;
> pct := p5;
> while (mux < target * m or pct < target * p5) do
>    n := n + 1;
>    System := Minimum(MaximumIID(A, n), Minimum(MaximumIID
>                      (B, 2), MaximumIID(C, 2)));
>    mux := Mean(System);
>    pct := IDF(System, 0.05);
> end do;
> print(n);
```

It turns out that five components of type A arranged in parallel produce the desired increase of 12% in both the mean and 5th percentile, leaving the designers to then conduct a cost-benefit analysis of the improved system. Further still, one can investigate the asymptotic benefit of even more components of type A in parallel. Let **target** be the desired increase in mean system lifetime. By changing the value of the variable **target** in the loop to 1.13, 1.14, 1.15, and 1.16, one finds that the number of type A components in parallel is required to be 6, 7, 20, and ∞ respectively. The asymptotic upper limits of the mean and the 5th percentile are approximately 90.61 and 41.66 respectively as the number of parallel type A components approaches infinity.

Clearly, without a fully enumerated bootstrap, the above iterative method of exploring system design would become much more computationally expensive. Also, multiple resampling error analyses would have to be conducted at each iteration. In these situations, increasing the bootstrap iteration value B is not as simple as it sounds, as one would have iteration within iteration. One can imagine even more complex optimization algorithms than those above and the benefits of the enumerated bootstrapping become more apparent. Finally, in further research, we have also investigated the cold start backup of components (statistically, distributions of convolutions) using enumerated bootstrapping. Comparing parallel structures of systems to cold start backup structures are now possible, each time getting an exact distribution for the system. In fact all sorts of algebraic operations are possible for other types of systems as well as statistics, many of which we are still investigating. Work on this is also being extended to k-out-of-n type systems as well as cold backup systems.

This section has proved that using bootstrapping on a parallel (or series) systems with statistically-identical components underestimates (or overestimates) the expected system lifetime. An expression for the bias for all values of n is closed-form. Exact bias results are calculated for two and three component systems. Fully enumerated bootstraps are possible, and an application of such bootstrapping is provided.

This completes our initial chapter on APPL applications to reliability and survival analysis. The next chapter concerns applications of APPL in time series analysis.

11

Symbolic ARMA Model Analysis

This chapter extends the APPL language to include the analysis of ARMA (autoregressive moving average) time series models. ARMA models provide a parsimonious and flexible mechanism for modeling the evolution of a time series. Some useful measures of these models (e.g., the autocorrelation function or the spectral density function) are oftentimes tedious to compute by hand, and APPL can help ease the computational burden.

Many problems in time series analysis rely on approximate values from Monte Carlo simulations or the central limit theorem rather than exact results. This chapter describes our time series extension to APPL which can compute autocorrelation and partial autocorrelation functions of ARMA models which would otherwise require extremely tedious pencil and paper work or simulation to find, along with several other tools for working with ARMA models. Simulation is not used in any of the results.

The time series extension to APPL provides procedures for ARMA models to:

- calculate an autocorrelation (TSCorrelation), including autocorrelations of models with heteroskedastic error terms,
- calculate a partial autocorrelation (TSPartialCorrelation),
- calculate a mean (TSMean),
- calculate a variance or covariance (TSVariance),
- plot an autocorrelation or partial correlation function (TSPlot),
- calculate a spectral density function (SDF),
- perform a unit roots analysis to determine whether or not an $AR(p)$ model is stationary or an $MA(q)$ model is invertible (UnitRoots),
- forecast an $AR(p)$ model and calculate a confidence interval (TSForecast),
- generate a realization of a process (TSRealization),
- perform an exploratory time series analysis (ETSA) which displays several of the previous commands at once.

© Springer International Publishing AG 2017

J.H. Drew et al., *Computational Probability*, International Series in Operations Research & Management Science 246,

DOI 10.1007/978-3-319-43323-3_11

The remainder of this chapter is organized as follows. Section 11.1 describes the ARMA model. Section 11.2 describes the data structure for storing ARMA models and explains the ways that the mean, variance, autocorrelation, partial autocorrelation and forecasts are computed. Section 11.3 illustrates how to use the time series extension through a series of examples.

11.1 ARMA Model Basics

An $AR(p)$ model has p autoregressive components, 0 moving average components and is defined by

$$Y_t = c + \phi_1 Y_{t-1} + \phi_2 Y_{t-2} + \cdots + \phi_p Y_{t-p} + \epsilon_t, \tag{11.1}$$

where Y_t is the value of the time series at time t, c is a real-valued constant, $\phi_1, \phi_2, \ldots, \phi_p$ are real-valued parameters, and $\epsilon_0, \epsilon_1, \epsilon_2, \ldots, \epsilon_t$ are error terms, typically mutually independent and normally distributed with a mean of 0. An $MA(q)$ model has 0 autoregressive components, q moving average components and is defined by

$$Y_t = c + \epsilon_t + \theta_1 \epsilon_{t-1} + \theta_2 \epsilon_{t-2} + \cdots + \theta_q \epsilon_{t-q}, \tag{11.2}$$

with variables defined similarly. An $ARMA(p, q)$ model is a combination of (11.1) and (11.2):

$$Y_t = c + \phi_1 Y_{t-1} + \phi_2 Y_{t-2} + \cdots + \phi_p Y_{t-p} + \epsilon_t + \theta_1 \epsilon_{t-1} + \theta_2 \epsilon_{t-2} + \cdots + \theta_q \epsilon_{t-q}. \tag{11.3}$$

See Box and Jenkins [13] for more details.

One aspect of analyzing ARMA (and other time series) models is finding autocorrelation functions, that is, the jth autocorrelation, denoted by ρ_j, is the correlation between the value of a time series Y_t at time t and its value Y_{t-j} at time $t - j$. For models in which the ϵ_t terms are iid, ρ_j is the same at any time t. The jth autocorrelation of $MA(q)$ models with $q < \infty$ is 0 when $j > q$ because Y_t and Y_{t-j} do not share terms. On the other hand, the jth autocorrelation of stationary $AR(p)$ models asymptotically approaches 0 as j increases because the value of Y_t affects all future values, but its impact decreases over time. Finding the exact autocorrelation function for an ARMA model can be difficult to do by hand, especially when error terms are not iid. Section 11.2 explains two ways that this process is automated with Maple and APPL.

APPL is essential for working with various distributions of error terms. For example, suppose that $\epsilon_t \sim$ exponential(4) and we want to find $E[\epsilon_t^2]$. APPL can find the exact solution quickly:

```
> X := ExponentialRV(4);
> g := [[x -> x ^ 2], [0, infinity]];
> U := Transform(X, g);
> Mean(U);
```

$$\frac{1}{8}$$

The Mean, Variance, and Transform procedures, as well as the data structure for probability distributions in APPL, are all used when working with ARMA models. This enables our software to use an arbitrary probability distribution for the error terms.

11.2 Implementation

The time series analysis extension to the APPL software consists of approximately 1000 lines of Maple code which allow a user to define an ARMA model and compute various measures associated with the model. The data structure for storing the ARMA model and methods for computing the autocorrelation and partial autocorrelation are described here.

Data Structure. One of the first steps in developing the time series extension is to define a data structure within Maple to store a times series model. A list-of-lists data structure was used to describe an ARMA model. This allows for both symbolic and numeric values for the parameters, and the error terms ϵ_t can have arbitrary probability distributions. The first sub-list contains the parameters $\phi_1, \phi_2, \ldots, \phi_p$ for the autoregressive portion of the model; the second sub-list contains the parameters $\theta_1, \theta_2, \ldots, \theta_q$ for the moving average portion of the model; the third element of the list contains the probability distribution of the error terms. For example, the statement

```
> X := [[0.2, 0.93], [1.4], NormalRV(0, sigma)];
```

sets the variable X to an ARMA$(2, 1)$ model with $\phi_1 = 0.2$, $\phi_2 = 0.93$, $\theta_1 = 1.4$, and $N(0, \sigma^2)$ error terms (Gaussian white noise). Procedures AR, MA, and ARMA have also been written to save on keystrokes. For example,

```
> Y := ARMA(1, 1);
```

sets the variable Y to a ARMA$(1, 1)$ model with arbitrary parameters ϕ_1, θ_1, and the default standard normal error terms (the default allows a realization to be computed). Other error terms can also be included using these procedures, for example,

```
> Z := MA([2, 3, 1 / 4], ExponentialRV(1));
```

sets the variable Z to an MA(3) model with parameters $\theta_1 = 2$, $\theta_2 = 3$, $\theta_3 = 1/4$, and unit exponential error terms. Time-dependent error terms can be introduced in the data structure by using the variable tau, the time series placeholder for t in the expression for the error-term distribution. Time-dependent errors are illustrated in Sect. 11.3.

Computing Autocorrelation by the Defining Formula. The autocorrelation between Y_t and Y_{t-j} can be calculated using the defining formula:

$$\rho_j = \frac{\text{Cov}(Y_t, Y_{t-j})}{\sigma_{Y_t}\sigma_{Y_{t-j}}} = \frac{E[(Y_t - E[Y_t])(Y_{t-j} - E[Y_{t-j}])]}{\sqrt{E[(Y_t - E[Y_t])^2]}\sqrt{E[(Y_{t-j} - E[Y_{t-j}])^2]}}. \tag{11.4}$$

For MA(q) models, this is a straightforward but tedious computation to perform by hand. Since for any random variables X_1, X_2 and constants a, b we have

$$E[aX_1 + bX_2] = aE[X_1] + bE[X_2], \tag{11.5}$$

then

$$
\begin{aligned}
E[Y_t] &= E[c + \epsilon_t + \theta_1\epsilon_{t-1} + \theta_2\epsilon_{t-2} + \cdots + \theta_q\epsilon_{t-q}] \\
&= c + E[\epsilon_t] + \theta_1 E[\epsilon_{t-1}] + \theta_2 E[\epsilon_{t-2}] + \cdots + \theta_q E[\epsilon_{t-q}]. \tag{11.6}
\end{aligned}
$$

Maple and APPL can then compute $E[Y_t]$ term-by-term, even for error term distributions that are non-normal or time-dependent (i.e., not iid). The computed values of $E[Y_t]$ and $E[Y_{t-j}]$ are then substituted into (11.4). The resulting expressions within each of the expected values are then expanded so that Maple and APPL can compute each expected value term-by-term once again using (11.5). With these results, Maple can compute the mean, variance, jth covariance, and jth autocorrelation for MA(q) models.

For stationary AR(p) and ARMA(p, q) models, this method will not work because computing $E[Y_t]$ requires computing $E[Y_{t-1}]$, which in turn requires computing $E[Y_{t-2}]$, etc. However, autocorrelations for either stationary or non-stationary AR(p) and ARMA(p, q) models over a finite time horizon can be computed using this method. For example, consider an AR(2) model at time 7,

$$Y_7 = c + \phi_1 Y_6 + \phi_2 Y_5 + \epsilon_7 \tag{11.7}$$

and let the two initial observations be $Y_0 = c + \epsilon_0$ and $Y_1 = c + \phi_1 Y_0 + \epsilon_1$. Substitute $c + \phi_1 Y_5 + \phi_2 Y_4 + \epsilon_6$ for Y_6 in (11.7), then substitute $c + \phi_1 Y_4 + \phi_2 Y_3 + \epsilon_5$ for $Y_5, \ldots, c + \epsilon_0$ for Y_0. The resulting Y_7 will not contain any Y_t terms and thus the jth autocorrelation of Y_7, $\rho_{j,7}$, can be computed using the method described above.

Computing Autocorrelation with the Yule–Walker Equations. Autocorrelation for a steady-state AR(p) model (i.e., a stationary process that has run for long enough that covariances are equal) cannot be computed using the above method. Instead, a well-known system of equations called the Yule–Walker equations describes the autocorrelations (provided that the error terms are iid). For an AR(p) model, the Yule–Walker equations are given in Hamilton [44, page 59]:

$$
\begin{aligned}
\rho_0 &= 1 \\
\rho_1 &= \phi_1\rho_0 + \phi_2\rho_{-1} + \cdots + \phi_p\rho_{1-p} \\
\rho_2 &= \phi_1\rho_1 + \phi_2\rho_0 + \cdots + \phi_p\rho_{2-p} \\
&\vdots \\
\rho_p &= \phi_1\rho_{p-1} + \phi_2\rho_{p-2} + \cdots + \phi_p\rho_0. \tag{11.8}
\end{aligned}
$$

Since for real-valued processes $\rho_1 = \rho_{-1}, \rho_2 = \rho_{-2}, \ldots$, we can solve for any ρ_j whenever the process is stationary. Covariances can be computed with the similar system of equations, where γ_j is the jth covariance:

$$\gamma_0 = \phi_1\gamma_1 + \phi_2\gamma_2 + \cdots + \phi_p\gamma_p + \sigma^2$$
$$\gamma_j = \phi_{j-1}\gamma_1 + \phi_{j-2}\gamma_2 + \cdots + \phi_p\gamma_{j-p} \qquad j = 1, 2, \ldots. \qquad (11.9)$$

For a stationary ARMA(p, q) process there is not an easily definable system of equations for computing the autocorrelations. However, a system can be generated in the following manner. Consider finding the equation for γ_j of an ARMA(p, q) model, $j \le q$. Observe that when $j > q$, then (11.9) describes γ_j. For simplicity, let $c = 0$ even though the results hold when this is not the case.

$$Y_t = \phi_1 Y_{t-1} + \phi_2 Y_{t-2} + \cdots + \phi_p Y_{t-p} + \epsilon_t + \theta_1 \epsilon_{t-1} + \theta_2 \epsilon_{t-2} + \cdots + \theta_q \epsilon_{t-q}$$
$$E[Y_t Y_{t-j}] = E[Y_{t-j}(\phi_1 Y_{t-1} + \phi_2 Y_{t-2} + \cdots + \phi_p Y_{t-p} + \epsilon_t + \theta_1 \epsilon_{t-1} + \theta_2 \epsilon_{t-2} + \cdots + \theta_q \epsilon_{t-q})]$$
$$\gamma_j = \phi_1 \gamma_{j-1} + \phi_2 \gamma_{j-2} + \cdots + \phi_p \gamma_{j-p} + E[Y_{t-j}(\epsilon_t + \theta_1 \epsilon_{t-1} + \theta_2 \epsilon_{t-2} + \cdots + \theta_q \epsilon_{t-q})].$$

Now define $\xi_k = E[Y_t \epsilon_{t-k}]$ and let $\theta_0 = 1$, giving us the following solvable system of equations (once the expectations ξ_k are computed):

$$\gamma_j = \phi_1 \gamma_{j-1} + \phi_2 \gamma_{j-2} + \cdots + \phi_p \gamma_{j-p} + \theta_j \xi_0 + \theta_{j+1}\xi_1 + \cdots + \theta_q \xi_{q-j} \qquad j \le q$$
$$\gamma_j = \phi_{j-1}\gamma_1 + \phi_{j-2}\gamma_2 + \cdots + \phi_p \gamma_{j-p} \qquad\qquad\qquad j > q.$$
$$(11.10)$$

The expectation ξ_k can be computed in the following way and then substituted into (11.10):

$$\xi_0 = E[\epsilon_t(\phi_1 Y_{t-1} + \phi_2 Y_{t-2} + \cdots + \phi_p Y_{t-p} + \epsilon_t + \theta_1 \epsilon_{t-1} + \theta_2 \epsilon_{t-2} + \cdots + \theta_q \epsilon_{t-q})]$$
$$= E[\epsilon_t^2] = \sigma^2$$
$$\xi_1 = E[\epsilon_{t-1}(\phi_1 Y_{t-1} + \phi_2 Y_{t-2} + \cdots + \phi_p Y_{t-p} + \epsilon_t + \theta_1 \epsilon_{t-1} + \theta_2 \epsilon_{t-2} + \cdots + \theta_q \epsilon_{t-q})]$$
$$= \phi_1 \xi_0 + \theta_1 \sigma^2$$
$$\xi_2 = E[\epsilon_{t-2}(\phi_1 Y_{t-1} + \phi_2 Y_{t-2} + \cdots + \phi_p Y_{t-p} + \epsilon_t + \theta_1 \epsilon_{t-1} + \theta_2 \epsilon_{t-2} + \cdots + \theta_q \epsilon_{t-q})]$$
$$= \phi_1 \xi_1 + \phi_2 \xi_0 + \theta_2 \sigma^2$$
$$\vdots$$
$$\xi_k = E[\epsilon_{t-k}(\phi_1 Y_{t-1} + \phi_2 Y_{t-2} + \cdots + \phi_p Y_{t-p} + \epsilon_t + \theta_1 \epsilon_{t-1} + \theta_2 \epsilon_{t-2} + \cdots + \theta_q \epsilon_{t-q})]$$
$$= \phi_1 \xi_{k-1} + \phi_2 \xi_{k-2} + \cdots + \phi_k \xi_0 + \theta_k \sigma^2.$$

Maple will solve (11.8)–(11.10) either with arbitrary parameters or with values that describe a stationary process.

Computing Partial Autocorrelation. Partial autocorrelations are computed by first finding autocorrelations with one of the above methods. Then the jth partial autocorrelation is [120, pages 579–580]

$$\alpha_j = \frac{\begin{vmatrix} \rho_0 & \rho_1 & \rho_2 & \cdots & \rho_{j-2} & \rho_1 \\ \rho_1 & \rho_0 & \rho_1 & \cdots & \rho_{j-3} & \rho_2 \\ \rho_2 & \rho_1 & \rho_0 & \cdots & \rho_{j-4} & \rho_3 \\ \vdots & \vdots & \vdots & \ddots & \vdots & \vdots \\ \rho_{j-1} & \rho_{j-2} & \rho_{j-3} & \cdots & \rho_1 & \rho_j \end{vmatrix}}{\begin{vmatrix} \rho_0 & \rho_1 & \rho_2 & \cdots & \rho_{j-2} & \rho_{j-1} \\ \rho_1 & \rho_0 & \rho_1 & \cdots & \rho_{j-3} & \rho_{j-2} \\ \rho_2 & \rho_1 & \rho_0 & \cdots & \rho_{j-4} & \rho_{j-3} \\ \vdots & \vdots & \vdots & \ddots & \vdots & \vdots \\ \rho_{j-1} & \rho_{j-2} & \rho_{j-3} & \cdots & \rho_1 & \rho_0 \end{vmatrix}}, \tag{11.11}$$

which can be evaluated using Maple's `LinearAlgebra` package.

Invertibility for MA(q) Models. MA(q) models as defined by (11.2) are invertible if they can be represented by an AR(∞) model. For models with iid error terms, this is true when the roots of the model's characteristic equation lie outside the unit circle in the complex plane. The characteristic equation of an MA(q) model is

$$1 + \theta_1 z + \theta_2 z^2 + \cdots + \theta_q z^q = 0. \tag{11.12}$$

However, non-invertible MA(q) models have an invertible representation with identical moments from Hamilton [44, page 68]. If $\lambda_1, \lambda_2, \ldots, \lambda_n$ are the roots of (11.12) inside the unit circle and $\lambda_{n+1}, \lambda_{n+2}, \ldots, \lambda_q$ are the roots of (11.12) outside the unit circle, then an invertible model with identical moments is

$$Y_t = c + \left(\prod_{i=1}^{n} (1 - \lambda_i L) \right) \left(\prod_{i=n+1}^{q} (1 - \lambda_i^{-1} L) \right) \epsilon_t', \tag{11.13}$$

where $V[\epsilon_t'] = \sigma^2 \lambda_1^{-2} \lambda_2^{-2} \cdots \lambda_n^{-2}$ and L is the lag or backshift operator, i.e., $\epsilon_t L = \epsilon_{t-1}$, $\epsilon_t L^2 = \epsilon_{t-2}$, etc.

Forecasting AR(p) Models. Consider an AR(p) model where the values of Y_t, Y_{t-1}, \ldots are known, but the values of Y_{t+1}, Y_{t+2}, \ldots are unknown, and we want to find the expected value and a confidence interval for Y_{t+s} for some positive integer s. Then,

$$Y_{t+s} = \phi_1 Y_{t+s-1} + \phi_2 Y_{t+s-2} + \cdots + \phi_p Y_{t+s-p} + \epsilon_{t+s}$$
$$Y_{t+s-1} = \phi_1 Y_{t+s-2} + \phi_2 Y_{t+s-3} + \cdots + \phi_p Y_{t+s-p-1} + \epsilon_{t+s-1}$$

$$\vdots$$

$$Y_{t+1} = \phi_1 Y_t + \phi_2 Y_{t-1} + \cdots + \phi_p Y_{t-p+1} + \epsilon_{t+1}.$$

Since Y_{t+1} is an error term plus a constant, then we can substitute Y_{t+1} into Y_{t+2}, then substitute Y_{t+2} into Y_{t+3}, \ldots, then substitute Y_{t+s-1} into

Y_{t+s}, resulting in an expression consisting of error terms and constants. The software assumes that the error terms are mutually independent and normally distributed, which implies that the value of Y_{t+s} is also normally distributed. Maple can then compute the mean and variance of Y_{t+s} and call APPL's IDF procedure to find values for a confidence interval.

11.3 Examples

This section consists of a series of examples that highlight the capability of the time series extension to the APPL language. Some of the long symbolic results may find uses in statistical software which currently use simulation to find measures of interest. Some examples contain graphical output, which is included in the text rather than referring to a figure.

Example 11.1. Find the lag 2 autocorrelation at time t of an MA(2) model with time-dependent error terms with $\epsilon_t \sim N\big(0, \ln(t)^2\big)$.

The MA(2) model is

$$Y_t = c + \epsilon_t + \theta_1 \epsilon_{t-1} + \theta_2 \epsilon_{t-2},$$

where $\epsilon_t \sim N\big(0, \ln(t)^2\big)$, $\epsilon_{t-1} \sim N\big(0, \ln(t-1)^2\big)$, etc. First we compute the second autocorrelation at time t, $\rho_{2,t}$, by hand and then we will compute $\rho_{2,t}$ using the time series extension.
Using the defining formula for correlation, we have

$$\rho_{2,t} = \frac{\mathrm{Cov}(Y_t, Y_{t-2})}{\sigma_{Y_t} \sigma_{Y_{t-2}}} = \frac{E[(Y_t - E[Y_t])(Y_{t-2} - E[Y_{t-2}])]}{\sqrt{E[(Y_t - E[Y_t])^2]}\sqrt{E[(Y_{t-2} - E[Y_{t-2}])^2]}}.$$

Notice that since all error terms have a mean of 0, then $E[Y_t] = c$ and $E[Y_{t-2}] = c$. Substituting in the expressions for Y_t and Y_{t-2} we have,

$$\rho_{2,t} = \frac{E[(\epsilon_t + \theta_1 \epsilon_{t-1} + \theta_2 \epsilon_{t-2})(\epsilon_{t-2} + \theta_1 \epsilon_{t-3} + \theta_2 \epsilon_{t-4})]}{\sqrt{E[(\epsilon_t + \theta_1 \epsilon_{t-1} + \theta_2 \epsilon_{t-2})^2]}\sqrt{E[(\epsilon_{t-2} + \theta_1 \epsilon_{t-3} + \theta_2 \epsilon_{t-4})^2]}}.$$

While the error term distributions are not identical, they are independent and have mean 0, meaning that for any $i \neq j$ we have $E[\epsilon_{t-i}\epsilon_{t-j}] = 0$. This allows us to simplify when expanding the polynomials inside the expected values. So we have,

$$\rho_{2,t} = \frac{E[\theta_2 \epsilon_{t-2}^2]}{\sqrt{E[\epsilon_t^2 + \theta_1^2 \epsilon_{t-1}^2 + \theta_2^2 \epsilon_{t-2}^2]}\sqrt{E[\epsilon_{t-2}^2 + \theta_1^2 \epsilon_{t-3}^2 + \theta_2^2 \epsilon_{t-4}^2]}}.$$

Each θ_i parameter is a constant (i.e., we can bring them outside of the expected value operator), so it remains to find $E[\epsilon_t^2], \ldots, E[\epsilon_{t-4}^2]$ to

complete our calculation of $\rho_{2,t}$. For any $k > 0$ and $\epsilon_k \sim N\big(0, \ln(k)^2\big)$ we can use the properties of the normal and chi square distributions to find $E[\epsilon_k^2]$. First, $\epsilon_k^2/\ln(k)^2 \sim \chi^2(1)$, which implies that $E[\epsilon_k^2/\ln(k)^2] = 1$ because the expected value of a chi square random variable is its degrees of freedom. Then, since $1/\ln(k)^2$ is a constant, we can bring it outside of the expected value operator and solve, giving us $E[\epsilon_k^2] = \ln(k)^2$. Using this result, we have:

$$
\rho_{2,t} = \frac{\theta_2 \ln(t-2)^2}{\sqrt{\ln(t)^2 + \theta_1^2 \ln(t-1)^2 + \theta_2^2 \ln(t-2)^2}\sqrt{\ln(t-2)^2 + \theta_1^2 \ln(t-3)^2 + \theta_2^2 \ln(t-4)^2}}.
$$

Now we will compute the same autocorrelation in just two commands using the time series extension.

```
> X := MA(2, NormalRV(0, ln(tau))):
> TSCorrelation(X, 2);
```

$$
\frac{\theta_2 \ln(t-2)^2}{\sqrt{\ln(t)^2 + \theta_1^2 \ln(t-1)^2 + \theta_2^2 \ln(t-2)^2}\sqrt{\ln(t-2)^2 + \theta_1^2 \ln(t-3)^2 + \theta_2^2 \ln(t-4)^2}}
$$

The first command assigns the time series list-of-lists data structure to X. The first argument of the MA() procedure is the order of the model and the second argument is the distribution of the error terms in APPL format (APPL takes standard deviation rather than variance for its probability distributions), with tau as the placeholder for t. The second command computes the second autocorrelation of the assigned model.

Example 11.2. Find the unit roots of an MA(3) model with parameters $\theta_1 = 0.7$, $\theta_2 = 0.9$, and $\theta_3 = -1.3$.

Since all MA(q) models are stationary, a unit root analysis of an MA(q) model tests for invertibility. If the model is not invertible, then an invertible representation will be found according to Eq. (11.13). A call to the procedure UnitRoots prints the values of the unit roots, plots the unit roots in the complex plane, and gives an invertible representation of the model with identical autocorrelation function.

```
> X := MA([0.7, 0.9, -1.3]):
> UnitRoots(X):
```

The roots are:
$$[-0.3730587892 + 0.6289673928I, -0.3730587892 - 0.6289673928I, 1.438425271]$$
Number of roots inside the unit circle: 2
Assuming error terms are i.i.d, then an invertible representation is:
$$\mathsf{Y_t} =$$
$$\mathsf{c} + \epsilon_t + 0.0509128847\epsilon_{t-1} + 0.0160683988\epsilon_{t-2} - 0.3717765894\epsilon_{t-3}$$

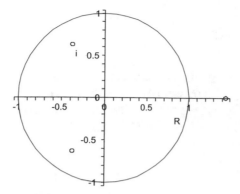

If the parameter values were given as exact values rather than decimals (and $q \leq 4$), then the invertible model would also have exact parameters. So if the model were instead input as

```
> X := MA([7 / 10, 9 / 10, -13 / 10]):
```

then the UnitRoots procedure prints the exact values of the three roots of the cubic polynomial

$$1 + \frac{7}{10}z + \frac{9}{10}z^2 - \frac{13}{10}z^3 = 0,$$

which are the real root

$$\frac{1}{78}\sqrt[3]{217836 + 780\sqrt{73329}} + \frac{236}{13\sqrt[3]{217836 + 780\sqrt{73329}}} + \frac{3}{13}$$

and the complex conjugates

$$-\frac{1}{156}\sqrt[3]{217836 + 780\sqrt{73329}} - \frac{118}{13\sqrt[3]{217836 + 780\sqrt{73329}}} + \frac{3}{13} \pm$$

$$\frac{i\sqrt{3}}{2}\left(\frac{1}{78}\sqrt[3]{217836 + 780\sqrt{73329}} - \frac{236}{13\sqrt[3]{217836 + 780\sqrt{73329}}}\right).$$

Example 11.3. Find the unit roots of the AR(3) model with $\phi_1 = 0.5$, $\phi_2 = 0.1$, and $\phi_3 = -0.3$.

Instead of testing for invertibility as in the previous example, unit roots analysis of $AR(p)$ and $ARMA(p, q)$ models tests for stationarity. $AR(p)$ models with iid error terms are stationary when the roots of $1 - \phi_1 z - \phi_2 z^2 - \cdots - \phi_p z^p = 0$ lie outside the unit circle.

```
> X := AR([0.5, 0.1, -0.3]):
> UnitRoots(X);
```

The roots are:
$$[1.034099254 + 0.9230480107I, 1.034099254 - 0.9230480107I, -1.734865174]$$
All roots lie outside the unit circle, so the ARMA process is stationary

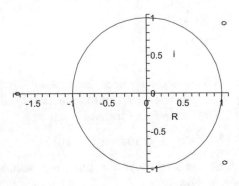

Example 11.4. Find the lag 1 autocorrelation of an MA(1) model with real-valued parameter θ_1 and exponential(4) error terms.

As expected, non-normal but iid error terms produce the same auto-correlation function as the case with normal error terms.

```
> X := MA(1, ExponentialRV(4)):
> TSCorrelation(X, 1);
```

$$\frac{\theta_1}{1 + \theta_1^2}$$

Example 11.5. Find the lag 2 autocorrelation for an AR(2) model with parameters $\phi_1 = 1/2$ and $\phi_2 = 1/4$ at time $t = 7$.

In this example, $\rho_{2,7}$ for $Y_7 = c + \phi_1 Y_6 + \phi_2 Y_5 + \epsilon_7$ of an AR(2) model is calculated using recursive substitution then taking of expected values as described in Sect. 11.2. Notice that the constant term, c, appears in the expression.

```
> X := AR([1 / 2, 1 / 4]):
> TSCorrelation(X, 2, 7);
```

$$\frac{1067\sqrt{30621}\sqrt{16384}}{31355904\sqrt{\frac{2209}{16384}c^2 + \frac{461}{256}}}$$

This can also be evaluated with arbitrary parameters, but the resulting expression for the second autocorrelation value is very long.

Example 11.6. Find the lag 2 autocorrelation of an AR(4) model with parameters ϕ_1, ϕ_2, ϕ_3, and ϕ_4 with the autocorrelation computed by the Yule–Walker equations.

We can compute ρ_2 for an AR(4) model that is in steady-state (i.e., stationary) by the Yule–Walker method described in Sect. 11.2. The optional "YW" flag tells Maple to find the correlation by solving the Yule–Walker equations rather than the defining formula.

```
X := AR(4):
TSCorrelation(X, 2, "YW");
```

$$-\frac{-\phi_4{}^2\phi_2 + \phi_4\phi_3{}^2 + \phi_1\phi_4\phi_3 - \phi_2{}^2\phi_4 + \phi_1{}^2 + \phi_2 + \phi_1\phi_3 - \phi_2{}^2}{-1 + \phi_4 + \phi_2 + \phi_4{}^2 - \phi_4{}^3 - \phi_4{}^2\phi_2 + \phi_1\phi_3 + \phi_3{}^2 + \phi_1{}^2\phi_4 + \phi_1\phi_4\phi_3}$$

Example 11.7. Find the lag 2 autocorrelation of an ARMA(3, 3) model with parameters $\phi_1 = 1/2, \phi_2 = 3/10, \phi_3 = 1/10, \theta_1 = 9/10, \theta_2 = 1/2$, and $\theta_3 = 3/10$ at time $t = 8$.

The lag 2 autocorrelation of Y_8 is computed as described in Sect. 11.2.

```
> X := ARMA([1 / 2, 3 / 10, 1 / 10],
            [9 / 10, 1 / 2, 3 / 10]):
> TSCorrelation(X, 2, 8);
```

$$\frac{3935943161}{31174413101120000000}\sqrt{97420040941}\sqrt{6400000000}\,\frac{1}{\sqrt{\frac{204404181}{16000000} + \frac{14874485521}{25600000000}c^2}}$$

This can be computed with arbitrary parameters but the resulting equation is very long. Again, notice that the constant term, c, appears in the expression.

Example 11.8. Find the lag 2 autocorrelation of an ARMA(3, 2) model with parameters $\phi_1 = 1/2$, $\phi_2 = 3/10$, $\phi_3 = 1/10$, $\theta_1 = 3/4$, and $\theta_2 = 1/4$ by solving the Yule–Walker equations.

The lag 2 autocorrelation of Y_t is computed as described in Sect. 11.2.

```
> X := ARMA([1 / 2, 3 / 10, 1 / 10], [3 / 4, 1 / 4]):
> TSCorrelation(X, 2, "YW");
```

$$\frac{1636}{1831}$$

Example 11.9. Find the lag 3 partial autocorrelation of an MA(2) model with parameters θ_1 and θ_2.

The lag 3 partial autocorrelation of an MA(2) model is computed as described in Sect. 11.2.

```
> TSPartialCorrelation(MA(2), 3);
```

$$\frac{\theta_1\,(1+\theta_2)\,(-2\,\theta_2 - 2\,\theta_2{}^3 + \theta_1{}^2 + \theta_2{}^2\theta_1{}^2 + \theta_2{}^2)}{1 + 2\,\theta_2{}^2 + \theta_1{}^6 + \theta_1{}^2 + \theta_1{}^4 + 2\,\theta_2{}^4 - 2\,\theta_2\theta_1{}^2 + \theta_2{}^2\theta_1{}^4 + 5\,\theta_2{}^2\theta_1{}^2 - 4\,\theta_2\theta_1{}^4 - 2\,\theta_2{}^3\theta_1{}^2 + \theta_2{}^6 + \theta_2{}^4\theta_1{}^2}$$

Example 11.10. Find the lag 3 partial autocorrelation of an MA(1) model with parameter θ_1 and error terms $\epsilon_t \sim N(0,t)$.

The lag 3 partial autocorrelation is computed as described in Sect. 11.2 and $\epsilon_t \sim N(0,t)$.

```
> X := MA(1, NormalRV(0, sqrt(tau))):
> TSPartialCorrelation(X, 3);
```

$$\frac{\theta_1{}^3\,(t-1)^3}{\sqrt{t + \theta_1{}^2 t - \theta_1{}^2}\,\sqrt{t - 1 + \theta_1{}^2 t - 2\,\theta_1{}^2}\,(t^2 - t + \theta_1{}^4 t^2 - 3\,\theta_1{}^4 t - \theta_1{}^2 + 2\,\theta_1{}^4)}$$

Example 11.11. Find the lag 2 partial autocorrelation of an AR(3) model with parameters $\phi_1 = 1/2$, $\phi_2 = 3/10$, and $\phi_3 = 1/10$ at time $t = 7$.

In this example, the first two autocorrelations of Y_7 are computed initially, then they are then used to find the lag 2 partial autocorrelation. All calculations are performed using the methods described in Sect. 11.2.

```
> TSPartialCorrelation(AR([1 / 2, 3 / 10, 1 / 10]), 2, 7);
```

$$\frac{1}{132731033}\left(\frac{22811879108847780543893}{33554432000000000000000}c^2 - \frac{58277673273951481}{83886080000000000000000}\right)$$
$$\sqrt{132731033}\sqrt{51200000}\sqrt{\frac{170119849}{256000000}c^2 + \frac{289249}{128000}} +$$
$$\frac{89625931935942986179113}{83886080000000000000000}\right)\sqrt{132731033}\sqrt{51200000}$$
$$\left(\frac{170119849}{256000000}c^2 + \frac{289249}{128000}\right)^{-1/2}$$
$$\left(\frac{5269767243516833}{13107200000000000}c^2 + \frac{2761515025601}{1000000000000}\right)^{-1}$$

Again, this partial autocorrelation can be computed with arbitrary parameters, but the resulting expression is very long. As in two of the previous examples, the constant term c appears in the expression.

Example 11.12. Find the lag 2 partial autocorrelation of an AR(3) model with parameters ϕ_1, ϕ_2, and ϕ_3 using the Yule–Walker equations.

By using the "YW" flag in TSPartialCorrelation, the computer finds the autocorrelations by Yule–Walker which it then uses to find the lag 2 partial autocorrelation as described in Sect. 11.2.

> TSPartialCorrelation(AR(3), 2, "YW");

$$-\frac{4\phi_3\phi_1\phi_2 + \phi_2\phi_3{}^2 - \phi_2 + 2\phi_2{}^2 - \phi_1\phi_2{}^2\phi_3 - \phi_2{}^3 + \phi_1{}^3\phi_3 + 2\phi_1{}^2\phi_3{}^2 + \phi_1{}^2\phi_2 + \phi_3{}^3\phi_1 - \phi_3\phi_1}{\phi_1{}^2\phi_3{}^2 + 2\phi_3{}^3\phi_1 - 2\phi_3\phi_1 + \phi_3{}^4 - 2\phi_3{}^2 + 2\phi_2\phi_3{}^2 + 1 - 2\phi_2 + \phi_2{}^2 - \phi_2{}^2\phi_3{}^2 - \phi_1{}^2}$$

Example 11.13. Plot the autocorrelation function of an ARMA(3, 3) model with parameters $\phi_1 = 0.4$, $\phi_2 = 0.2$, $\phi_3 = -0.3$, $\theta_1 = 0.9$, $\theta_2 = 0.3$, and $\theta_3 = -0.2$ using the Yule–Walker equations.

TSPlot will find several autocorrelations or partial autocorrelations (by using the optional "partial" flag as a third parameter) and then plot them. In this example, the "YW" flag is used to tell Maple to find autocorrelations by solving the Yule–Walker equations as described in Sect. 11.2.

> X := ARMA([0.4, 0.2, -0.3], [0.9, 0.3, -0.2]):
> TSPlot(X, "YW");

Example 11.14. Give a point estimate and 95% confidence interval for a forecasted value of an AR(3) model with parameters $\phi_1 = 0.5$, $\phi_2 = 0.1$, and $\phi_3 = -0.3$, five time periods into the future.

In this example an AR(3) model is forecast five time periods into the future as described in Sect. 11.2. The last three known values of the model are [-5, -1, 4] , with -5 being the most recent. Such small data sets may be used, for example, for the price of a stock shortly after its launch or for space missions where there is limited data on past missions. The fourth (optional) argument in TSForecast contains the lower and upper percentiles on the confidence interval. The default is a symmetric 90% confidence level. This example plots a symmetric 95% confidence interval.

```
> X := AR([0.5, 0.1, -0.3], NormalRV(0, 4)):
> TSForecast(X, 5, [-5, -1, 4], [.025, .975]);
```

Assuming error terms are normal for this forecast.

Expected value and confidence interval are:
$$[0.36950, 12.78829339, -12.04929337]$$

After the three known values are plotted, the next five expected values and confidence interval bounds are plotted.

Example 11.15. Plot a realization of an ARMA(3, 3) model with parameters $\phi_1 = 1/2$, $\phi_2 = 3/10$, $\phi_3 = 1/10$, $\theta_1 = 1/10$, $\theta_2 = 1/2$, and $\theta_3 = 7/10$.

TSRealization will sample normally distributed error terms with the Box–Muller transformation and then plot a realization for the given model. If AR parameters are given, then the system warms up before plotting. This example plots a realization of 75 values (the default is 50 values).

```
> X := ARMA([1 / 2, 3 / 10, 1 / 10], [1 / 10, 1 / 2, 7 / 10]):
> TSRealization(X, 75);
```

This realization shows the evolution of one instantiation of 75 values, Y_1, Y_2, \ldots, Y_{75} from this particular ARMA(3, 3) model.

Example 11.16. Execute the APPL **ETSA** (Exploratory Time Series Analysis) procedure for an MA(3) model with parameters $\theta_1 = 1/5$, $\theta_2 = -7/5$, and $\theta_3 = 11/5$.

The **ETSA()** procedure calls autocorrelation, partial autocorrelation, unit roots analysis, spectral density function, and realization procedures. This calculates various quantities associated with the MA(3) model. It also produces plots that might be of interest to a modeler in order to gain some intuition associated with the time series model.

```
> X := MA([1 / 5, -7 / 5, 11 / 5]):
> ETSA(X);
```

Time series model:
$$Y_t = c + \epsilon_t + \tfrac{1}{5}\epsilon_{t-1} - \tfrac{7}{5}\epsilon_{t-2} + \tfrac{11}{5}\epsilon_{t-3}$$

Unit Roots Analysis:
The roots are:

$$\left[-\tfrac{1}{33}(8171 + 33\sqrt{61305})^{(1/3)} - \frac{16}{33(8171+33\sqrt{61305})^{(1/3)}} + \tfrac{7}{33}, \tfrac{1}{66}(8171 + \right.$$

$$33\sqrt{61305})^{(1/3)} + \frac{8}{33(8171+33\sqrt{61305})^{(1/3)}} + \tfrac{7}{33}$$

$$-\tfrac{1}{2}I\sqrt{3}\left(-\tfrac{1}{33}(8171 + 33\sqrt{61305})^{(1/3)} + \right.$$

$$\left. \frac{16}{33(8171+33\sqrt{61305})^{(1/3)}} \right), \tfrac{1}{66}(8171 + 33\sqrt{61305})^{(1/3)}$$

$$+\frac{8}{33(8171+33\sqrt{61305})^{(1/3)}}+\frac{7}{33}-\frac{1}{2}I\sqrt{3}\bigg(-\frac{1}{33}(8171+33\sqrt{61305})^{(1/3)}+$$

$$\frac{16}{33(8171+33\sqrt{61305})^{(1/3)}}\bigg)\bigg]$$

All roots are inside the unit circle.

Assuming error terms are i.i.d., then an invertible representation is:

$$Y_t = c+\epsilon_t-\frac{7}{11}\epsilon_{t-1}+\frac{1}{11}\epsilon_{t-2}+\bigg(\frac{1}{1089}\sqrt{61305}+\frac{4096}{35937(8171+33\sqrt{61305})}+\frac{8164}{35937}\bigg)\epsilon_{t-3}$$

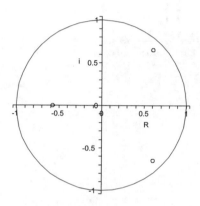

Spectral Density Function:

$$f(\omega)=\frac{\frac{196}{25}-\frac{158}{25}\cos(\omega)-\frac{48}{25}\cos(2\omega)+\frac{22}{5}\cos(3\omega)}{\pi}\qquad 0<\omega<\pi$$

First Realization: *Second Realization:*

12

Stochastic Simulation

This chapter considers applications from stochastic simulation. Section 12.1 considers tests of randomness. Section 12.2 investigates the use of computational probability in input modeling. Section 12.3 contains a development of an algorithm to find the distribution of the Kolmogorov–Smirnov goodness-of-fit test statistic in the all-parameters-known case.

12.1 Tests of Randomness

Since random numbers play a pivotal role in Monte Carlo simulation, discrete-event simulation, and bootstrapping, having random number generators with good statistical properties is essential. Some of the most common types of random number generators are in the class of Lehmer generators. These generators are based on Lehmer's algorithm [77] which defines an iterative equation that can be used to produce a stream of random numbers. A Lehmer generator requires three input parameters: m, a, and z_0. The *modulus* parameter m is a fixed large prime integer. The *multiplier* parameter a is a fixed positive integer less than m. The *initial seed* z_0 can be any positive integer less than m and can vary from experiment to experiment. Lehmer's algorithm produces integers in the range $[1, m - 1]$ via the iterative equation $z_{i+1} = az_i \bmod m$ for $i = 0, 1, \ldots$. For proper choices of a and m (see Banks et al. [7], Fishman [33], or Law [70]), the period length of the generator is $m - 1$. To convert the z_i's into $U(0, 1)$ pseudo-random variates, referred to here as simply "random numbers," each is divided by the modulus m, $x_i = z_i/m$, yielding a rational number in the set $\{\frac{1}{m}, \frac{2}{m}, \ldots, \frac{m-1}{m}\}$.

It is known that the numbers produced by a Lehmer generator are not truly random, but with carefully chosen m and a it is possible to generate output that will be "random enough" from a statistical point of view. Lehmer generators are also considered "good" generators because their output can be

© Springer International Publishing AG 2017
J.H. Drew et al., *Computational Probability*, International Series
in Operations Research & Management Science 246,
DOI 10.1007/978-3-319-43323-3_12

replicated, they're portable, efficient, and thoroughly documented. However, a disturbing amount of regularity inherent in the output of these generators was discovered by Marsaglia [83]. He observed a lattice structure when consecutive random numbers were plotted as overlapping ordered n-tuples [i.e, $(x_0, x_1, \ldots, x_{n-1}), (x_1, x_2, \ldots, x_n), \ldots$]. We consider two types of testing for random numbers in this section. The first is the more traditional Kolmogorov–Smirnov goodness-of-fit test and the second is based on the distance between consecutive random number pairs.

Kolmogorov–Smirnov Goodness-of-Fit Test. Consider the use of the Kolmogorov–Smirnov test for assessing model adequacy (goodness of fit) for a Lehmer generator. Assume that $m = 2^{31} - 1 = 2147483647$, $a = 7^5 = 16807$, and $z_0 = 987654321$, which is a popular random number generator described in Park and Miller [97]. The first five random numbers generated are

$$\frac{1605065384}{2147483647} \qquad \frac{1791818921}{2147483647} \qquad \frac{937423366}{2147483647}$$

$$\frac{1334477970}{2147483647} \qquad \frac{252032522}{2147483647}$$

or, approximately

$$0.7474168 \qquad 0.8343807 \qquad 0.4365218$$

$$0.6214147 \qquad 0.1173618.$$

Since these five data values are being evaluated for their uniformity, there should be a reasonable match between their empirical CDF and the CDF for a $U(0, 1)$ random variable. If we let the Maple list `Sample` contain the five random numbers generated above, then the APPL statements required to plot these two functions over the interval $(0, 1)$, shown in Fig. 12.1, are

```
> n := 5;
> a := 7 ^ 5;
> seed := 987654321;
> m := 2 ^ 31 - 1;
> Sample := [];
> for j from 1 to n do
>    seed := a * seed mod m:
>    Sample := [op(Sample), seed / m]:
> od;
> U := UniformRV(0, 1);
> PlotEmpVsFittedCDF(U, Sample, [], 0, 1);
```

The five parameters that are required by the APPL plotting procedure `PlotEmpVsFittedCDF` are the random variable whose CDF is to be plotted, the data values [held in a list], the parameters associated with the random variable [empty in this case of $U(0, 1)$ random variable], and the optional lower and upper plotting limits.

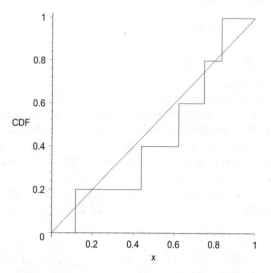

Fig. 12.1. The empirical CDF of `Sample` and the theoretical $U(0,1)$ CDF

Let $F(x)$ be the hypothesized CDF and $F_5(x)$ be the empirical CDF. To determine the Kolmogorov–Smirnov test statistic,

$$D_5 = \sup_x |F(x) - F_5(x)|,$$

which measures the largest vertical distance between the two CDFs, requires the following additional command:

```
> TestStat := KSTest(U, Sample, []);
```

The approximate value of the test statistic for the five random numbers is 0.2365, which occurs just to the left of the random number 0.4365.

Since large values of the test statistic indicate a poor fit, and the CDF $F_{D_5}(x)$ of the test statistic under the null hypothesis is (see Sect. 12.3)

$$F_{D_5}(x) = \begin{cases} 0 & y < \frac{1}{10} \\ \frac{24}{625}(10x-1)^5 & \frac{1}{10} \le y < \frac{1}{5} \\ -288x^4 + 240x^3 - \frac{1464}{25}x^2 + \frac{672}{125}x - \frac{96}{625} & \frac{1}{5} \le y < \frac{3}{10} \\ 160x^5 - 240x^4 + \frac{424}{5}x^3 + 12x^2 - \frac{168}{25}x + \frac{336}{625} & \frac{3}{10} \le y < \frac{2}{5} \\ -20x^5 + 74x^4 - \frac{456}{5}x^3 + \frac{224}{5}x^2 - \frac{728}{125}x & \frac{2}{5} \le y < \frac{1}{2} \\ 12x^5 - 6x^4 - \frac{56}{5}x^3 + \frac{24}{5}x^2 + \frac{522}{125}x - 1 & \frac{1}{2} \le y < \frac{3}{5} \\ -20y^6 + 32y^5 - \frac{185}{9}y^3 + \frac{175}{36}y^2 + \frac{3371}{648}y - 1 & \frac{1}{2} \le y < \frac{3}{5} \\ -8x^5 + 22x^4 - \frac{92}{5}x^3 + \frac{12}{25}x^2 + \frac{738}{125}x - 1 & \frac{3}{5} \le y < \frac{4}{5} \\ 2x^5 - 10x^4 + 20x^3 - 20x^2 + 10x - 1 & \frac{4}{5} \le y < 1 \\ 1 & y \ge 1, \end{cases}$$

the p-value for this particular test can be found with the additional APPL statement

```
> p := SF(KSRV(5), TestStat);
```

which yields $p \cong 0.8838$.

If this process is repeated for a total of 1000 groups of nonoverlapping consecutive sets of five random number pairs, the empirical CDF of the Kolmogorov–Smirnov statistics should be close to the theoretical from APPL if the random number generator is valid. Figure 12.2 is a plot of the empirical CDF of the 1000 Kolmogorov–Smirnov statistics versus the theoretical Kolmogorov–Smirnov CDF with $n = 5$. The empirical CDF lies slightly above the theoretical. If this experiment were performed repeatedly, the empirical CDFs should fluctuate around the theoretical CDF for a good random number generator.

Fig. 12.2. Empirical CDF of 1000 Kolmogorov–Smirnov test statistics and the theoretical Kolmogorov–Smirnov CDF for $n = 5$

Distance Between Consecutive Random Number Pairs. While it is true that the lattice structure discovered by Marsaglia [83] does not appear to be random at all, it may be the case that there is still a degree of randomness embedded within it. This randomness is not to be found in the structure of the lattice, however, but in the order in which the points are generated. More specifically, the empirical distribution of the distance between consecutive random number pairs should be close to the theoretical distance between consecutive random number pairs for a good random number generator. Duggan

et al. [24] have developed a test of randomness based on the distance between consecutive random number pairs. We consider only the use of APPL to find the distribution of the distance between two of these pairs here. The derivations and analysis are an extension of the work by L'Ecuyer et al. [65] who test for excessive clustering or repulsion of random points in space, where points X_i are non-overlapping k-dimensional vectors of successive output values of a random number generator. They consider the L_p-distances between all pairs of points generated X_i and X_j, whereas we only consider distances between consecutively generated points X_i and X_{i+1}, but treat both the overlapping and non-overlapping cases. Knuth [63] discusses specific tests that have traditionally been used to check the randomness of sequences, but all differ from our method. An ideal random number generator will have a close match between the empirical and theoretical distribution of the distance between consecutive random number pairs.

A random number generator should produce a stream of numbers that appears to be random despite the fact that they are generated by a deterministic function. Even though Marsaglia discovered that a plot of the n-tuples constructed from consecutive random numbers results in a lattice structure, the order in which the points are generated will affect the generator's statistical properties. That is, we hope to observe that points generally are not generated in order along a hyperplane nor is there necessarily a regular pattern of jumping between hyperplanes. With this in mind, we compare the distribution of the distances between consecutive points (ordered pairs) and the distribution of the distances between truly random points on the unit square. The assumption being made is that generators that yield a distribution of distances similar to the purely random points will be better random number generators.

When considering the distance between consecutive pairs of random numbers, the first question that must be addressed is whether to consider overlapping or non-overlapping ordered pairs. We choose the simpler case here, the non-overlapping ordered pairs (x_i, x_{i+1}) and (x_{i+2}, x_{i+3}), because APPL can be used to find the theoretical distance between the points in two-dimensional space. When working with non-overlapping pairs, it must be noted that analysis is only performed on half of the possible ordered pairs at a time and then merged together at the end. The two sets of ordered pairs that are examined are

$$\{(x_0, x_1), (x_2, x_3), (x_4, x_5), \ldots, (x_{m-3}, x_{m-2})\}$$

and

$$\{(x_1, x_2), (x_3, x_4), (x_5, x_6), \ldots, (x_{m-2}, x_{m-1})\},$$

where the first set consists of all ordered pairs with the first coordinate having an even index and the second set consists of all ordered pairs with the first coordinate having an odd index. The distances are calculated between consecutive points in the same set and then combined to form a single distribution.

Under the assumption that the pairs contain non-overlapping coordinates, the calculation of the theoretical distribution of the distances between

consecutive pairs of randomly generated numbers is straightforward. If we assume that X_1, X_2, X_3, and X_4 are iid $U(0,1)$ random variables, then we must simply find the distribution of the distance between (X_1, X_2) and (X_3, X_4) given by

$$D = \sqrt{(X_1 - X_3)^2 + (X_2 - X_4)^2}.$$

Since D represents a distance between two points, it is obvious that $D > 0$ (the inequality is strict because the Lehmer algorithm cannot generate two identical pairs unless $m = 3$). Furthermore, since the longest distance possible is across the diagonal of the unit square, $D < \sqrt{2}$ (the inequality is also strict because the Lehmer algorithm cannot generate the numbers 0 or 1).

APPL provides an easy way to find the distribution of D which is a rather cumbersome hand calculation. The APPL code below calculates the PDF of D.

```
> U1 := UniformRV(0,1);
> U2 := UniformRV(0,1);
> V1 := Difference(U1, U2);
> g1 := [[x -> x ^ 2, x -> x ^ 2], [-infinity, 0, infinity]];
> V2 := Transform(V1, g1);
> V3 := Convolution(V2, V2);
> g2 := [[x -> sqrt(x)], [0, 2]];
> V4 := Transform(V3, g2);
```

The theoretical PDF of D is returned as

$$f(x) = \begin{cases} 2x(x^2 - 4x + \pi) & 0 < x \le 1 \\ \dfrac{-2x\left(2\sqrt{x^2-1}+4-4x^2+2\sqrt{x^2-1}\arcsin\left(\frac{x^2-2}{x^2}\right)+x^2\sqrt{x^2-1}\right)}{\sqrt{x^2-1}} & 1 < x < \sqrt{2}, \end{cases}$$

which is a continuous function because $\lim_{x\uparrow 1} f(x) = \lim_{x\downarrow 1} f(x) = 2(\pi - 3)$. The graph in Fig. 12.3 is a plot of the PDF.

The corresponding CDF can be found with the additional CDF(V4); statement, yielding

$$F(x) = \begin{cases} \frac{1}{2}x^4 - \frac{8}{3}x^3 + \pi x^2 & 0 < x \le 1 \\ \dfrac{\sqrt{x^2-1}\left(2-12x^2\arcsin\left(\frac{x^2-2}{x^2}\right)-3x^4-12x^2\right)+24x^2-24+16(x^2-1)^2}{6\sqrt{x^2-1}} & 1 < x < \sqrt{2}. \end{cases}$$

The distribution of D can then be used in the following fashion. Generate many non-overlapping pairs of random numbers, then calculate and store the distance between consecutive pairs. Perform a goodness-of-fit test to see if the distances provide a reasonable fit to the data values. If so, then the random number generator can be considered adequate. Details are given in Duggan et al. [24]. Khoshkenara and Mahloojia [61] consider tests for randomness based on the Manhattan distance between consecutive random numbers pairs.

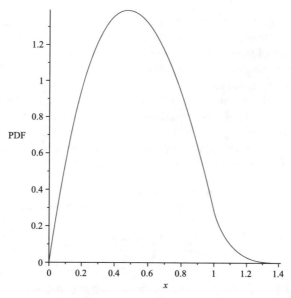

Fig. 12.3. The PDF of D

12.2 Input Modeling

Stochastic simulation models have stochastic elements that mimic the probabilistic nature of the system under consideration. A close match between the input model and the true underlying probabilistic mechanism associated with the system is required for successful "input modeling." The general question considered by input modeling is how best to model an element (e.g., arrival process, service times) in a discrete-event simulation given a data set collected on the element of interest. Oftentimes this involves fitting standard univariate parametric probability distributions to data sets. For brevity, it is assumed that data is available on the aspect of the simulation of interest. It is also assumed that raw data is available, as opposed to censored data, grouped data, or summary statistics. This example-driven section examines some applications of APPL for input modeling. Most simulation texts (e.g., Law [70]) have a broader treatment of input modeling than presented here. Nelson and Yamnitsky [91] survey advanced techniques.

Input modeling is typically performed using an input modeling package. These packages fit several distributions to a data set, then determine the distribution with the best fit by comparing goodness-of-fit statistics. But what if an appropriate input model is not included in one of these packages? The modeler must resort to deriving the appropriate estimators by hand for the appropriate input model. The purpose of this section is to investigate the use of APPL for input modeling. APPL allows an analyst to specify a standard or nonstandard distribution for an input model, and have the derivations performed

automatically. Input modeling serves as an excellent arena for illustrating the applicability and usefulness of APPL. Besides including pre-defined distributions for over 45 different types of continuous and discrete random variables and over 30 procedures for manipulating random variables (e.g., `Convolution`, `Transform`), APPL contains input modeling procedures for parameter estimation, plotting empirical and fitted CDFs, and performing goodness-of-fit tests. Using examples, we illustrate its utility for input modeling.

Both APPL and Maple can be adapted for use in input modeling. This section gives several examples from Evans and Leemis [28] of cases where a symbolic language is of use in analyzing a data set.

Example 12.1. One of the tools for selecting a suitable input model is a plot of the coefficient of variation ($\gamma = \sigma/\mu$) versus the skewness

$$\gamma_3 = E\left[\left(\frac{X - \mu}{\sigma}\right)^3\right]$$

for several popular input distributions. After constructing this plot, the sample coefficient of variation and the sample skewness can be plotted for a particular data set or data sets to determine an appropriate distribution for modeling the data. The Maple code that produces the plot in Fig. 12.4 for the Weibull, gamma, log normal, and log logistic distributions uses the additional APPL procedures `CoefOfVar` and `Skewness`. The statements necessary to plot the gamma distribution's coefficient of variation versus skewness are shown below. The plots for the other distributions are calculated similarly. The Maple statement used to display all four plots in one graphic is also provided.

```
> unassign('kappa');
> lambda := 1;
> X := GammaRV(lambda, kappa);
> c := CoefOfVar(X);
> s := Skewness(X);
> GammaPlot := plot([c, s, kappa = 0 .. 750],
      labels = [gamma, gamma_3]):
      .
      .
      .
> plots[display]({GammaPlot, WeibullPlot,
      LogNormalPlot, LogLogisticPlot},
      scaling = unconstrained);
```

The Maple `unassign` command is used to treat κ as a symbolic parameter. Vargo et al. [114] include an updated version of Fig. 12.4.

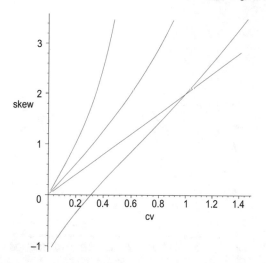

Fig. 12.4. Coefficient of variation γ vs. skewness γ_3 for the gamma, Weibull, log normal, and log logistic distributions

Example 12.2. The following $n = 23$ ball bearing failure times (in millions of revolutions) are collected to determine an input model in a stochastic simulation:

105.84	28.92	98.64	55.56	128.04	45.60
67.80	105.12	48.48	51.84	173.40	51.96
54.12	68.64	93.12	68.88	84.12	68.64
41.52	127.92	42.12	17.88	33.00.	

The original source for this data set is Lieblein and Zelen [79]. First, consider fitting an exponential distribution to this data set using maximum likelihood. The data set for the service times, `BallBearing`, is a pre-defined list in APPL. The APPL `MLE` procedure returns the maximum likelihood estimators as a list. Its arguments are the model, the data, and the parameters to be estimated. The APPL statements

```
> unassign('lambda');
> X := ExponentialRV(lambda);
> lamhat := MLE(X, BallBearing, [lambda]);
```

return $\hat{\lambda} \cong 0.0138$ as the maximum likelihood estimator. The additional APPL command

```
> PlotEmpVsFittedCDF(X, BallBearing,
                [lambda = lamhat[1]], 0, 180);
```

produces a plot of the empirical and fitted CDFs on one set of axes, as seen in Fig. 12.5.

Fig. 12.5. Empirical and fitted exponential CDFs for the ball bearing data set

In order to assess the model adequacy, either a formal goodness-of-fit test can be performed, or goodness-of-fit statistics can be compared for competing models. The Kolmogorov–Smirnov test statistic, for example, can be computed with the additional APPL statement

```
> KSTest(X, BallBearing, [lambda = lamhat[1]]);
```

which returns 0.3068, indicating a rather poor fit.

As an alternative, one might consider fitting the *reciprocal* of an exponential random variable to the service times, as in the following example.

Example 12.3. Fit the reciprocal of an exponential random variable to the service times in the previous example.

The APPL statements required to find the distribution of the reciprocal of an exponential random variable and find the MLE for the unknown parameter are

```
> unassign('lambda');
> X := ExponentialRV(lambda);
> g := [[x -> 1 / x], [0, infinity]];
> Y := Transform(X, g);
> lamhat := MLE(Y, BallBearing, [lambda]);
```

which derive the PDF of Y to be

$$f_Y(y) = \frac{\lambda}{y^2} e^{-\lambda/y} \qquad\qquad y > 0$$

and calculate the MLE $\hat{\lambda} \cong 55.06$. The function **g** is used to find the distribution of $Y = g(X) = 1/X$. As can be seen in Fig. 12.6, the reciprocal of the exponential also provides a poor fit to the ball bearing data.

Fig. 12.6. Empirical and reciprocal exponential fitted CDFs for the ball bearing data set

Neither the exponential model nor its reciprocal are appropriate for modeling the service times. It might be appropriate to consider two-parameter distributions as potential models, as shown in the next example.

Example 12.4. Fit the inverse Gaussian and Weibull distributions to the data set. Again using the APPL procedures MLE and KSTest,

```
> unassign('lambda');
> X := InverseGaussianRV(lambda, mu);
> hat := MLE(X, BallBearing, [lambda, mu]);
> KSValue := KSTest(X, BallBearing,
                [lambda = hat[1], mu = hat[2]]);
```

yields an improved fit with $\hat{\lambda} \cong 231.67$, $\hat{\mu} \cong 72.22$, and a Kolmogorov–Smirnov test statistic of 0.088. The MLE procedure is able to return the appropriate values because the maximum likelihood estimators can be expressed in closed form for this particular distribution. Unfortunately, the statements

```
> unassign('lambda'); unassign('kappa');
> Y := WeibullRV(lambda, kappa);
> hat := MLE(Y, BallBearing, [lambda, kappa]);
```

fail to return the MLEs in APPL. The Maple numerical equation solving procedure `fsolve` is not clever enough to exploit some of the structure in the score vector that is necessary to find the MLEs. Therefore a special routine, `MLEWeibull`, has been written that computes MLEs for the Weibull distribution.

To summarize the ball bearing data set examples, Table 12.1 contains the Kolmogorov–Smirnov test statistic values for various distributions that were fit to the data via maximum likelihood estimation.

Table 12.1. Kolmogorov–Smirnov test statistics for various distributions

Model	Test statistic
Exponential	0.307
Reciprocal of exponential	0.306
Weibull	0.151
Gamma	0.123
Arctangent	0.093
Log normal	0.090
Inverse Gaussian	0.088

Another difficulty that can present itself in input modeling is the presence of censoring. A right-censored data set, for example, often occurs in reliability and biostatistical applications. Examples likely to arise in discrete-event input modeling situations include machine failure times (when some machines have not yet failed) and the analysis of rare events.

Example 12.5. Consider the problem of determining an input model for the remission time for the treatment group in the study concerning the drug 6-MP given by Gehan [35]. Letting an asterisk denote a right-censored observation, the remission times (in weeks) are

$$6 \quad 6 \quad 6 \quad 6^* \quad 7 \quad 9^* \quad 10 \quad 10^* \quad 11^* \quad 13 \quad 16$$
$$17^* \quad 19^* \quad 20^* \quad 22 \quad 23 \quad 25^* \quad 32^* \quad 32^* \quad 34^* \quad 35^*.$$

Both `MP6` and `MP6Censor` are pre-defined lists in APPL. `MP6` is simply the 21 data values given above, and `MP6Censor` is the list

```
[1, 1, 1, 0, 1, 0, 1, 0, 0, 1, 1, 0, 0, 0, 1, 1, 0, 0, 0, 0, 0]
```

where 0 represents a right-censored observation and 1 represents an uncensored observation. The statements used to determine the MLE for an exponential distribution are

```
> unassign('lambda');
> X := ExponentialRV(lambda);
> hat := MLE(X, MP6, [lambda], MP6Censor);
```

The code yields the MLE for the failure rate as $\hat{\lambda} = 9/359$ remissions per week. Similarly, the statements

```
> unassign('lambda'); unassign('kappa');
> Y := WeibullRV(lambda, kappa);
> hat := MLEWeibull(MP6, MP6Censor);
```

yield the MLE estimates $\hat{\lambda} \cong 0.03$ and $\hat{\kappa} \cong 1.35$ for the Weibull distribution. The Kaplan–Meier product-limit survivor function estimate for the MP6 data set, along with the fitted Weibull survivor function are plotted in Fig. 12.7 using the additional APPL statement

```
> PlotEmpVsFittedSF(Y, MP6,
                    [lambda = hat[1], kappa = hat[2]],
                    MP6Censor, 0, 23);
```

The downward steps in the estimated survivor function occur only at observed remission times. The six arguments required by the PlotEmpVsFitted procedure are the random variable whose SF is to be plotted, the data values in a list, the parameters associated with the random variable, the right-censoring vector in a list, and the lower and upper plotting limits. Note that the product-limit estimator cuts off after the largest observed remission time per the convention given in Lawless [71].

Fig. 12.7. Product-limit survivor function estimate and fitted Weibull survivor function for the 6-MP treatment group

All of the input modeling examples thus far have been limited to continuous data. The next example fits the geometric distribution as a model for daily demand at a vending machine.

Example 12.6. A vending machine has capacity for 24 cans of "Purple Passion" grape drink. The machine is restocked to capacity every day at noon. Restocking time is negligible. The last 5 days have produced the following Purple Passion sales:

$$14 \quad 24 \quad 18 \quad 20 \quad 24.$$

The *demand* for Purple Passion at this particular vending machine can be estimated from the data by treating the 24-can sales figures as *right-censored* demand observations. If demand has the geometric distribution, with probability function

$$f(t) = p(1-p)^t \qquad t = 0, 1, 2, \ldots,$$

find the MLE \hat{p}.

As discussed earlier, many procedures, like MLE, are able to handle discrete distributions. Since the pre-defined geometric distribution in APPL is parameterized for $t = 1, 2, \ldots$, we need to define a geometric random variable with the different parameterization (used above) in the list-of-sublists data structure. The statements

```
> X := [[x -> p * (1 - p) ^ x],
         [0 .. infinity],
         ["Discrete", "PDF"]];
> PurplePass := [14, 24, 18, 20, 24];
> PurplePassCensor := [1, 0, 1, 1, 0];
> MLE(X, PurplePass, [p], PurplePassCensor);
```

yield $\hat{p} = 3/103$. Model adequacy is not considered for this particular example.

All previous examples have considered time-independent observations. There are occasions when a series of event times may be time dependent, and a more complicated input model may be appropriate.

Example 12.7. Ignoring preventive maintenance, ten odometer readings (from a certain model of car) associated with failures appearing over the first 100,000 miles, reading row-wise, are

12942	17256	23254	28489	45561
85603	88143	94078	98231	99900.

Consider fitting a nonhomogeneous Poisson process to the above data set, where the ending time of the observation interval is assumed to

be 100,000 miles. The data can be approximated by a power law process (i.e., the intensity function has the same parametric form as the hazard function for a Weibull random variable). The following APPL statements, including the additional procedure MLENHPP, return $\hat{\lambda} \cong 0.000058$ and $\hat{\kappa} \cong 1.30$.

```
> unassign('lambda'); unassign('kappa');
> CarFailures := [12942, 17256, 23254, 28489, 45561,
                  85603, 88143, 94078, 98231, 99900];
> X := WeibullRV(lambda, kappa);
> hat := MLENHPP(X, CarFailures, [lambda, kappa]);
```

Every example considered thus far has used maximum likelihood to estimate the unknown parameters. APPL includes the procedure MOM for computing the method of moments estimators.

Example 12.8. Hurricanes typically strike the eastern and southern coastal regions of the United States, although they occasionally sweep inland before completely dissipating. The U.S. Weather Bureau reported that during the period from 1900 to 1969 a total of 36 hurricanes moved as far as the Appalachian Mountains. The maximum 24 hour precipitation levels (measured in inches) recorded from those 36 storms during the time they were over the mountains are [69, pages 319–322]

31.00	2.82	3.98	4.02	9.50	4.50
11.40	10.71	6.31	4.95	5.64	5.51
13.40	9.72	6.47	10.16	4.21	11.60
4.75	6.85	6.25	3.42	11.80	0.80
3.69	3.10	22.22	7.43	5.00	4.58
4.46	8.00	3.73	3.50	6.20	0.67.

A histogram of the data, which can be plotted in Maple, suggests that the random variable X, which is the maximum 24 hour precipitation, might be well approximated by the gamma distribution. The following APPL code finds the method of moments estimates for the parameters λ and κ, where Hurricane is the above data set pre-defined in APPL:

```
> X := GammaRV(lambda, kappa);
> hat := MOM(X, Hurricane, [lambda, kappa]);
```

The resulting estimates for the parameters are $\hat{\lambda} = \frac{954000}{4252153} \cong 0.224$ and $\hat{\kappa} = \frac{6952275}{4252153} \cong 1.64$.

12.3 Kolmogorov–Smirnov Goodness-of-Fit Test

In this final section, we present an algorithm for computing the CDF of the Kolmogorov–Smirnov test statistic D_n in the all-parameters-known case. Birnbaum [12] gives an n-fold integral for the CDF of the test statistic which yields

a function defined in a piecewise fashion, where each piece is a polynomial of degree n. Unfortunately, it is difficult to determine the appropriate limits of integration for computing these polynomials. Our algorithm performs the required integrations in a manner that avoids calculating the same integrals repeatedly, resulting in shorter computation time. It can be used to compute the entire CDF or just a portion of the CDF, which is more efficient for finding a critical value or a p-value associated with a hypothesis test. To date, critical tables have been approximated by various techniques including asymptotic approximations, recursive formulas, and Monte Carlo simulation. Our approach yields exact critical values and significance levels. The algorithm has been implemented in APPL.

The Kolmogorov–Smirnov (KS) test statistic D_n is defined by

$$D_n = \sup_x |F(x) - F_n(x)|,$$

where n is the sample size, $F(x)$ is a hypothesized CDF with fixed parameters, and $F_n(x)$, also called the empirical distribution function, is a step-function that increases by $1/n$ at each data value. This statistic has been used for goodness-of-fit testing for continuous populations for decades. The KS test's appeal is its straightforward computation of the test statistic and the distribution-free characteristic of D_n. Its drawback is that its CDF under the null hypothesis is difficult to determine, leaving one to calculate critical values with various approximation methods.

Birnbaum [12] gives the CDF of $D_n - \frac{1}{2n}$ as

$$\Pr\left(D_n < \frac{1}{2n} + v\right) = n! \int_{\frac{1}{2n}-v}^{\frac{1}{2n}+v} \int_{\frac{3}{2n}-v}^{\frac{3}{2n}+v} \cdots \int_{\frac{2n-1}{2n}-v}^{\frac{2n-1}{2n}+v} g(u_1, u_2, \ldots, u_n)$$

$$du_n \ldots du_2\, du_1$$

for $0 \le v \le \frac{2n-1}{2n}$, where

$$g(u_1, u_2, \ldots, u_n) = 1$$

for $0 \le u_1 \le u_2 \le \cdots \le u_n \le 1$, and is 0 otherwise. Equivalently,

$$\Pr\left(D_n < \frac{1}{2n} + v\right) = n! \int_{\max(0,\frac{1}{2n}-v)}^{\min(1,\frac{1}{2n}+v)} \int_{\max(u_1,\frac{3}{2n}-v)}^{\min(1,\frac{3}{2n}+v)} \cdots \int_{\max(u_{n-1},\frac{2n-1}{2n}-v)}^{\min(1,\frac{2n-1}{2n}+v)} 1$$

$$du_n \ldots du_2\, du_1.$$

Birnbaum's rather harmless-looking integral is tedious to compute by hand, even for small values of n, due to the complexity of the region where g is nonzero. Evaluating his expression requires calculating many n-fold integrals whose limits are determined by considering a carefully chosen partition of the

support of D_n. The difficult part of the process is setting the appropriate limits on these integrals. The final expression for the KS CDF is a piecewise polynomial function in which each polynomial has degree n.

The literature available on the KS statistic is extensive. Stephens' article (Chap. 4 of D'Agostino and Stephens [19]) contains comprehensive coverage on the use of the KS statistic, as well as other statistics based on the empirical distribution function. He calculates the power of these goodness-of-fit tests. Johnson et al. [55, page 460] consider this source to be so complete that they have deleted KS discussions in their second edition and refer the reader to that compendium instead. For computing critical points of the KS distribution, we find five-digit accuracy as early as 1956 in Miller [87]. Miller relies on asymptotic results that converge fairly quickly to produce these estimates. Birnbaum's article [12] also explains how various sets of recursive formulas can be used to calculate certain critical points to reasonable levels of accuracy. Law [70, pages 351–356] has a well-written introduction to the KS test and indicates that critical points can only be calculated via computer techniques. There appears to be no source that produces exact distribution functions for any distribution where $n > 3$ in the literature. Birnbaum [12, page 144] gives the CDF of $D_n - \frac{1}{2n}$ for $n = 2$ and $n = 3$. Knuth [63] provides a functional form for the CDFs for the two statistics D_n^+ and D_n^-, but does not provide the CDF for $D_n = \max\{D_n^+, D_n^-\}$, a harder problem given the dependence between the two random variables. As a consequence of the apparent complexity of the required integration and lack of literature on exact distributions, we believed early on that a computational algebra system, such as Maple, would be necessary to compute the polynomials needed for the distribution of the KS test statistic.

Computing the Distribution of D_n. When $0 < v < \frac{1}{2n}$, Birnbaum's integral is easy to compute because none of the intervals of integration overlap. Additionally, these intervals are all wholly contained within the interval from 0 to 1. For example, u_1 varies over the interval from $\frac{1}{2n} - v$ to $\frac{1}{2n} + v$, and u_2 varies over the interval from $\frac{3}{2n} - v$ to $\frac{3}{2n} + v$. The interval for u_1 begins to the right of 0 and ends before the interval for u_2 begins. Thus, when $0 < v < \frac{1}{2n}$, the limits of integration of Birnbaum's integral guarantee that $0 \leq u_1 \leq u_2 \leq \cdots \leq u_n \leq 1$. Because the u_i's are constrained in this way, we may replace Birnbaum's integrand $g(u_1, u_2, \ldots, u_n)$ with 1, and our computation of the KS CDF requires only a single n-fold integral:

$$\Pr\left(D_n < \frac{1}{2n} + v\right) = n! \int_{\frac{1}{2n}-v}^{\frac{1}{2n}+v} \int_{\frac{3}{2n}-v}^{\frac{3}{2n}+v} \cdots \int_{\frac{2n-1}{2n}-v}^{\frac{2n-1}{2n}+v} 1 \, du_n \ldots du_2 \, du_1$$

$$= n!(2v)^n \qquad 0 < v < \frac{1}{2n}.$$

When v has a fixed value greater than $\frac{1}{2n}$, it is also desirable to replace Birnbaum's integrand by 1. In order to permit this replacement, we must only allow integration over that region of n-dimensional space for which

$0 \leq u_1 \leq u_2 \leq \cdots \leq u_n \leq 1$. Since the intervals of integration specified in Birnbaum's integral for different u_i's can overlap, the smallest allowable value for any u_i is influenced by all u_j's with subscripts less than i that can take on values in u_i's interval of integration. This overlapping requires partitioning the interval from $\frac{1}{2n} - v$ to $\frac{2n-1}{2n} + v$ into subintervals (which we will henceforth refer to as u-subintervals), with the first u-subinterval starting at $\frac{1}{2n} - v$ and a new u-subinterval beginning whenever the left endpoint of one of Birnbaum's intervals of integration is encountered. When any u_i lies in a u-subinterval that consists entirely of negative values, Birnbaum's integrand is zero. For this reason, only u-subintervals having a positive right endpoint contribute to the KS CDF.

Of course, the number of u-subintervals having a positive right endpoint depends on the value of v. For small values of v (i.e., $0 < v < \frac{1}{2n}$), $n + 1$ u-subintervals have a positive right endpoint since all n of Birnbaum's lower limits of integration are positive. For large values of v (i.e., $\frac{2n-3}{2n} < v < \frac{2n-1}{2n}$), only two u-subintervals have a positive right endpoint since only u_n's lower limit of integration is positive. Intermediate values of v produce all possibilities from 2 to $n + 1$ u-subintervals having a positive right endpoint.

Because different values of v can yield different numbers of u-subintervals having a positive right endpoint, the interval $0 < v < \frac{2n-1}{2n}$ must be subdivided at the following values of v: $\frac{1}{2n}, \frac{3}{2n}, \frac{5}{2n}, \ldots, \frac{2n-3}{2n}$. When the values of v remain within one of the resulting subintervals for v, the number of u-subintervals having a positive right endpoint will remain fixed.

Another complication arises because it is necessary to know, for a fixed value of v and for the u-subintervals produced by this value, which variables of integration u_i can take on values in each of the u-subintervals of $[0, 1]$. The previous subdivision of the values of v is not fine enough to allow unambiguous specification of the range for each u_i. When $i < j$, u_i and u_j have overlapping intervals of integration if the upper integration limit for u_i exceeds the lower integration limit for u_j, i.e., $\frac{2i-1}{2n} + v \geq \frac{2j-1}{2n} - v$. As a result, as v increases from 0 to $\frac{2n-1}{2n}$, new overlaps take place when v equals $\frac{1}{2n}, \frac{2}{2n}, \frac{3}{2n}, \ldots, \frac{n-1}{2n}$. The interval $0 < v < \frac{2n-1}{2n}$ must be divided into subintervals at these values of v as well as at the values of v listed previously. We will henceforth refer to the resulting subintervals as v-subintervals.

Indicator matrices will be used to summarize the interrelationships between the possible values for the variables of integration u_1, u_2, \ldots, u_n. For a fixed n and for values of v in the kth v-subinterval of $\left(0, \frac{2n-1}{2n}\right)$, the indicator matrix A_k will show, by the presence of a 1 or a 0 in row i and column j, whether or not u_i can take on values in the jth u-subinterval. Finally, by defining paths through these indicator matrices, we will produce a complete set of n-fold integrals which satisfy the requirement $0 \leq u_1 \leq u_2 \leq \cdots \leq u_n \leq 1$, and which produce the KS CDF when summed.

The algorithm for computing the CDF of D_n is divided into four phases. In Phase 1, an appropriate partition of the support of $D_n - \frac{1}{2n}$ is determined. In Phase 2, we define matrices A_k that are instrumental in determining the

limits of integration in the n-fold integrals. In Phase 3, these integrals and their associated limits are computed. To take advantage of the similarities in the limits of integration of these n-fold integrals, they are grouped for efficiency and all evaluated to the same level of integration before proceeding to the next level. Finally, Phase 4 consists of translating the support of $D_n - \frac{1}{2n}$ to the support of D_n. Thus, the input to the algorithm is a positive integer n and the output is the piecewise CDF of D_n.

Phase 1: Partition the Support of $D_n - \frac{1}{2n}$. The endpoints of the segments that define the support of the KS test statistic can be determined from the limits of integration given by Birnbaum. Using Birnbaum's formula, the *baseline* lower and upper limits of integration associated with $v = 0$ are

$$\left\{ \frac{1}{2n}, \frac{3}{2n}, \frac{5}{2n}, \ldots, \frac{2n-1}{2n} \right\}.$$

As v increases, the support of $D_n - \frac{1}{2n}$ is broken into disjoint v-subintervals. The endpoints of these support v-subintervals consist of $v = 0$ and the values of v for which the endpoints of the n intervals of integration either:

- equal 0 or 1, *or*
- coincide.

The 0 or 1 cases occur at v values in the set

$$\left\{ \frac{1}{2n}, \frac{3}{2n}, \frac{5}{2n}, \ldots, \frac{2n-1}{2n} \right\}.$$

The set of values of v for which the endpoints of different intervals of integration coincide are

$$\left\{ \frac{1}{2n}, \frac{2}{2n}, \frac{3}{2n}, \ldots, \frac{n-1}{2n} \right\}.$$

Thus, the union of these two sets and 0 comprise the endpoints of the v-subintervals of the support of $D_n - \frac{1}{2n}$.

The first phase of the algorithm given at the end of this section computes the above endpoints $v_0 = 0, v_1, v_2, \ldots, v_m$, where m is the number of v-subintervals on which the CDF of $D_n - \frac{1}{2n}$ is defined. For any n,

$$m = \left\lceil \frac{3n}{2} \right\rceil - 1.$$

Phase 2: Define the A Matrices. At this point in the algorithm, v_0, v_1, \ldots, v_m are the endpoints of the m v-subintervals on which the CDF of $D_n - \frac{1}{2n}$ is defined.

Two book-keeping steps are needed in this phase of the algorithm. They are:

1. Define c_1, c_2, \ldots, c_m as the midpoints of the v-subintervals of support for $D_n - \frac{1}{2n}$.

2. Define x_1, x_2, \ldots, x_n as the midpoints of the intervals of integration for u_1, u_2, \ldots, u_n in Birnbaum's n-fold integral. Thus

$$x_i = \frac{2i - 1}{2n} \qquad i = 1, 2, \ldots, n.$$

Part of the algorithm involves defining the $n \times n$ indicator matrices A_1, A_2, \ldots, A_m corresponding to the v-subintervals which form the support of $D_n - \frac{1}{2n}$. The rows of an A matrix refer to the variables of integration u_1, u_2, \ldots, u_n. The columns of an A matrix refer to the u-subintervals, with the jth column corresponding to the interval from $x_j - v$ to $x_{j+1} - v$, except for the nth column, which corresponds to the interval from $x_n - v$ to $x_n + v$. If the (i, j) element of an A matrix equals 1, then the range of u_i includes at least part of the jth u-subinterval.

Since the (n, n) element of each A matrix is the only element in the nth row that is 1, the integrals that need to be computed for each segment of support of $D_n - \frac{1}{2n}$ can be visualized as a path in A, consisting of a sequence of moves from the (n, n) position of A to a nonzero element in the first row of A. All moves in the A matrix from row i to row $i - 1$ (for $i = n, n - 1, \ldots, 2$) require that the following conditions are met:

1. The move must be from one of the 1's in row i to one of the 1's in row $i - 1$.
2. The move is either directly vertical or vertical and one or more spaces to the left.

That is, if the move begins at a 1 element in column j of row i, it must end at a 1 element in column 1 through j of row $i - 1$.

Example 12.9. For $n = 3$ and $k = 3$, the CDF is defined on the v-subinterval $\frac{2}{6} < v < \frac{3}{6}$ and has the following A_3 matrix:

$$A_3 = \begin{bmatrix} 1 & 1 & 1 \\ 0 & 1 & 1 \\ 0 & 0 & 1 \end{bmatrix}.$$

There are five different paths from the $(3, 3)$ element to the top row of the matrix. In the five matrices below, these five paths to the top row of the A matrix are displayed by using boldface type on the path. Next to each matrix is the corresponding triple integral that the path represents.

$$\begin{bmatrix} \mathbf{1} & 1 & 1 \\ 0 & \mathbf{1} & 1 \\ 0 & 0 & \mathbf{1} \end{bmatrix} \qquad \int_0^{\frac{3}{6}-v} \int_{\frac{3}{6}-v}^{\frac{5}{6}-v} \int_{\frac{5}{6}-v}^1 1 \, du_3 \, du_2 \, du_1$$

$$\begin{bmatrix} 1 & \mathbf{1} & 1 \\ 0 & \mathbf{1} & 1 \\ 0 & 0 & \mathbf{1} \end{bmatrix} \qquad \int_{\frac{3}{6}-v}^{\frac{5}{6}-v} \int_{u_1}^{\frac{5}{6}-v} \int_{\frac{5}{6}-v}^1 1 \, du_3 \, du_2 \, du_1$$

$$\begin{bmatrix} \mathbf{1} & \mathbf{1} & 1 \\ 0 & 1 & 1 \\ 0 & 0 & 1 \end{bmatrix} \qquad \int_0^{\frac{3}{6}-v} \int_{\frac{5}{6}-v}^{\frac{3}{6}+v} \int_{u_2}^1 1 \, du_3 \, du_2 \, du_1$$

$$\begin{bmatrix} \mathbf{1} & \mathbf{1} & 1 \\ 0 & 1 & 1 \\ 0 & 0 & 1 \end{bmatrix} \qquad \int_{\frac{3}{6}-v}^{\frac{5}{6}-v} \int_{\frac{5}{6}-v}^{\frac{3}{6}+v} \int_{u_2}^1 1 \, du_3 \, du_2 \, du_1$$

$$\begin{bmatrix} \mathbf{1} & \mathbf{1} & 1 \\ 0 & 1 & 1 \\ 0 & 0 & 1 \end{bmatrix} \qquad \int_{\frac{5}{6}-v}^{\frac{1}{6}+v} \int_{u_1}^{\frac{3}{6}+v} \int_{u_2}^1 1 \, du_3 \, du_2 \, du_1$$

When these five integrals are summed, Birnbaum's expression reduces to

$$\Pr\left(D_3 < \frac{1}{6} + v\right) = -4v^3 + \frac{11}{3}v - \frac{11}{27} \qquad \frac{2}{6} < v < \frac{3}{6}.$$

Note that the innermost integral

$$\int_{u_2}^1 1 \, du_3$$

on the third, fourth, and fifth triple integrals are identical. The number of identical integrals of this type grows rapidly as n increases. The algorithm we have developed avoids recalculation of duplicate integrals. Continuing in this fashion for the other A matrices (i.e., A_1, A_2, and A_4), yields the CDF

$$\Pr\left(D_3 < \frac{1}{6} + v\right) = \begin{cases} 48v^3 & 0 < v < \frac{1}{6} \\ -12v^3 + 8v^2 + v - \frac{1}{9} & \frac{1}{6} < v < \frac{2}{6} \\ -4v^3 + \frac{11}{3}v - \frac{11}{27} & \frac{2}{6} < v < \frac{3}{6} \\ 2v^3 - 5v^2 + \frac{25}{6}v - \frac{17}{108} & \frac{3}{6} < v < \frac{5}{6}. \end{cases}$$

Note that Birnbaum [12] contains a sign error in the fourth v-subinterval of support.

Table 12.2 shows how rapidly the number of n-fold integrations required to compute the CDF of $D_n - \frac{1}{2n}$ grows with n. Note that the five triple integrals in the previous example corresponds to $n = 3$ and the third v-subinterval ($k = 3$), shown in boldface in the table. Note that the fifth column in Table 12.2 is the product of the first and fourth columns.

Example 12.10. The method for determining the integers in the third column of the Table 12.2 is somewhat analogous to determining the entries in a row of Pascal's triangle by using the entries of the previous row. Beginning at each nonzero A matrix element in rows 2 through

Table 12.2. Computational requirements for computing the D_n CDF for small n

n	m	Number of n-fold integrals required for each v-subinterval	Total number of n-fold integrals required	Total number of single integrals required
1	1	1	1	1
2	2	1, 2	3	6
3	4	1, 4, **5**, 3	13	39
4	5	1, 8, 13, 9, 4	35	140
5	7	1, 16, 34, 27, 28, 14, 5	125	625
6	8	1, 32, 89, 81, 89, 48, 20, 6	366	2196

n, there are a certain number of paths whose moves to the top row of the matrix satisfy the two conditions stated previously. We will assign such elements a number p (for paths) that denotes the number of paths from that element to the top of the matrix. Thus, if $p_{2,2} = 2$ (as is the case for the (2, 2) element of the A_3 matrix for $n = 3$ and $k = 3$ in Example 12.9), then no matter how we reached this element of the A matrix, there are only two possible paths remaining to the top. The $p_{i,j}$ values are assigned as follows. Only A matrix elements with value 1 are assigned a $p_{i,j}$ value. If the jth entry in the first row of A is nonzero, assign $p_{1,j} = 1$. For the second and subsequent rows $(i = 2, 3, \ldots, n)$, $p_{i,j} = \sum_{q=1}^{j} p_{i-1,q}$. Upon completion, $p_{n,n}$ represents the total number of paths from the lower right corner of the matrix to the top row of the matrix. Consider, for example, the case of $n = 5$ and $k = 4$. The A_4 matrix with superscripts denoting the $p_{i,j}$ values is

$$A_4 = \begin{bmatrix} 0 & 1^1 & 1^1 & 1^1 & 0 \\ 0 & 1^1 & 1^2 & 1^3 & 1^3 \\ 0 & 0 & 1^3 & 1^6 & 1^9 \\ 0 & 0 & 0 & 1^9 & 1^{18} \\ 0 & 0 & 0 & 0 & 1^{27} \end{bmatrix}.$$

Here we see that $p_{5,5} = 27$, meaning there are 27 possible paths to the top row of A_4, consistent with the $n = 5$ and $k = 4$ element in Table 12.2.

Phase 3: Set Limits on the Appropriate Integrals. The indicator matrix A shows, by the presence of a 1 in row i and column j, for $j < n$, that u_i can assume values from the maximum of $x_j - v$ and 0 to the minimum of $x_{j+1} - v$ and $x_i + v$. The presence of a 1 in row i and column n, means that u_i assumes values from $x_n - v$ to the minimum of $x_i + v$ and 1. Each path in A represents a set of allowable intervals of integration for the variables u_1 to u_n in a particular n-fold integral. For a particular path, the limits of

integration for each u_i are those given above that correspond to entry $a_{i,j}$ of A in row i that lies on the path, with one exception: if both $a_{i,j}$ and $a_{i-1,j}$ are on the same path, the lower limit of integration for u_i must be u_{i-1}. This exception occurs because the path allows u_i and u_{i-1} to assume values in the same u-subinterval. In such a case, the lower limit of integration for u_i must be u_{i-1} in order to reflect the requirement $u_{i-1} \leq u_i$.

For each path through the matrix A, and for each nonzero entry $a_{i,j}$ on that path, a single integration with respect to u_i must be performed. If $a_{i-1,j}$ is on the same path as $a_{i,j}$, then the lower limit of integration will be the variable u_{i-1}. If $a_{i-1,j}$ is not on the same path as $a_{i,j}$, then the lower limit of integration will be a fixed number: the maximum of $x_j - v$ and 0. Thus, for each path which passes through $a_{i,j}$, either a variable or fixed lower limit of integration might be appropriate.

Example 12.11. Consider again the case of $n = 5$ and the fourth v-subinterval ($k = 4$). The A_4 matrix shown below has its 1's replaced with \mathcal{F}, \mathcal{V}, or \mathcal{B}, indicating whether a fixed-limit integral, a variable-limit integral, or both need to be computed for each entry.

$$A_4 = \begin{bmatrix} 0 & \mathcal{F} & \mathcal{F} & \mathcal{F} & 0 \\ 0 & \mathcal{V} & \mathcal{B} & \mathcal{B} & \mathcal{F} \\ 0 & 0 & \mathcal{B} & \mathcal{B} & \mathcal{B} \\ 0 & 0 & 0 & \mathcal{B} & \mathcal{B} \\ 0 & 0 & 0 & 0 & \mathcal{B} \end{bmatrix}.$$

In general, when the A matrix contains a zero, neither the fixed nor variable lower limits need to be computed. Now consider the elements of the A matrix that contain a 1. The positions associated with the first 1 in each column of A require only a fixed lower limit to be calculated. The positions below the first 1 in the first nonzero column require only a variable lower limit to be calculated. All other positions in the A matrix associated with 1 elements require both a fixed and a variable lower limit to be calculated. Table 12.3 shows the computational efficiency of performing the integrations for various values of n by using the algorithm.

The number of single integrals calculated by the algorithm for any n is no more than

$$\frac{n(n+1)}{2} \cdot 2 \cdot \left(\left\lceil \frac{3n}{2} \right\rceil - 2 \right) \cong \frac{3}{2} n^3,$$

where the first factor corresponds to the maximum number of 1's in any A matrix, the second factor accounts for computing both the fixed and variable matrices (described subsequently), and the last factor is one less than m since no integration is required for the first v-subinterval. Due to savings associated with 0 elements in the A matrix and the fact that both fixed and variable matrix elements do not need to be calculated in all cases, this order n^3 expression is closely approximated by $0.80n^3$ for $n = 20$, 30, and 50.

Table 12.3. Computational efficiency associated with using the F and V arrays

n	m	Total number of single integrals required (Table 12.2)	Total number of single integrals required (algorithm)
1	1	1	0
2	2	6	4
3	4	39	23
4	5	140	48
5	7	625	108
6	8	2196	170
10	14	442540	800
15	22	318612735	2793
20	29	232093052440	6400
30	44	136617382366486800	21550
50	74	6105637995599418238684313815O	99500

Table 12.4 lists the coefficients of the polynomials that define the CDF of $D_n - \frac{1}{2n}$ as computed by the APPL procedure KSRV whose algorithm is given at the end of this section, for $n = 1, 2, \ldots, 6$. The ability to store fractions exactly is of particular benefit for these calculations.

Rather than detail every aspect of the logic of Phase 3 of the algorithm, we illustrate the evolution of the $n \times n$ matrices F (for fixed limits) and V (for variable limits) in a row-by-row fashion for a particular combination of n and k.

Example 12.12. Consider again the case of $n = 3$ and $k = 3$, which corresponds to the v-subinterval $\frac{2}{6} < v < \frac{3}{6}$. The x_i values associated with $n = 3$ are

$$x_1 = \frac{1}{6} \qquad x_2 = \frac{3}{6} \qquad x_3 = \frac{5}{6}.$$

The v_i values associated with $n = 3$ are $v_0 = 0$,

$$v_1 = \frac{1}{6} \qquad v_2 = \frac{2}{6} \qquad v_3 = \frac{3}{6} \qquad v_4 = \frac{5}{6}.$$

The centers of the v-subintervals are

$$c_1 = \frac{1}{12} \qquad c_2 = \frac{3}{12} \qquad c_3 = \frac{5}{12} \qquad c_4 = \frac{8}{12}.$$

The A_3 matrix is

$$A_3 = \begin{bmatrix} 1 & 1 & 1 \\ 0 & 1 & 1 \\ 0 & 0 & 1 \end{bmatrix}.$$

Table 12.4. CDFs of $D_n - \frac{1}{2n}$ for $n = 1, 2, \ldots, 6$

n	m	k	Coefficients of CDF polynomials	Subinterval
1	1	1	$2, 0$	$0 < v < \frac{1}{2}$
2	2	1	$8, 0, 0$	$0 < v < \frac{1}{4}$
		2	$-2, 3, -\frac{1}{8}$	$\frac{1}{4} < v < \frac{3}{4}$
3	4	1	$48, 0, 0, 0$	$0 < v < \frac{1}{6}$
		2	$-12, 8, 1, -\frac{1}{9}$	$\frac{1}{6} < v < \frac{2}{6}$
		3	$-4, 0, \frac{11}{3}, -\frac{11}{27}$	$\frac{2}{6} < v < \frac{3}{6}$
		4	$2, -5, \frac{25}{6}, -\frac{17}{108}$	$\frac{3}{6} < v < \frac{5}{6}$
4	5	1	$384, 0, 0, 0, 0$	$0 < v < \frac{1}{8}$
		2	$-48, 0, 15, -\frac{9}{8}, \frac{3}{256}$	$\frac{1}{8} < v < \frac{2}{8}$
		3	$16, -40, 21, -\frac{5}{8}, -\frac{29}{256}$	$\frac{2}{8} < v < \frac{3}{8}$
		4	$6, -7, -\frac{27}{16}, \frac{293}{64}, -\frac{853}{2048}$	$\frac{3}{8} < v < \frac{5}{8}$
		5	$-2, 7, -\frac{147}{16}, \frac{343}{64}, -\frac{353}{2048}$	$\frac{5}{8} < v < \frac{7}{8}$
5	7	1	$3840, 0, 0, 0, 0, 0$	$0 < v < \frac{1}{10}$
		2	$0, -288, \frac{624}{5}, -\frac{96}{25}, -\frac{36}{125}, \frac{6}{625}$	$\frac{1}{10} < v < \frac{2}{10}$
		3	$160, -160, \frac{24}{5}, \frac{616}{25}, -\frac{332}{125}, \frac{6}{125}$	$\frac{2}{10} < v < \frac{3}{10}$
		4	$-20, 64, -\frac{318}{5}, \frac{542}{25}, \frac{343}{500}, -\frac{273}{1250}$	$\frac{3}{10} < v < \frac{4}{10}$
		5	$12, 0, -\frac{62}{5}, \frac{6}{5}, \frac{2391}{500}, -\frac{3413}{6250}$	$\frac{4}{10} < v < \frac{5}{10}$
		6	$-8, 18, -\frac{52}{5}, -\frac{19}{5}, \frac{1838}{250}, -\frac{10527}{25000}$	$\frac{5}{10} < v < \frac{7}{10}$
		7	$2, -9, \frac{81}{5}, -\frac{729}{50}, \frac{6561}{1000}, -\frac{9049}{50000}$	$\frac{7}{10} < v < \frac{9}{10}$
6	8	1	$46080, 0, 0, 0, 0, 0, 0$	$0 < v < \frac{1}{12}$
		2	$2880, -3360, 660, 60, -\frac{95}{12}, \frac{5}{24}, -\frac{5}{5184}$	$\frac{1}{12} < v < \frac{2}{12}$
		3	$320, 480, -700, \frac{5620}{27}, -\frac{125}{36}, -\frac{275}{216}, \frac{2195}{46656}$	$\frac{2}{12} < v < \frac{3}{12}$
		4	$-280, 420, -\frac{335}{2}, \frac{1675}{54}, \frac{26005}{864}, -\frac{3125}{1728}, \frac{20645}{373248}$	$\frac{3}{12} < v < \frac{4}{12}$
		5	$104, -188, \frac{1235}{6}, -\frac{7435}{54}, \frac{36245}{864}, -\frac{7327}{5184}, \frac{69797}{373248}$	$\frac{4}{12} < v < \frac{5}{12}$
		6	$-20, 22, \frac{45}{4}, -\frac{2005}{108}, \frac{185}{1728}, \frac{57971}{10368}, -\frac{406469}{746496}$	$\frac{5}{12} < v < \frac{7}{12}$
		7	$10, -33, \frac{925}{24}, -\frac{3065}{216}, \frac{22175}{3456}, -\frac{134807}{20736}, \frac{632863}{1492992}$	$\frac{7}{12} < v < \frac{9}{12}$
		8	$-2, 11, -\frac{605}{24}, \frac{6655}{216}, -\frac{73205}{3456}, \frac{161051}{20736}, -\frac{278569}{1492992}$	$\frac{9}{12} < v < \frac{11}{12}$

The F and V matrices are computed simultaneously, row by row, starting with the last row and ending with the first. Thus, all of the elements of the third row of F and of V are computed before computing the elements of the second row is begun. In general, integrals that are calculated in a particular row become integrand candidates for the integration to be carried out in the row immediately above. The F and V matrices are designed to store the intermediate results

of integration of the n-fold integrals so that the necessary inner integration operations are performed only once. Entries in the ith rows of F and V each result from $n+1-i$ single integrations, one integration corresponding to the ith row and one integration for every lower row. The F matrix stores the results of all integration in which the last integral had a fixed lower limit and the V matrix stores the results of all integration in which the last integral had a variable lower limit. In this example, for the inner-most integration variable u_3, there is only one possible fixed integration to compute and one possible variable integration to compute on the third u-subinterval. These computations are seen in the (3, 3) element of the respective matrices:

$$
F = \begin{bmatrix} F_{1,1} & F_{1,2} & F_{1,3} \\ 0 & F_{2,2} & F_{2,3} \\ 0 & 0 & \int_{\frac{5}{6}-v}^{1} 1 du_3 \end{bmatrix}, \qquad V = \begin{bmatrix} V_{1,1} & V_{1,2} & V_{1,3} \\ 0 & V_{2,2} & V_{2,3} \\ 0 & 0 & \int_{u_2}^{1} 1 du_3 \end{bmatrix}.
$$

The second variable, u_2, has four possible combinations of limits and integrands. Thus, there are four storage elements for u_2, found in the second rows of the F and V matrices. Elements (2, 2) and (2, 3) of the F matrix store integration results that have all fixed lower-limits of integration, but the integrands are fixed and variable, respectively. Similarly, elements (2, 2) and (2, 3) of the V matrix are for the cases when the lower-limit of integration contains the u_1 variable, while the integrands are fixed and variable, respectively. $F_{2,2}$ is computed to cover the case in which u_2 varies over the second u-subinterval and u_3 varies over the third u-subinterval. $F_{2,3}$ covers the case in which both u_2 and u_3 vary over the third u-subinterval, so u_3 must have u_2 as its variable lower limit; hence the integrand is $V_{3,3}$, not $F_{3,3}$. Similar cases are covered by $V_{2,2}$ and $V_{2,3}$, except that variable lower limits are used in anticipation of the fact that u_1 and u_2 can both vary over the same u-subinterval, making it necessary to have u_1 as the (variable) lower limit for u_2. The integrations in the second row of each matrix are shown below (note, the integration for the (3, 3) element, previously discussed, has been carried out):

$$
F = \begin{bmatrix} F_{1,1} & F_{1,2} & F_{1,3} \\ 0 & \int_{\frac{1}{2}-v}^{\frac{5}{6}-v} F_{3,3}\, du_2 & \int_{\frac{5}{6}-v}^{\frac{1}{2}+v} V_{3,3}\, du_2 \\ 0 & 0 & \frac{1}{6}+v \end{bmatrix},
$$

$$
V = \begin{bmatrix} V_{1,1} & V_{1,2} & V_{1,3} \\ 0 & \int_{u_1}^{\frac{5}{6}-v} F_{3,3}\, du_2 & \int_{u_1}^{\frac{1}{2}+v} V_{3,3}\, du_2 \\ 0 & 0 & 1-u_2 \end{bmatrix}.
$$

For the third and last variable, u_1, only fixed-limit integration takes place so only the F matrix gets updated. The (1, 1) element covers the

case in which u_1 varies over the first u-subinterval and u_2 varies over the second and third u-subinterval. The $(1, 2)$ element covers the case in which u_1 varies over the second u-subinterval and u_2 varies over the second and third u-subinterval. The $(1, 3)$ element allows both u_1 and u_2 to vary over the third u-subinterval. The integrals in this first row are shown below (note again, all previously discussed integration has been carried out in the second and third rows of F and V):

$$
F = \begin{bmatrix} \int_0^{\frac{1}{2}-v} (F_{2,2} + F_{2,3})\, du_1 & \int_{\frac{1}{2}-v}^{\frac{5}{6}-v} (V_{2,2} + F_{2,3})\, du_1 & \int_{\frac{5}{6}-v}^{\frac{1}{6}+v} V_{2,3}\, du_1 \\ 0 & \frac{1}{3}v + \frac{1}{18} & -\frac{1}{9} + \frac{2}{3}v \\ 0 & 0 & \frac{1}{6} + v \end{bmatrix},
$$

$$
V = \begin{bmatrix} V_{1,1} & V_{1,2} & V_{1,3} \\ 0 & -v^2 + \frac{2}{3}v + \frac{5}{36} - vu_1 - \frac{1}{6}u_1 & \frac{1}{2}v^2 + \frac{1}{2}v + \frac{3}{8} + \frac{1}{2}u_1^2 - u_1 \\ 0 & 0 & 1 - u_2 \end{bmatrix}.
$$

The completely evaluated F matrix is given by

$$
F = \begin{bmatrix} -v^2 + \frac{5}{9}v - \frac{1}{36} & -\frac{1}{36} + \frac{5}{18}v & -\frac{2}{3}v^3 + v^2 - \frac{2}{9}v - \frac{1}{81} \\ 0 & -\frac{1}{3}v + \frac{1}{18} & -\frac{1}{9} + \frac{2}{3}v \\ 0 & 0 & \frac{1}{6} + v \end{bmatrix}.
$$

Finally, the CDF on the third v-subinterval is the sum of the elements in the first row of the F matrix, all multiplied by $3! = 6$.

Phase 4: Shift the Distribution. At this point, the CDF is computed in the form $\Pr(D_n < v + \frac{1}{2n})$. Now we convert the distribution into the more usable form $\Pr(D_n < y) = F_{D_n}(y)$ by making the substitution $y = v + \frac{1}{2n}$ in both the polynomials and the v-subintervals of the CDF. Specifically, we add $\frac{1}{2n}$ to each endpoint of the v-subintervals and we substitute $(y - \frac{1}{2n})$ for v in the CDF polynomials. We then take these two lists of numbers and polynomials respectively, simplify them (using Maple's `simplify()` command), and create the CDF representation in the usual list-of-sublists representation for distributions. Specifically, we can now verify critical values for the distribution, but more importantly, we can calculate exact significance levels of any given statistic. An example of the shifted distribution for $n = 6$ is

$$
F(y) = \begin{cases} 0 & y < \frac{1}{12} \\ 46080\,y^6 - 23040\,y^5 + 4800\,y^4 - \frac{1600}{3}y^3 + \frac{100}{3}y^2 - \frac{10}{9}y + \frac{5}{324} & \frac{1}{12} \le y < \frac{1}{6} \\ 2880\,y^6 - 4800\,y^5 + 2360\,y^4 - \frac{1280}{3}y^3 + \frac{235}{9}y^2 + \frac{10}{27}y - \frac{5}{81} & \frac{1}{6} \le y < \frac{1}{4} \\ 320\,y^6 + 320\,y^5 - \frac{2600}{3}y^4 + \frac{4240}{9}y^3 - \frac{785}{9}y^2 + \frac{145}{27}y - \frac{35}{1296} & \frac{1}{4} \le y < \frac{1}{3} \\ -280\,y^6 + 560\,y^5 - \frac{1115}{3}y^4 + \frac{515}{9}y^3 + \frac{1525}{54}y^2 - \frac{565}{81}y + \frac{5}{16} & \frac{1}{3} \le y < \frac{5}{12} \\ 104\,y^6 - 240\,y^5 + 295\,y^4 - \frac{1985}{9}y^3 + \frac{775}{9}y^2 - \frac{7645}{648}y + \frac{5}{16} & \frac{5}{12} \le y < \frac{1}{2} \\ -20\,y^6 + 32\,y^5 - \frac{185}{9}y^3 + \frac{175}{36}y^2 + \frac{3371}{648}y - 1 & \frac{1}{2} \le y < \frac{2}{3} \\ 10\,y^6 - 38\,y^5 + \frac{160}{3}y^4 - \frac{265}{9}y^3 - \frac{115}{108}y^2 + \frac{4651}{648}y - 1 & \frac{2}{3} \le y < \frac{5}{6} \\ -2\,y^6 + 12\,y^5 - 30\,y^4 + 40\,y^3 - 30\,y^2 + 12\,y - 1 & \frac{5}{6} \le y < 1 \\ 1 & y \ge 1. \end{cases}
$$

The CDF and PDF of D_6 are shown in Figs. 12.8 and 12.9. The APPL commands used to produce these plots are

```
> Y := KSRV(6);
> PlotDist(CDF(Y));
> PlotDist(PDF(Y));
```

Fig. 12.8. The CDF of the D_6 random variable

We have noted that there is a discontinuity in the PDF at $y = \frac{1}{6}$, and more generally, at $y = \frac{1}{n}$. This corresponds to the value $v = \frac{1}{2n}$ in Birnbaum's original integral. This is the smallest value of v for which the n-dimensional hypercube centered at $\left(\frac{1}{2n}, \frac{3}{2n}, \ldots, \frac{2n-1}{2n}\right)$ makes contact with the boundary of the region in n-space that satisfies $0 \leq u_1 \leq u_2 \leq \cdots \leq u_n \leq 1$.

Critical values associated with the distribution of D_n are determined with the APPL IDF procedure. Arbitrary accuracy can be achieved by using the Maple `Digits` command to define the precision. We have verified critical values by comparing them with those given in Owen [95] for $\alpha = 0.1$, 0.05, and 0.01 and for $n = 1, 2, \ldots, 10$ and those values match ours exactly, except that ours are not limited to five decimal places.

In summary, we have developed and implemented an algorithm for computing the CDF of the KS test statistic in the all-parameters-known case for a positive integer parameter n. Computing the CDF provides a challenging calculus problem, even for small values of n. The CDF can be constructed from a set of piecewise degree-n polynomials. The algorithm can be used to plot the entire CDF or to just find particular fractiles or probabilities associated with the distribution. Further details are given in Drew et al. [23]. An

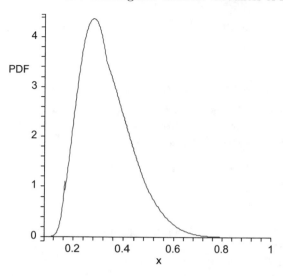

Fig. 12.9. The PDF of the D_6 random variable. Note the discontinuity at $y = 1/6$

extension to the parameters-estimated-from-data case for sampling from an exponential population and a small sample size is given in Evans et al. [31]. We conclude this section by presenting the algorithm for KSRV, which returns the CDF of the KS test statistic D_n in the all-parameters-known case.

Procedure KSRV: Compute the CDF of the Kolmogorov–Smirnov test statistic.

Input: A positive integer n, the sample size associated with the KS test statistic when all parameters are known.

Output: The CDF of the KS test statistic when all parameters are known is returned. The random variable will be in the usual list-of-sublists format.

Algorithm:

Phase 1: Determine the endpoints of the segments that define the support of the KS test statistic.

Dimension $v\left[0..\left\lceil\frac{3n}{2}\right\rceil - 1\right]$ v holds the partition values for the support of D_n
$m \leftarrow 0$
$v_m \leftarrow 0$
$g \leftarrow \frac{1}{2n}$ g is the gap between adjacent integration limits when $v = 0$
For $i \leftarrow 1$ to $n - 1$
 $m \leftarrow m + 1$
 $v_m \leftarrow i \cdot g$

For $j \leftarrow 2 \cdot \lfloor \frac{n}{2} \rfloor + 1$ to $2n - 1$ by 2 lower limit is n or $n + 1$, whichever is odd
 $m \leftarrow m + 1$
 $v_m \leftarrow j \cdot g$

Phase 2: This phase of the algorithm has three parts to it.
1. Define c_1, c_2, \ldots, c_m as the midpoints of the support intervals
2. Define x_1, x_2, \ldots, x_n as the limits of integration when $v = 0$
3. Define the $n \times n$ indicator matrices A_1, A_2, \ldots, A_m.

Dimension $c[m]$ define vector c of midpoints of the support subintervals
For $k \leftarrow 1$ to m
 $c_k \leftarrow (v_{k-1} + v_k)/2$

Dimension $x[n]$ define vector x of centers of integration limits
For $i \leftarrow 1$ to n
 $x_i \leftarrow (2i - 1) \cdot g$

Dimension $A[n, n]$

For $i \leftarrow 2$ to n
 For $j \leftarrow 1$ to $i - 1$
 $A_{ij} \leftarrow 0$ zero out lower triangle of A

For $k \leftarrow 1$ to m loop through all subintervals of support
 For $i \leftarrow 1$ to n
 For $j \leftarrow i$ to n
 $A_{ij} \leftarrow 0$ zero out upper triangle of A
 $z \leftarrow \max\{\lfloor nc_k - \frac{1}{2} \rfloor, 0\}$ $z \equiv$ number of leading zero columns in A
 $l \leftarrow \min\{\lceil 2nc_k \rceil, n\}$ $l \equiv$ number of u-subint. intersecting $(\frac{1}{2n} - v, \frac{1}{2n} + v)$
 For $i \leftarrow 1$ to n
 For $j \leftarrow \max\{i, z + 1\}$ to $\min\{n, i + l - 1\}$
 $A_{ij} \leftarrow 1$ place 1's in A

Phase 3: Compute the fixed and variable limit matrices F and V and combine to give the piecewise polynomial CDF.

Dimension $P[m]$ nth order polynomials for CDF segments param. by v
Dimension $F[n, n]$ array of fixed integrals parameterized by v
Dimension $V[n, n]$ array of variable integrals parameterized by v
Real $S(v)$ a function of v, used as a running sum $S(v)$ is the sum of
 all F elements one row down, starting one element to the right

$P_1(v) \leftarrow n!(2v)^n$ $\left[P_1(v) \leftarrow n! \int_{x_1-v}^{x_1+v} \int_{x_2-v}^{x_2+v} \cdots \int_{x_n-v}^{x_n+v} 1 \, du_n \cdots du_2 \, du_1 \right]$
For $k \leftarrow 2$ to m loop through all subintervals of support
 $z \leftarrow \max\{\lfloor nc_k - \frac{1}{2} \rfloor, 0\}$

$l \leftarrow \min\{\lceil 2nc_k \rceil, n\}$

$F_{n,n}(v) \leftarrow \int_{x_n - v}^{1} 1 \, du_n$

$V_{n,n}(v) \leftarrow \int_{u_{n-1}}^{1} 1 \, du_n$

For $i \leftarrow n - 1$ to 2 by -1

 If $i + l > n$ then

 $S(v) \leftarrow 0$

 else

 $S(v) \leftarrow F_{i+1,i+l}(v)$

 If $i + l > n + 1$ then

 $F_{i,n}(v) \leftarrow \int_{x_n - v}^{\lfloor x_i + c_k \rfloor + (x_i + v)(1 - \lfloor x_i + c_k \rfloor)} V_{i+1,n}(v) \, du_i$

 $V_{i,n}(v) \leftarrow \int_{u_{i-1}}^{\lfloor x_i + c_k \rfloor + (x_i + v)(1 - \lfloor x_i + c_k \rfloor)} V_{i+1,n}(v) \, du_i$

 If $i + l = n + 1$ then

 $F_{i,n}(v) \leftarrow \int_{x_n - v}^{x_i + v} V_{i+1,n}(v) \, du_i$

 If $i + l < n + 1$ then

 $F_{i,i+l-1}(v) \leftarrow \int_{x_{i+l-1} - v}^{x_i + v} \{V_{i+1,i+l-1}(v) + S(v)\} \, du_i$

 $S(v) \leftarrow S(v) + F_{i+1,\min\{i+l-1,n\}}(v)$

 For $j \leftarrow \min\{n-1, i+l-2\}$ to $\max\{i+1, z+2\}$ by -1

 $F_{i,j}(v) \leftarrow \int_{x_j - v}^{x_{j+1} - v} \{V_{i+1,j}(v) + S(v)\} \, du_i$

 $V_{i,j}(v) \leftarrow \int_{u_{i-1}}^{x_{j+1} - v} \{V_{i+1,j}(v) + S(v)\} \, du_i$

 $S(v) \leftarrow S(v) + F_{i+1,j}(v)$

 If $z + 1 \leq i$ then

 $V_{i,i}(v) \leftarrow \int_{u_{i-1}}^{x_{i+1} - v} S(v) \, du_i$

 If $z + 1 > i$ then

 $V_{i,z+1}(v) \leftarrow \int_{u_{i-1}}^{x_{z+2} - v} \{V_{i+1,z+1}(v) + S(v)\} \, du_i$

 If $z + 1 < i$ then

 $F_{i,i}(v) \leftarrow \int_{x_i - v}^{x_{i+1} - v} S(v) \, du_i$

If $l = n$ then rows $2, 3, \ldots, n$ of A matrix completed at this point

 $S(v) \leftarrow 0$

 $F_{1,n}(v) \leftarrow \int_{x_n - v}^{x_1 + v} V_{2,n}(v) \, du_1$

else

 $S(v) \leftarrow F_{2,l+1}(v)$

If $l < n$ then

 $F_{1,l}(v) \leftarrow \int_{x_l - v}^{x_1 + v} \{V_{2,l}(v) + S(v)\} \, du_1$

$S(v) \leftarrow S(v) + F_{2,l}(v)$

For $j \leftarrow \min\{n-1, l-1\}$ to $\max\{2, z+1\}$ by -1

 $F_{1,j}(v) \leftarrow \int_{(x_j - v)\lceil x_j - c_k \rceil}^{x_{j+1} - v} \{V_{2,j}(v) + S(v)\} \, du_1$

 $S(v) \leftarrow S(v) + F_{2,j}(v)$

If $z = 0$ then

 $F_{1,1}(v) \leftarrow \int_{0}^{x_2 - v} S(v) \, du_1$

$P_k(v) \leftarrow 0$ For $j \leftarrow z + 1$ to l

$$P_k(v) \leftarrow P_k(v) + F_{1,j}(v)$$
$$P_k(v) \leftarrow n! P_k(v)$$

Phase 4: Shift the distribution to the form of $\Pr(D_n < y)$ by substituting $y = v + \frac{1}{2n}$.

$y_0 \leftarrow v_0 + \frac{1}{2n}$

For $k \leftarrow 1$ to m loop through all subintervals of support

 $y_i \leftarrow v_k + \frac{1}{2n}$

 $F_k(y) \leftarrow P_k(y - \frac{1}{2n})$

Return: y_0, y_1, \ldots, y_m and $F_1(y), F_2(y), \ldots, F_m(y)$.

13

Transient Queueing Analysis

An APPL extension that computes the exact distribution of the nth customer's sojourn time in an $M/M/s$ queue with k customers initially present is derived in this chapter. Algorithms for computing the covariance between sojourn times for an $M/M/1$ queue with k customers present at time zero are also developed. Maple computer code is developed to implement the transient queue analysis for many system measures of performance without regard to traffic intensity (i.e., the system may be unstable with traffic intensity greater than one).

13.1 Introduction

Many traditional simulation studies analyze queueing systems in steady-state, requiring appropriate warm-up periods and associated long simulation runs. However, in many cases the system being modeled never reaches steady-state; thus, steady-state simulation results do not accurately portray the system behavior. The ability to analyze transient results associated with such models is often complicated by intractable theory, leaving simulation as the only method for analysis. Further complicating the transient analysis is the effect of initial conditions [60]. Since steady-state results depend on running the system long enough to eliminate the impact of initial conditions, these steady-state results reveal nothing about the transient behavior of the queueing system. Our purpose here is to combine new and existing results in transient queueing analysis with a symbolic engine in computational probability.

There are many classes of queueing systems in which a transient analysis is required, e.g., service businesses often model queues that never reach equilibrium. Recognizing the need to develop theory for transient results, as opposed

© Springer International Publishing AG 2017
J.H. Drew et al., *Computational Probability*, International Series
in Operations Research & Management Science 246,
DOI 10.1007/978-3-319-43323-3_13

to steady-state results, has resulted in a wide literature in this area. Initial work in transient analysis ironically appeared as an attempt to measure when a system achieved equilibrium. Law [66] notes the consequences of failing to adequately account for the initial transient period, leading to Gafarian et al. [34] outlining a comprehensive framework for the initial transient problem. Morisaku [90] addresses the time to equilibrium in simulations modeling the $M/M/1$ queue and provides schematics for the transition probabilities given $k \geq 0$ customers initially present at time $t = 0$. Grassmann [40] compares three methods for finding transient solutions in Markovian queueing systems, Runge–Kutta, Liou's method, and randomization, where randomization is shown as superior for large sparse transition matrices. Pegden and Rosenshine [101] provide a closed-form expression for the probability of exactly i arrivals and j servicings over a time horizon of length t in an $M/M/1$ queue starting empty and idle, allowing the calculation of certain performance measures for a specified time period. Odoni and Roth [94] take an empirical approach to compare observed and predicted transient state queue length for the $M/M/1$ queue, noting that for small values of t the expected queue length is strongly influenced by initial conditions, and provide a good approximation for an upper bound of the time to steady-state. Kelton and Law [60] consider the $M/M/s$ $(s \geq 1)$ queue and provide expressions to calculate the probabilities of having up to $n + k$ customers in the system upon the arrival of the nth customer, where k is the number of customers in the system at time $t = 0$. They then apply these calculations to a variety of measures of performance with implications to convergence on steady-state delays and offer methods for choosing queue initialization in simulation. Much of the work in this chapter is motivated by their results. Kelton [59] extends the previous work by considering $M/E_m/1$ and $E_m/M/1$ queues. Parthasarathy [99] provides a transient solution for the probability that there are n customers in the system at time t for an $M/M/1$ queue. Abate and Whitt [1] use Laplace transformations to analyze some transient results of interest in the $M/M/1$ queue. Leguesdron et al. [76] provide transient probabilities for the $M/M/1$ queue by inverting the generating function of the uniformized Markov chain describing the $M/M/1$ process. Transient distributions of cumulative reward are calculated by e Silva et al. [25] using uniformization, where the distribution of cumulative reward is over a finite interval with reward rates represented by Markov model states. Grassmann [41] investigates warm-up periods in simulation and shows that in many cases these warm-up periods should not be used, especially if the simulation begins in a high probability state. In this chapter we will focus on the transient analysis of the $M/M/1$ and the more general $M/M/s$ queues, specifically on the distribution of the nth customer's sojourn time, which is the sum of the nth customer's delay time and service time. Almost all the references above address measures of performance specified over a finite time interval and are the results of numerical work, which is in stark contrast to the measures proposed here, which are based on the exact distribution of specific customers. Rather than arriving at the results numerically, a computer algebra system is utilized offering exact measures of performance based on a given number of customers.

The $M/M/s$ queue is defined in Sect. 13.2 for a positive integer s, and a method is given for calculating the probability distribution of the number of customers an arriving customer sees upon arrival to an $M/M/s$ queue. Section 13.3 describes how the sojourn time distribution is calculated for a given customer in an $M/M/s$ queue with k customers initially present in the system, $k \geq 0$. Section 13.4 includes examples using the implemented procedures to calculate exact sojourn time distributions, related measures of performance, and graphical illustrations for varying parameters such as traffic intensity and number of customers in the system. Section 13.5 offers two approaches for calculating the covariance and correlation among customers in an $M/M/1$ queue. Section 13.6 extends the covariance and correlation calculations by automating the process of finding the joint probability distribution function of the sojourn times of two customers, and provides the exact covariance and correlation calculations for varying traffic intensities. Commented code is available for all computations conducted here.

13.2 Basics of the $M/M/s$ Queue

The $M/M/s$ queue is governed by iid exponential interarrival times (the arrival stream is a Poisson process) with arrival rate λ, and iid exponential service times among s identical servers, each with service rate μ. The interarrival times and the service times are mutually independent. The traffic intensity of the system is $\rho = \lambda/(s\mu)$. The system consists of a single queue with customers waiting to be serviced by one of the s identical parallel servers. If an arriving customer finds at least one idle server, the customer immediately proceeds to service; otherwise the customer joins the single queue of those waiting for service in a first-come, first-served manner. To achieve classic steady-state results, the traffic intensity must satisfy $\rho < 1$. This critical assumption is not required in transient analysis, described here, because the system of interest never reaches equilibrium.

Let $P_k(n, i)$ be the probability that upon the arrival of the nth customer there are i customers in the system including the nth customer (in queue or in service), given k customers are present at time $t = 0$. Using propositions provided by Kelton and Law [60], reprinted here for completeness (proofs are available in the reference), and a recursion algorithm, $P_k(n, i)$ for $i = 1, 2, \ldots, n + k$ can be computed. Using these probabilities, it is possible to find the distribution of the sojourn time for the nth customer in an $M/M/s$ queue, given k customers are present at time $t = 0$. Proposition 1 addresses the case of no exits prior to the nth customer's arrival, given $k \geq 1$. Proposition 2 is identical to Proposition 1 except that the system is empty and idle at $t = 0$ (i.e., $k = 0$). Proposition 3 addresses the case that the first customer finds $i - 1$ other customers present for $k > 0$. Proposition 4 is the more general case that customer $n \geq 2$ finds i other customers present, given $k \geq 0$.

Proposition 1. If $k \geq 1$, then for $n \geq 1$,

$$P_k(n, k+n) = \begin{cases} [\rho/(\rho+1)]^n & k \geq s \\ \rho^n / \prod_{j=1}^{n} [\rho + (k+j-1)/s] & k+n \leq s \\ \rho^n / \left[(\rho+1)^{n-s+k} \prod_{j=1}^{s-k} [\rho + (k+j-1)/s] \right] & k < s < k+n. \end{cases}$$

Proposition 2. For $n \geq 1$,

$$P_0(n, n) = \begin{cases} \rho^n / \prod_{j=1}^{n} [\rho + (j-1) s] & n \leq s \\ \rho^n / \left[(\rho+1)^{n-s} \prod_{j=1}^{s} [\rho + (j-1)/s] \right] & n > s. \end{cases}$$

Proposition 3. If $k \geq 1$, then for $2 \leq i \leq k$,

$$P_k(1, i) = \begin{cases} \{\rho/[\rho+(i-1)/s]\} \cdot \\ \qquad \prod_{j=1}^{k-i+1} \{1 - \rho/[\rho+(k-j+1)/s]\} & k \leq s \\ \rho/(\rho+1)^{k-i+2} & k > s \text{ and } i > s \\ \{\rho/[(\rho+1)^{k-s+1}[\rho+(i-1)/s]]\} \cdot \\ \qquad \prod_{j=1}^{s-i} \{1 - \rho/[\rho+(s-j)/s]\} & i \leq s < k. \end{cases}$$

Proposition 4. For $n \geq 2$, and $2 \leq i \leq k+n-1$,

$$P_k(n, i) = \begin{cases} [\rho/(\rho+1)] \sum_{j=i-1}^{k+n-1} [1/(\rho+1)]^{j-i+1} P_k(n-1, j) & i > s \\ \{\rho/[\rho+(i-1)/s]\} \cdot \\ \quad \left\{ \sum_{j=i-1}^{s-1} \left[\prod_{h=1}^{j-i+1} \{1 - \rho/[\rho+(j-h+1)/s]\} \right] \cdot \right. \\ \quad P_k(n-1, j) + \left[\prod_{h=1}^{s-i} \{1 - \rho/[\rho+(s-h)/s]\} \right] \cdot \\ \quad \left. \sum_{j=s}^{k+n-1} [1/(\rho+1)]^{j-s+1} P_k(n-1, j) \right\} & i \leq s. \end{cases}$$

Using these four propositions, $P_k(n, 1)$ is calculated by subtracting the complementary probability from one. These results can be generated by invoking the Maple procedure MMsQprob(n, k, s), where

- n is the index of the customer of interest,
- k is the number of customers in the system at time $t = 0$,
- s is the number of identical parallel servers.

The procedure is written in Maple and uses APPL. We choose to calculate the distribution of the sojourn time because it is a purely continuous random variable enabling us to exploit associated procedures in APPL.

13.3 Creating the Sojourn Time Distribution

Once the necessary $P_k(n, i)$ probabilities are calculated, for $i = 1, 2, \ldots, n+k$, the exact sojourn time distribution for the nth customer can be calculated.

We define X_n as the number of customers, including customer n, in the system upon the arrival of the nth customer. The possible values of X_n can vary from a minimum of 1, which occurs when customer n arrives to an empty queue with idle servers, to a maximum of $n + k$, which occurs when 0 exits occur prior to customer n's arrival, matching the possible values for i in the expression $P_k(n, i)$ above. The mathematical derivations for both the $M/M/1$ and $M/M/s$ queues make extensive use of the memoryless property, permitting the construction of the distribution of T_n, the sojourn time of customer n. We present each case separately below.

Distribution of T_n for the $M/M/1$ Queue. For an $M/M/1$ queue starting empty and idle, the delay time of the first customer is zero because the customer proceeds directly to service upon arrival. Therefore, the first customer has an exponential(μ) sojourn time distribution. Conditioning on customer 1's service time, one can calculate the probabilities of customer 2 arriving before and after customer 1 finishes service. These well-known results [49, 62, 118] are

$$\Pr(Y < X) = \frac{\mu}{\lambda + \mu}, \qquad \Pr(X < Y) = \frac{\lambda}{\lambda + \mu},$$

where X is an exponential(λ) interarrival time and Y is an exponential(μ) service time. The first probability represents customer 2 proceeding directly to service, in which case his sojourn time is simply his service time, which is exponential(μ). The second probability represents the likelihood that customer 2 will delay prior to service. Using the memoryless property, customer 2 delays an exponential(μ) time before being serviced in an additional exponential(μ) time. Using these two probabilities, it is easy to see that customer 2's sojourn time distribution is a mixture, where the mix probabilities are the $P_0(n, i)$'s and the distributions are determined by the orderings of delays and services potentially encountered. It is well known that for X_1, X_2, \ldots, X_n iid exponential(λ) random variables that

$$\sum_{i=1}^{n} X_i \sim \text{Erlang}(\lambda, n). \tag{13.1}$$

Using this result, the $M/M/1$ queue sojourn time distribution for $k = 0$ initial customers generalizes very elegantly to include $k > 0$, as indicated in Table 13.1. Line i of the table occurs with probability $P_k(n, i)$ and lists the distribution of the sojourn time for the nth customer, conditioned on i customers being in the system upon his arrival.

Let $g_i(t)$ be the PDF of an Erlang(μ, i) random variable. Using the conditional sojourn time distributions for $i = 1, 2, \ldots, n + k$ potential customers in the system, each with probability $P_k(n, i)$, the PDF for the nth customer's sojourn time T_n is the mixture

$$f_n(t) = \sum_{i=1}^{n+k} P_k(n, i) g_i(t) \qquad t > 0. \tag{13.2}$$

Table 13.1. Conditional sojourn time distributions for the $M/M/1$ queue

X_n	Delay distribution	Service time distribution	Conditional sojourn time distribution
1	0	exponential(μ)	exponential(μ)
2	exponential(μ)	exponential(μ)	Erlang(μ, 2)
3	Erlang(μ, 2)	exponential(μ)	Erlang(μ, 3)
4	Erlang(μ, 3)	exponential(μ)	Erlang(μ, 4)
\vdots	\vdots	\vdots	\vdots
$n+k$	Erlang(μ, $n+k-1$)	exponential(μ)	Erlang(μ, $n+k$)

This result is simple in the $M/M/1$ case because we can take advantage of (13.1), resulting in a mixture of $n+k$ Erlang distributions.

Distribution of T_n for the $M/M/s$ Queue. Given $s > 1$ parallel identical servers, the nth customer's sojourn time distribution is still a mixture of $n+k$ conditional sojourn time distributions. However, each distribution might be more complicated. For illustration, consider an $M/M/3$ queue starting empty and idle with exponential(λ) arrivals and three identical exponential(μ) servers. It is clear that for customers 1, 2, and 3, the sojourn time is exponential(μ) because all three customers proceed directly to service. Therefore, in the general case, for the number of customers in the system including customer n, which we defined as X_n, when $X_n \leq s$ the conditional sojourn time distribution is exponential(μ). However, if $X_n > s$, then the nth customer experiences a delay while observing $X_n - s$ service completions. When $s > 1$ and $X_n > s$, the service time distribution observed by customers in queue is exponential with rate $s\mu$.

Using this result, it is apparent that the delay time for the nth customer is the sum of $X_n - s$ independent exponential($s\mu$) random variables, and using (13.1) is Erlang($s\mu, X_n - s$). To calculate the nth customer's sojourn time for a particular value of X_n, we sum his delay time and his service time. Table 13.2 shows the distributions conditioned on the number of customers X_n encountered by customer n (including himself) for the $M/M/3$ queue, given $k = 0$ customers present at time $t = 0$. The APPL procedure `Convolution` calculates the distribution of a sum of independent random variables. We use the symbol \oplus to denote convolution.

Since X_n represents the number of customers in the system upon arrival of the nth customer, including himself, the first row in Table 13.2 corresponds to customer n arriving to an empty system and the last row corresponds to no service completions prior to customer n's arrival. The general form for the $M/M/s$ sojourn time PDF is identical to (13.2), however, in the $M/M/s$ case each $g_i(t)$ can potentially require an additional step to calculate the distribution of a sum of random variables.

Table 13.2. Conditional sojourn time distributions for the $M/M/3$ queue

X_n	Delay distribution	Service time distribution	Conditional sojourn time distribution
1	0	exponential(μ)	exponential(μ)
2	0	exponential(μ)	exponential(μ)
3	0	exponential(μ)	exponential(μ)
4	exponential(3μ)	exponential(μ)	exponential$(3\mu) \oplus$ exponential(μ)
5	Erlang$(3\mu, 2)$	exponential(μ)	Erlang$(3\mu, 2) \oplus$ exponential(μ)
\vdots	\vdots	\vdots	\vdots
n	Erlang$(3\mu, n-3)$	exponential(μ)	Erlang$(3\mu, n-3) \oplus$ exponential(μ)

13.4 Transient Analysis Applications

It is apparent that calculating (13.2) for large n is tedious. Kelton and Law [60] acknowledge the computational difficulty in achieving the $P_k(n, i)$ probabilities alone. Conducting the added steps of up to $n - s$ convolutions for the $M/M/s$ queue and then mixing the resulting conditional distributions with the appropriate probabilities can be complicated to implement. APPL provides the underlying computational engine to achieve exact results for such problems. The APPL procedure Queue(X, Y, n, k, s) returns the exact sojourn time distribution for customer n. Queue calls MMsQprob(n, k, s), which uses recursion to calculate the necessary $P_k(n, i)$ probabilities. APPL is capable of symbolic results, as illustrated in Examples 13.1 and 13.2.

Example 13.1. Consider an $M/M/1$ queue with arrival rate λ and service rate μ starting empty and idle at time $t = 0$. For the fourth customer, calculate the probabilities $P_0(4, i)$ for $i = 1, 2, 3, 4$.

The APPL command MMsQprob(4, 0, 1) returns the exact symbolic probabilities

$$P_0(4, 1) = \frac{5\rho^2 + 4\rho + 1}{(\rho + 1)^5},$$

$$P_0(4, 2) = \frac{\rho\left(5\rho^2 + 4\rho + 1\right)}{(\rho + 1)^5},$$

$$P_0(4, 3) = \frac{\rho^2\left(3\rho + 1\right)}{(\rho + 1)^4},$$

$$P_0(4, 4) = \frac{\rho^3}{(\rho + 1)^3},$$

where $\rho = \lambda/\mu$ is the traffic intensity. It is easy to verify that for any $\rho > 0$, $\sum_{i=1}^{4} P_0(4, i) = 1$, as required. For example, a simple substitution letting $\rho = 9/10$ yields

$$P_0(4, 1) = \frac{865000}{2476099}, P_0(4, 2) = \frac{778500}{2476099}, P_0(4, 3) = \frac{29970}{130321}, P_0(4, 4) = \frac{729}{6859}.$$

Example 13.2. For the queue described in Example 13.1, calculate the fourth customer's sojourn time distribution, mean sojourn time and sojourn time variance.

The APPL statements

```
> X := ExponentialRV(lambda);
> Y := ExponentialRV(mu);
> T := Queue(X, Y, 4, 0, 1);
> Mean(T);
> Variance(T);
```

calculate the desired results. The first two lines define the interarrival and service time distributions, while the third line calculates the fourth customer's sojourn time distribution. The last two lines are self explanatory. The resulting PDF is

$$f_4(t) = \frac{1}{6(\lambda+\mu)^5}\mu^4 e^{-\mu t}\left(30\lambda^2 + 30\lambda^3 t + 24\lambda\mu + 24\lambda^2\mu t + 6\mu^2+\right.$$

$$\left. 6\mu^2\lambda t + 9t^2\lambda^4 + 12t^2\lambda^3\mu + 3t^2\lambda^2\mu^2 + t^3\lambda^5 + 2t^3\lambda^4\mu + t^3\lambda^3\mu^2\right)$$

for $t > 0$. Using $f_4(t)$ above, the Mean and Variance commands return

$$E[T_4] = \frac{\mu^5 + 6\lambda\mu^4 + 26\mu^2\lambda^3 + 16\mu^3\lambda^2 + 17\mu\lambda^4 + 4\lambda^5}{\mu(\lambda+\mu)^5}$$

and

$$\text{Var}[T_4] = \left(181\mu^2\lambda^8 + 484\mu^3\lambda^7 + 816\mu^4\lambda^6 + 868\mu^5\lambda^5 + 574\mu^6\lambda^4 +\right.$$
$$\left. 244\mu^7\lambda^3 + 40\mu\lambda^9 + 68\mu^8\lambda^2 + 12\mu^9\lambda + \mu^{10} + 4\lambda^{10}\right)/\left(\mu^2(\lambda+\mu)^{10}\right).$$

Substituting $\lambda = 1$ and $\mu = 10/9$, for example, the results simplify to

$$f_4(t) = \frac{5000}{66854673}e^{-10/9t}\left(361t^3 + 2109t^2 + 5190t + 5190\right) \qquad t > 0,$$

$$E[T_4] = \frac{23323347}{12380495} \cong 1.8839,$$

and

$$\text{Var}[T_4] = \frac{383506725720906}{153276656445025} \cong 2.5021.$$

The CPU time associated with these two examples was negligible. Examples 13.1 and 13.2 represent simple applications of these procedures that circumvent time intensive hand-calculations. They serve only as indications of more challenging problems solvable using these procedures.

Example 13.3. Calculate the mean sojourn time of the 9th customer in an $M/M/2$ queue with arrival rate $\lambda = 1$, service rate $\mu = 9/20$, ($\rho = 10/9$), and $k = 3$ customers initially present.

The mean can be calculated in a single APPL statement by embedding the function calls

```
> Mean(Queue(ExponentialRV(1), ExponentialRV(9 / 20), 9, 3, 2));
```

which yields

$$E[T_9] = \frac{888596294932695525996949989160095340}{156268000432990839922685380588656603}$$

or, to 4 digits, $E[T_9] = 5.6864$.

Being able to represent the sojourn time distribution for the nth customer in closed form also provides valuable information on asymptotic behavior for queueing systems, including steady-state convergence rates for different initial conditions. Figure 13.1 shows the mean sojourn time for customer $n = 1, 2, \ldots, 120$ in an $M/M/1$ queue with $\lambda = 1$, $\mu = 10/9$, and $\rho = 9/10$ for several values of k. The points that are plotted have been connected by lines. As expected, despite the initial condition, all cases appear to move toward the steady-state value of 9 with increasing n. The horizontal axis is only limited to $n = 120$ for display purposes and in fact, identical computations were carried out for $n > 300$ customers to verify convergence. However, as shown in the cases where $k = 6$ and $k = 10$, the convergence to steady-state is not always monotone. Additionally, in testing various traffic intensities, the rate of convergence to steady-state increases rapidly with decreasing traffic intensity for varying values of k.

APPL also has the ability to calculate the closed-form CDF for the nth customer's sojourn time, permitting CDF comparisons for varying n as well as distribution percentiles for a given customer. The procedure call CDF(T) returns the exact CDF for customer 4 (from Example 13.1). Figure 13.2 displays the sojourn time CDF for varying n with fixed $k = 0$ and $\rho = 9/10$. The differences in CDFs across n correspond to the increasing mean attributed

Fig. 13.1. $M/M/1$ mean sojourn time for $\rho = 9/10$ given k at $t = 0$

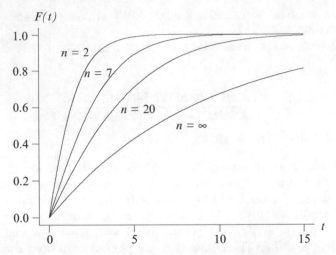

Fig. 13.2. $M/M/1$ sojourn time CDFs for various n given $\rho = 9/10$ and $k = 0$

to the delays experienced by successive customers, e.g., customer 2 has delay time zero or exponential(μ) whereas the nth customer (for $n > 2$) faces a finite mixture of n potential delay distributions. The CDF associated with $n = \infty$ corresponds to the steady-state distribution of the sojourn time, which is exponentially distributed with a mean of 9 [62].

Varying k for an $M/M/1$ queue also provides another basis for comparison of CDFs. Figure 13.3 fixes $n = 2$, $\rho = 9/10$, and plots the resulting CDFs across k. Kelton and Law [60] make a similar comparison using convergence to steady state delay time. Using the CDF for multiple values of k allows direct comparison of sojourn time percentiles for customer n. As depicted, the sojourn time CDF for customer 2 is extremely sensitive to the initial condition k. As an illustration, the 80th percentiles for $k = 0, 3, 6$ are

$$F_2^{-1}(0.80) \cong \begin{cases} 1.935 & k = 0 \\ 4.432 & k = 3 \\ 7.510 & k = 6. \end{cases}$$

These percentiles are achieved using the APPL statements

```
> X := ExponentialRV(1);
> Y := ExponentialRV(10 / 9);
> Z := Queue(X, Y, 2, k, 1);
> IDF(Z, 0.8);
```

for $k = 0, 3, 6$. The last statement, IDF(Z, 0.8), numerically solves $F_Z(z) = 0.80$ on the interval $(0, \infty)$.

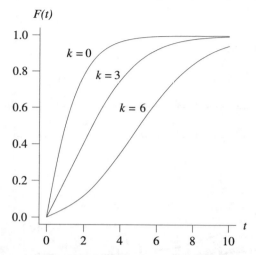

Fig. 13.3. $M/M/1$ sojourn time CDFs for customer $n = 2$ for various k given $\rho = 9/10$

Given the complete specification of the sojourn time distribution, one can use APPL to calculate not only the mean but also the second, third, and fourth moments for customer n. This is especially valuable for steady-state analysis. It is common in simulation to verify attainment of steady-state behavior by examining the mean delay or mean sojourn time. Though some literature exists on estimating transient mean and variance, we are not aware of any literature addressing higher moments. Literature addressing the second moment seems mostly focused on variance estimation and not necessarily convergence. Therefore, even when the first moment might acceptably approximate the steady state value, there is reason for further analysis of higher moments. For example, Fig. 13.4 displays the first four moments of the sojourn time for customer n in an $M/M/1$ queue, where $\lambda = 1$, $\mu = 2$, $\rho = 1/2$, with the initial condition $k = 0, 4, 8$.

The APPL code used to calculate the values plotted in Fig. 13.4 is

```
> X := ExponentialRV(1);
> Y := ExponentialRV(2);
> for i from 2 to 60 by 1 do
>    T := Queue(X, Y, i, k, 1):
>    print(i, evalf(Mean(T)), evalf(Variance(T)),
>              evalf(Skewness(T)), evalf(Kurtosis(T))):
> od:
```

for $k = 0, 4, 8$. The vertical dashed lines give the smallest customer number for which all three of the transient values are within 1% of the steady state value. The relatively low traffic intensity $\rho = 1/2$ was selected purposely to allow quick convergence and easy visual inspection. Even with this somewhat

Fig. 13.4. First four moments of the $M/M/1$ sojourn time for customers 2 through 60 for $\rho = 1/2$ and $k = 0, 4, 8$. The arrival rate is $\lambda = 1$ and the service rate is $\mu = 2$, resulting in steady-state values for the four measures of performance of $1, 1, 2, 9$

low traffic intensity, it is apparent that the higher moments converge more slowly than the lower moments. In other scenarios where $\rho > 1/2$, the higher moments exhibit an even slower convergence. Each vertical dashed line in Fig. 13.4 was triggered by the $k = 8$ curve, suggesting that the moments are more sensitive to a heavily pre-loaded system. For the cases $k = 0, 4, 8$, the customer numbers for which the transient results were within 1% of the steady state values are listed in Table 13.3. To verify the initial-condition effect on the convergence rate of the first four moments, k was increasingly incremented beyond eight and displayed a further slowing of convergence.

Table 13.3. Smallest customer number where the sojourn time transient result is within 1% of steady state for an $M/M/1$ queue with $k = 0, 4, 8$ and $\rho = 1/2$

	$k = 0$	$k = 4$	$k = 8$
$E[T_n]$	19	21	36
$\sqrt{\mathrm{Var}[T_n]}$	27	29	46
$E\left[((T_n - \mu)/\sigma)^3\right]$	28	29	50
$E\left[((T_n - \mu)/\sigma)^4\right]$	34	35	56

13.5 Covariance and Correlation in the $M/M/1$ Queue

The dependence exhibited in sojourn times of successive customers is one reason for the difficulty in calculating interval estimators for queue measures of performance. In the simplest case, consider an initially empty and idle $M/M/1$ queue with interarrival and service rates λ and μ. Our desire is to calculate the covariance between the sojourn times of customers 1 and 2. We outline two approaches to modeling the events in the queue, which will be helpful in the presentation of the analytic result.

Discrete-Event Simulation Approaches. As previously discussed, customer 1 proceeds directly to service and two cases exist for customer 2. In the first case, customer 2 proceeds directly to service. In the second case, he delays until customer 1's departure. Both cases are shown in Fig. 13.5. This section introduces two approaches for generating the first two customers' sojourn times.

The first approach is a standard discrete-event simulation model. Without loss of generality, assume that customer 1 arrives at time 0. In the next-event approach, customer 1's service time is distributed according to the service distribution, exponential(μ), and an arrival time, a_2, for customer 2 is distributed according to the time-between-arrivals distribution, exponential(λ). If the arrival occurs after customer 1's service completion (case 1), then customer 2 also has an independent exponential(μ) service time, which in this

Fig. 13.5. Standard discrete-event simulation approach for cases 1 and 2

case is equal to his sojourn time T_2. In the second case in which customer 2's arrival time occurs before customer 1's completion of service ($a_2 < T_1$), customer 2 delays for $T_1 - a_2$ time units. We then add the exponential(μ) service time to the delay time to calculate T_2.

We define the gap occurring in case 2 as, $Y = T_1 - a_2$. It can be shown analytically that $Y \sim$ exponential(μ) by computing the distribution of the difference $T_1 - A_2$, where A_2 is the random arrival time of the second customer, which is exponential(λ), and then truncating the result on the left at zero. (Alternately, it can be reasoned that $Y \sim$ exponential(μ) by the memoryless property for the exponential distribution because the remaining service time for customer 1 after customer 2's arrival has the same distribution as an unconditional service time.) Therefore, by using (13.1), in case 2 the sojourn time for customer 2 is distributed Erlang($\mu, 2$).

The second approach is a conditional discrete-event simulation model, where the initial event, whose occurrence time is denoted as E_1 in Fig. 13.6, is either a completion of service for customer 1 with probability $\mu/(\lambda + \mu)$ or the arrival of customer 2 with probability $\lambda/(\lambda + \mu)$. Since E_1 is the minimum of the arrival time of customer 2 and service time of customer 1, $E_1 \sim$ exponential($\lambda + \mu$).

Covariance Calculations. One way to calculate the exact covariance of the sojourn times between customers 1 and 2 requires their joint probability density function, $f_{T_1,T_2}(t_1, t_2)$. The method used here for computing the joint probability density function uses Theorem 13.1.

Theorem 13.1. Let $X_1 \sim$ exponential(λ_1), $X_2 \sim$ exponential(λ_2), and $X_3 \sim$ exponential(λ_3) be independent random variables. The joint probability density function of $(T_1, T_2) = (X_1 + X_2, X_1 + X_3)$ is

$$f_{T_1,T_2}(t_1, t_2) = \begin{cases} \dfrac{\lambda_1\lambda_2\lambda_3 \left(e^{\lambda_1 t_1} - e^{(\lambda_2+\lambda_3)t_1}\right) e^{-\lambda_1 t_1 - \lambda_2 t_1 - \lambda_3 t_2}}{\lambda_1 - \lambda_2 - \lambda_3} & 0 < t_1 < t_2 \\[4mm] \dfrac{\lambda_1\lambda_2\lambda_3 \left(e^{\lambda_1 t_2} - e^{(\lambda_2+\lambda_3)t_2}\right) e^{-\lambda_2 t_1 - \lambda_1 t_2 - \lambda_3 t_2}}{\lambda_1 - \lambda_2 - \lambda_3} & 0 < t_2 < t_1. \end{cases}$$

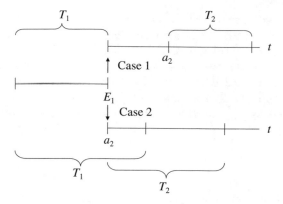

Fig. 13.6. Conditional discrete-event simulation approach for cases 1 and 2

Proof. The joint CDF of T_1 and T_2 is

$$
\begin{aligned}
F_{T_1,T_2}(t_1,t_2) &= \Pr\left(T_1 \le t_1, T_2 \le t_2\right) \\
&= \Pr\left(X_1 + X_2 \le t_1, X_1 + X_3 \le t_2\right) \\
&= \Pr\left(X_2 \le t_1 - X_1, X_3 \le t_2 - X_1\right) \\
&= \int_0^{\min\{t_1,t_2\}} \Pr\left(X_2 \le t_1 - x_1, X_3 \le t_2 - x_1 | X_1 = x_1\right) f_{X_1}(x_1)\, dx_1 \\
&= \int_0^{\min\{t_1,t_2\}} \Pr\left(X_2 \le t_1 - x_1 | X_1 = x_1\right) \cdot \\
&\qquad\qquad \Pr\left(X_3 \le t_2 - x_1 | X_1 = x_1\right) f_{X_1}(x_1)\, dx_1 \\
&= \int_0^{\min\{t_1,t_2\}} \left(1 - e^{-\lambda_2(t_1 - x_1)}\right)\left(1 - e^{-\lambda_3(t_2 - x_1)}\right) \lambda_1 e^{-\lambda_1 x_1}\, dx_1 \\
&= \begin{cases}
\int_0^{t_1} \left(1 - e^{-\lambda_2(t_1 - x_1)}\right)\left(1 - e^{-\lambda_3(t_2 - x_1)}\right) \lambda_1 e^{-\lambda_1 x_1}\, dx_1 & 0 < t_1 < t_2 \\
\int_0^{t_2} \left(1 - e^{-\lambda_2(t_1 - x_1)}\right)\left(1 - e^{-\lambda_3(t_2 - x_1)}\right) \lambda_1 e^{-\lambda_1 x_1}\, dx_1 & 0 < t_2 < t_1.
\end{cases}
\end{aligned}
$$

After evaluating the integrals and differentiating, $f_{T_1,T_2}(t_1,t_2)$ is

$$
f_{T_1,T_2}(t_1,t_2) = \begin{cases}
\dfrac{\lambda_1 \lambda_2 \lambda_3 \left(e^{\lambda_1 t_1} - e^{(\lambda_2+\lambda_3)t_1}\right) e^{-\lambda_1 t_1 - \lambda_2 t_1 - \lambda_3 t_2}}{\lambda_1 - \lambda_2 - \lambda_3} & 0 < t_1 < t_2 \\[2ex]
\dfrac{\lambda_1 \lambda_2 \lambda_3 \left(e^{\lambda_1 t_2} - e^{(\lambda_2+\lambda_3)t_2}\right) e^{-\lambda_2 t_1 - \lambda_1 t_2 - \lambda_3 t_2}}{\lambda_1 - \lambda_2 - \lambda_3} & 0 < t_2 < t_1,
\end{cases}
$$

which completes the proof. □

 Theorem 13.1 provides the joint PDF of the first two sojourn times for case 2, which must be weighted appropriately by the probability that the arrival of customer 2 occurs prior to customer 1's completion of service, or $\lambda/(\lambda + \mu)$. Using the conditional discrete-event model, case 1 consists

of independent sojourn times, so the joint density can be written as the product of the densities of the sojourn times $T_1 \sim$ exponential$(\lambda + \mu)$ and $T_2 \sim$ exponential(μ), and weighted by $\mu/(\lambda + \mu)$. The resulting joint density is a mixture of the two possible cases displayed in Fig. 13.6. We apply Theorem 13.1 to case 2 because of the dependence that occurs due to the overlap of the sojourn times. Figure 13.7 depicts the relationships between the sojourn times T_1, T_2 and the random variables X_1, X_2, and X_3 defined in Theorem 13.1.

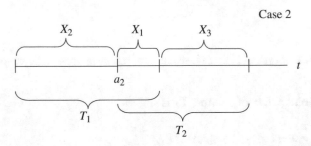

Fig. 13.7. Case 2 for Theorem 13.1 with $X_1 \sim$ exponential(λ_1), $X_2 \sim$ exponential(λ_2), and $X_3 \sim$ exponential(λ_3)

Substituting $\lambda_1 = \mu$, $\lambda_2 = \lambda + \mu$, and $\lambda_3 = \mu$ into the mixture of cases 1 and 2 yields the joint PDF of T_1 and T_2 as

$$f_{T_1,T_2}(t_1,t_2) = \begin{cases} \dfrac{\mu^2 \left(\lambda e^{-\mu t_2} + \mu e^{-\lambda t_1 - \mu t_1 - \mu t_2} \right)}{\lambda + \mu} & 0 < t_1 \leq t_2 \\ \dfrac{\mu^2 \left(\lambda e^{-\lambda t_1 - \mu t_1 + \lambda t_2} + \mu e^{-\lambda t_1 - \mu t_1 - \mu t_2} \right)}{\lambda + \mu} & 0 < t_2 < t_1. \end{cases}$$

(13.3)

Using this joint PDF, the covariance between the sojourn times of customers 1 and 2 is calculated to be

$$\text{Cov}(T_1, T_2) = \frac{\lambda(\lambda + 2\mu)}{(\lambda + \mu)^2 \mu^2}.$$

Substituting $\lambda = 1$ and $\mu = 2$, for example, produces

$$\text{Cov}(T_1, T_2) = \frac{5}{36} \cong 0.1389.$$

We now use the results of Theorem 13.1 in Example 13.4.

Example 13.4. Let T_1 and T_2 be the sojourn times for customers 1 and 2, respectively, in an initially empty and idle $M/M/1$ queue with exponential($\lambda = 1$) times between arrivals and exponential($\mu = 2$) service times. Find the distribution of the sample mean

$$\bar{T} = (T_1 + T_2)/2,$$

as well as $E[\bar{T}]$ and $\text{Var}[\bar{T}]$.

Applying Eq. (13.3) with $\lambda = 1$ and $\mu = 2$, the joint PDF of T_1 and T_2 is

$$f_{T_1,T_2}(t_1,t_2) = \begin{cases} \dfrac{8}{3}e^{-3t_1-2t_2} + \dfrac{4}{3}e^{-2t_2} & 0 < t_1 \leq t_2 \\[2mm] \dfrac{8}{3}e^{-3t_1-2t_2} + \dfrac{4}{3}e^{-3t_1+t_2} & 0 < t_2 < t_1. \end{cases}$$

Define the transformation

$$U = \bar{T} = (T_1 + T_2)/2 \qquad \text{and} \qquad V = (T_1 - T_2)/2$$

with inverse

$$T_1 = U + V \qquad \text{and} \qquad T_2 = U - V.$$

It can be shown that the functions U and V define a one-to-one transformation; thus, using the bivariate transformation technique described in Hogg et al. [52], the joint PDF of U and V is

$$f_{U,V}(u,v) = \begin{cases} f_{T_1,T_2}(u+v, u-v)|J| & -u \leq v < 0 \\ f_{T_1,T_2}(u+v, u-v)|J| & 0 < v < u, \end{cases}$$

where J is the Jacobian of the inverse transformation defined as

$$J = \begin{vmatrix} \dfrac{\partial t_1}{\partial u} & \dfrac{\partial t_1}{\partial v} \\[2mm] \dfrac{\partial t_2}{\partial u} & \dfrac{\partial t_2}{\partial v} \end{vmatrix} = \begin{vmatrix} 1 & 1 \\ 1 & -1 \end{vmatrix} = -2.$$

Substituting $t_1 = u + v$, $t_2 = u - v$, $J = -2$ and integrating out the dummy transformation variable v, the resulting PDF of $U = \bar{T}$ is

$$f_U(u) = 4e^{-4u} + 2e^{-2u} - 6e^{-6u} \qquad u > 0.$$

The mean of U is

$$E[U] = \int_0^\infty u \cdot f_U(u)\,du$$

$$= \int_0^\infty u \cdot \left(4e^{-4u} + 2e^{-2u} - 6e^{-6u}\right)du$$

$$= \frac{7}{12}.$$

Likewise, the variance of U using $\text{Var}[U] = E[U^2] - (E[U])^2$, where

$$
\begin{aligned}
E[U^2] &= \int_0^\infty u^2 \cdot f_U(u) du \\
&= \int_0^\infty u^2 \cdot \left(4e^{-4u} + 2e^{-2u} - 6e^{-6u}\right) du \\
&= \frac{41}{72},
\end{aligned}
$$

results in

$$
\text{Var}[U] = \frac{41}{72} - \left[\frac{7}{12}\right]^2 = \frac{11}{48}.
$$

Using the `Queue(X, Y, n, k, s)` procedure for customers 1 and 2, the mean sojourn times are $E[T_1] = 1/2$ and $E[T_2] = 2/3$, and the corresponding variances are $\text{Var}[T_1] = 1/4$ and $\text{Var}[T_2] = 7/18$. The covariance of sojourn times T_1 and T_2 was calculated to be $\text{Cov}(T_1, T_2) = 5/36$. Therefore, the mean sojourn time for customers 1 and 2 is

$$
E\left[\frac{T_1 + T_2}{2}\right] = \frac{E[T_1] + E[T_2]}{2} = \frac{7}{12},
$$

and the variance is

$$
\text{Var}\left[\frac{T_1 + T_2}{2}\right] = \frac{\text{Var}[T_1] + \text{Var}[T_2] + 2\text{Cov}(T_1, T_2)}{4} = \frac{11}{48},
$$

supporting the distribution of $U = \bar{T}$ given above.

Proceeding in this manner, we now derive similar expressions for the first three customers arriving to an initially empty and idle $M/M/1$ queue. We could use first principles to derive the trivariate PDF $f_{T_1, T_2, T_3}(t_1, t_2, t_3)$; however, since covariance only occurs between two customers, it is easier to calculate each respective paired joint distribution for covariance calculations. When considering $n = 3$ customers, there are five possible arrival and departure orderings. In general, for n customers, the number of ways arrivals and departures can occur is given by the nth Catalan number (see Stanley [111]), which is

$$
C_n = \frac{(2n)!}{n!(n+1)!}.
$$

Figure 13.8 shows the five possible arrangements for $n = 3$ customers along with the sojourn times T_1, T_2, and T_3 for each, with the arrival and completion times for the ith customer denoted by a_i and c_i respectively. The vertical arrows at event times represent service completions (pointing up) or arrivals (pointing down). This competing-event approach parallels the second discrete-event simulation approach described previously.

Using the same conditioning approach as in the proof of Theorem 13.1, the joint PDFs for each of the pairs (T_1, T_2), (T_1, T_3), and (T_2, T_3) in each of the five cases can be determined and then mixed to achieve the three

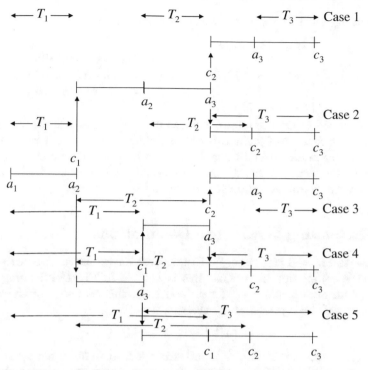

Fig. 13.8. Five cases for $n = 3$ customers' sojourn times in an $M/M/1$ queue

associated joint PDFs. The mixture probabilities are calculated by multiplying the appropriate number of competing arrivals (with probability $\lambda/(\lambda+\mu)$) or service completions (with probability $\mu/(\lambda+\mu)$). For example, in case 1 shown in Fig. 13.8, there are two instances with competing risks, both of which result in a service completion, thus the probability of this case is $\mu^2/(\lambda+\mu)^2$. Using these joint densities, the symmetric variance–covariance matrix for the first $n = 3$ customer sojourn times

$$\Sigma = \begin{bmatrix} \sigma_1^2 & \sigma_{12} & \sigma_{13} \\ \sigma_{12} & \sigma_2^2 & \sigma_{23} \\ \sigma_{13} & \sigma_{23} & \sigma_3^2 \end{bmatrix}$$

is

$$\Sigma = \begin{bmatrix} \dfrac{1}{\mu^2} & \dfrac{\lambda(2\mu+\lambda)}{(\lambda+\mu)^2\mu^2} & \dfrac{\lambda^2(\lambda^2+4\lambda\mu+5\mu^2)}{(\lambda+\mu)^4\mu^2} \\[2ex] \bullet & \dfrac{2\lambda^2+4\lambda\mu+\mu^2}{(\lambda+\mu)^2\mu^2} & \dfrac{\lambda(2\lambda^2+8\lambda^2\mu+11\lambda\mu^2+2\mu^3)}{(\lambda+\mu)^4\mu^2} \\[2ex] \bullet & \bullet & \dfrac{3\lambda^6+18\lambda^5\mu+45\lambda^4\mu^2+54\lambda^3\mu^3+30\lambda^2\mu^4+8\lambda\mu^5+\mu^6}{(\lambda+\mu)^6\mu^2} \end{bmatrix}.$$

Substituting $\lambda = 1$ and $\mu = 2$, for example, results in

$$\Sigma = \begin{bmatrix} 1/4 \; 5/36 & 29/324 \\ \bullet \quad 7/18 & 13/54 \\ \bullet \quad \bullet & 1451/2916 \end{bmatrix} \cong \begin{bmatrix} 0.2500 \; 0.1389 \; 0.0895 \\ \bullet \quad 0.3889 \; 0.2407 \\ \bullet \quad \bullet \quad 0.4976 \end{bmatrix}.$$

The sojourn time variance increases with customer number down the diagonal of the matrix because of the nature of the queueing process, where the sojourn time distribution for each additional customer is dependent on all the previous customers. On the other hand, the off-diagonal covariance entries in each row decrease with customer separation, for example $\sigma_{13} < \sigma_{12}$.

13.6 Extending Covariance Calculations

Consider the $n = 3$ case where all three customers arrive prior to the first customer's completion of service (this is Case 5 in Fig. 13.8). Using a 1 to represent an arrival and a -1 for a departure, this sequence of arrivals and departures can be represented by the vector

$$\begin{bmatrix} 1 \; 1 \; 1 \; -1 \; -1 \; -1 \end{bmatrix}.$$

Figure 13.9 depicts this case as a path from left to right in the figure, where moving up and right indicates an arrival and moving down and right indicates a service completion. Horizontal moves are not permitted. Each of the five possible sequences of arrivals and departures for $n = 3$, shown in Fig. 13.8, can be depicted by a specific path from the bottom left node to the bottom right node. The number in the system is depicted by the height of each node in the figure.

Ruskey and Williams [107] present an elegant algorithm that generates all such paths of arrival and service completions for a given number of customers n. The algorithm is based on a simple iterative successor rule that uses

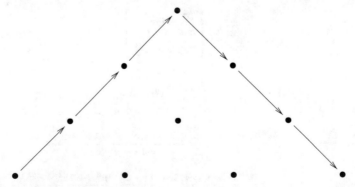

Fig. 13.9. Path for case 5 of $n = 3$ customers arrival and departure pattern in an $M/M/1$ queue

prefix shifts to exhaust the possible arrival and service completion scenarios. In Fig. 13.9 these are the $6!/(3!4!) = 5$ paths that can be drawn from the bottom left node to the bottom right node. The algorithm is "loopless" in that it requires a constant amount of computation in transforming the current case to its successor. Define the case matrix C with dimension $(2n)!/(n!(n+1)!)$ by $2n$ as the exhaustive list of possible arrival and service completion scenarios for n customers. To initiate the matrix the first row of C is

$$C_1 = \begin{bmatrix} 1 & -1 & 1 & 1 & -1 & -1 \end{bmatrix}.$$

The first row is always the ordered string created by an arrival, a service completion, $n-1$ arrivals, and $n-1$ service completions. The iterative successor rule described by Ruskey and Williams [107] is: "Locate the leftmost $[-1, 1]$ and suppose its 1 is in position k. If the $(k+1)$st prefix shift is valid (a possible arrival/service completion sequence), then it is the successor; if it is not valid then the kth prefix shift is the successor." The $(k + 1)$st prefix shift for the sequence

$$B = \{B_1, B_2, \ldots, B_{k-1}, B_k, B_{k+1}, \ldots, B_{2n}\}$$

is defined to be

$$B = \{B_1, B_{k+1}, B_2, \ldots, B_{k-1}, B_k, B_{k+2}, \ldots, B_{2n}\},$$

that is, the $(k+1)$st element of the sequence is shifted into the second position and the relative order of the other elements is left unchanged. The length of the sequence is always $2n$ because the number of arrivals and departures is balanced at n each. An example of an invalid sequence is

$$\begin{bmatrix} 1 & -1 & -1 & 1 & 1 & -1 \end{bmatrix}$$

because the second service completion occurs prior to the second arrival. For $n = 3$, the case matrix C is

$$C = \begin{bmatrix} 1 & -1 & 1 & 1 & -1 & -1 \\ 1 & 1 & -1 & 1 & -1 & -1 \\ 1 & -1 & 1 & -1 & 1 & -1 \\ 1 & 1 & -1 & -1 & 1 & -1 \\ 1 & 1 & 1 & -1 & -1 & -1 \end{bmatrix}.$$

(Note that the order of the five rows does not match the order of the cases in Fig. 13.8.)

Figure 13.10 further categorizes each segment of the path based on whether there exists a competing risk (competing event) in which case the distribution of the time until the next event (either an arrival or a completion) is given by

$$\min\{\text{exponential}(\lambda), \text{ exponential}(\mu)\} \sim \text{exponential}(\lambda + \mu),$$

where the time between arrivals is distributed as exponential(λ) and the service time distribution is exponential(μ).

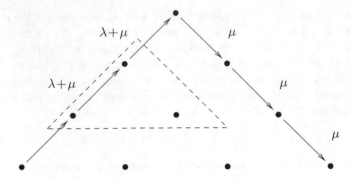

Fig. 13.10. Path segment distributions for case 5 for $n = 3$ customers

Competing risks can only occur along path segments originating inside the dashed triangle shown in Fig. 13.10. These path segments are exponential($\lambda + \mu$) distributed and are correspondingly labeled $\lambda + \mu$. Once all customers have arrived, the only possible events are service completions; thus each path segment along the rightmost edge of Fig. 13.10 is distributed exponential(μ) and labeled μ. If the path of interest reaches a node at the bottom of the figure, the queueing system empties and the next event must be an arrival, which occurs in an exponential(λ) time into the future. While the system is empty, none of the customers' sojourn times are affected, therefore waiting for the next arrival does not impact customer sojourn time distribution. The interior triangle in the path diagram also provides a method to calculate the probability of all possible paths. For path segments originating inside the triangle, a move right and up occurs with probability $\lambda/(\lambda + \mu)$ and a move right and down occurs with probability $\mu/(\lambda + \mu)$. For the particular path shown in Fig. 13.10 there are two segments originating inside the triangle, both of which are right and up, representing two successive arrivals. Thus, the probability of this case is

$$\frac{\lambda}{\lambda + \mu} \cdot \frac{\lambda}{\lambda + \mu} = \frac{\lambda^2}{(\lambda + \mu)^2}.$$

In order to capture the structure of the segment distributions for a given path, represented as a row of the case matrix C, another vector of length $2n-1$ is created, in which each entry corresponds to the sojourn time distribution for a particular segment. There are three possible entries in this vector:

1. exponential($\lambda + \mu$), which is indicated by a 1
2. exponential(μ), which is indicated by a 2
3. no distribution as a result of an emptied system, which is depicted as a 0.

The vector is of length $2n - 1$ since the first customer's arrival time can be ignored as it does not affect sojourn time. For the particular path shown in Fig. 13.10 the corresponding segment distribution vector is

$$\begin{bmatrix} 1 & 1 & 2 & 2 & 2 \end{bmatrix}.$$

Define the new matrix C' with dimension $(2n)!/(n!(n+1)!)$ by $2n-1$ as the segment distribution matrix for each case in C. For $n=3$, the matrix C' is

$$C' = \begin{bmatrix} 1 & 0 & 1 & 2 & 2 \\ 1 & 1 & 1 & 2 & 2 \\ 1 & 0 & 1 & 0 & 2 \\ 1 & 1 & 1 & 0 & 2 \\ 1 & 1 & 2 & 2 & 2 \end{bmatrix}.$$

The two vectors, which are each the fifth row of the corresponding matrices

$$C_5 = \begin{bmatrix} 1 & 1 & 1 & -1 & -1 & -1 \end{bmatrix} \quad \text{and} \quad C_5' = \begin{bmatrix} 1 & 1 & 2 & 2 & 2 \end{bmatrix}$$

contain the information necessary to calculate the contribution of Case 5 to the joint PDF for the sojourn times of any two customers. Denote C_l as the lth row vector of case matrix C, and define the 2×2 matrix R_l with elements

$$R_l = \begin{bmatrix} r_{is} & r_{if} \\ r_{js} & r_{jf} \end{bmatrix}$$

where r_{is} and r_{if} are the start and finish indices for customer i in row l of the case matrix C. Define r_{js} and r_{jf} similarly for customer j. Using C_5 above, for customers $i=1$ and $j=3$,

$$R_5 = \begin{bmatrix} 1 & 4 \\ 3 & 6 \end{bmatrix}.$$

Customer 1's arrival is the first event to occur. Customer 1's departure is the fourth event to occur. Customer 3's arrival is the third event to occur. Customer 3's departure is the sixth event to occur.

The R_l matrix provides two critical pieces of information. First, for the given case l, if $r_{if} < r_{js}$ then the sojourn times for customers i and j do not overlap because customer i departs prior to customer j's arrival. Since in each specific case the sojourn time for each customer is comprised of a uniquely determined sequence of independent time segments, consistent of either service completions distributed exponential(μ) or interarrival times distributed exponential($\lambda + \mu$), and the sequences for customers i and j have no time segments in common, the sojourn times for customers i and j are independent. Therefore, if $r_{if} < r_{js}$, the contribution of case l to the joint PDF is created by simply multiplying the sojourn time PDFs for customers i and j. Second, by computing $r_{if} - r_{is}$ and $r_{jf} - r_{js}$ and then indexing across C_l', the appropriate segment distributions can be combined to form the joint sojourn time PDF for customers i and j.

When $r_{if} > r_{js}$ the joint probability distribution is calculated by conditioning in a similar fashion to the proof of Theorem 13.1. However, it is first

necessary to find the independent and overlapping segments for the customers of interest. For the arrival and service completion scenario described by C_5, Fig. 13.11 shows sojourn times T_1 and T_3 for customers 1 and 3. The independent portion of customer 1's sojourn time consists of the two exponential($\lambda + \mu$) segments. The independent portion of customer 3's sojourn time consists of the two exponential(μ) segments shown on the right side of Fig. 13.11. The dependent (overlap) portion between customers 1 and 3 consists of the single exponential(μ) segment falling within the dashed vertical lines. Using C_5' and R_5, these segments can be determined without reference to Fig. 13.11, as follows: Given $r_{1f} > r_{3s}$, that is customer 3 arrives prior to customer 1 completing service, the independent portions of customer 1's sojourn time distribution are found by (a) calculating $r_{3s} - r_{1s} = 3 - 1 = 2$ and then (b) collecting the elements in C_5' beginning at index $r_{1s} = 1$ and indexing $r_{3s} - r_{1s} - 1 = 1$ additional element of the vector. For $C_5' = \begin{bmatrix} 1\ 1\ 2\ 2\ 2 \end{bmatrix}$, the first two entries, c_{51}' and c_{52}' correspond to the two exponential($\lambda + \mu$) segments. Likewise, customer 3's independent sojourn time segments are found by (a) calculating $r_{3f} - r_{1f} = 6 - 4 = 2$, and then (b) collecting the elements in C_5' beginning at index $r_{1f} = 4$ and indexing $r_{3f} - r_{1f} - 1 = 1$ additional element of the vector. This amounts to the two exponential(μ) segments in elements four and five of C_5'. The dependent portion is identified by starting at the element $r_{3s} = 3$ and indexing $r_{if} - r_{3s} - 1 = 0$ additional elements, the third element of C_5', a single exponential(μ) segment.

In this case, calculating the joint PDF is straightforward because the independent portions for each customer are iid exponential random variables. Defining the independent cumulative distribution function portions for customers 1 and 3 as $X_1 \sim \text{Erlang}(\lambda + \mu, 2)$ and $X_3 \sim \text{Erlang}(\mu, 2)$ respectively,

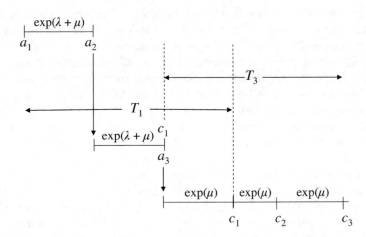

Fig. 13.11. Sojourn time segments for customers 1 and 3 in case 5 for $n = 3$ customers

and the dependent (overlap) random variable as $W \sim$ exponential(μ), the contribution of Case 5 to the joint CDF of $(T_1, T_3) = (X_1 + W, X_3 + W)$, conditioning on the dependent distribution segment W, is

$$
\begin{aligned}
F_{T_1,T_3}(t_1,t_3) &= \Pr(T_1 \leq t_1, T_3 \leq t_3) \\
&= \Pr(X_1 + W \leq t_1, X_3 + W \leq t_3) \\
&= \Pr(X_1 \leq t_1 - W, X_3 \leq t_3 - W) \\
&= \int_0^{\min\{t_1,t_3\}} \Pr(X_1 \leq t_1 - w, X_3 \leq t_3 - w | W = w) f_W(w)\, dw \\
&= \int_0^{\min\{t_1,t_3\}} \Pr(X_1 \leq t_1 - w | W = w) \cdot \\
&\qquad\qquad \Pr(X_3 \leq t_3 - w | W = w) f_W(w)\, dw \\
&= \int_0^{\min\{t_1,t_3\}} F_{X_1}(t_1 - w) F_{X_3}(t_3 - w) \mu e^{-\mu w}\, dw \\
&= \begin{cases} \int_0^{t_1} F_{X_1}(t_1 - w) F_{X_3}(t_3 - w) \mu e^{-\mu w}\, dw & 0 < t_1 < t_3 \\ \int_0^{t_3} F_{X_1}(t_1 - w) F_{X_3}(t_3 - w) \mu e^{-\mu w}\, dw & 0 < t_3 < t_1. \end{cases}
\end{aligned}
$$

Since closed-form versions of $F_{X_1}(t_1 - w)$ and $F_{X_3}(t_3 - w)$ are available, Maple is capable of evaluating this expression, though for large n it can be time consuming.

When the independent distribution segments are not iid exponential random variables, the calculation is more problematic because we can no longer use Eq. (13.1) to easily express $F_{X_1}(t_1 - w)$ and $F_{X_3}(t_3 - w)$. Convolution is required, and though capable, Maple, and subsequently APPL, tend to slow down significantly with increasing n. To overcome this shortfall, consider Theorem 13.2, which appears to be a faster approach than the two suggested in Hagwood [43].

Theorem 13.2. If $S_1 \sim$ Erlang(λ_1, m) and $S_2 \sim$ Erlang(λ_2, n) are independent random variables, then the PDF of $Y = S_1 + S_2$ is

$$
f_Y(y) = \left[\frac{\lambda_1^m \lambda_2^n e^{-\lambda_2 y}}{(m-1)!(n-1)!} \sum_{x=0}^{n-1} \left\{ (-1)^x \binom{n-1}{x} y^{n-1-x} e^{(\lambda_2 - \lambda_1)s} \cdot \right.\right.
$$
$$
\left.\left. \sum_{r=0}^{m-1+x} (-1)^r \frac{(m-1+x)! s^{m-1+x-r}}{(m-1+x-r)!(\lambda_2 - \lambda_1)^{r+1}} \right\} \right]_{s=0}^{y} \qquad y > 0.
$$

Proof. Since S_1 and S_2 are independent, the PDF of $Y = S_1 + S_2$ using convolution and the binomial theorem is

$$f_Y(y) = \int_0^y f_{S_1}(s) f_{S_2}(y-s) \, ds$$

$$= \int_0^y \frac{\lambda_1(\lambda_1 s)^{m-1} e^{-\lambda_1 s}}{(m-1)!} \cdot \frac{\lambda_2(\lambda_2(y-s))^{n-1} e^{-\lambda_2(y-s)}}{(n-1)!} \, ds$$

$$= \frac{\lambda_1^m \lambda_2^n}{(m-1)!(n-1)!} \int_0^y s^{m-1} e^{-\lambda_1 s} (y-s)^{n-1} e^{-\lambda_2(y-s)} \, ds$$

$$= \frac{\lambda_1^m \lambda_2^n e^{-\lambda_2 y}}{(m-1)!(n-1)!} \int_0^y s^{m-1} (y-s)^{n-1} e^{s(\lambda_2-\lambda_1)} \, ds$$

$$= \frac{\lambda_1^m \lambda_2^n e^{-\lambda_2 y}}{(m-1)!(n-1)!} \int_0^y s^{m-1} \left(\sum_{x=0}^{n-1} \binom{n-1}{x} y^{n-1-x} (-s)^x \right) e^{s(\lambda_2-\lambda_1)} \, ds$$

$$= \frac{\lambda_1^m \lambda_2^n e^{-\lambda_2 y}}{(m-1)!(n-1)!} \sum_{x=0}^{n-1} \left\{ (-1)^x \binom{n-1}{x} y^{n-1-x} \int_0^y s^{m-1+x} e^{s(\lambda_2-\lambda_1)} \, ds \right\}$$

$$= \left[\frac{\lambda_1^m \lambda_2^n e^{-\lambda_2 y}}{(m-1)!(n-1)!} \sum_{x=0}^{n-1} \left\{ (-1)^x \binom{n-1}{x} y^{n-1-x} e^{(\lambda_2-\lambda_1)s} \right. \right.$$
$$\left. \left. \sum_{r=0}^{m-1+x} (-1)^r \frac{(m-1+x)! s^{m-1+x-r}}{(m-1+x-r)!(\lambda_2-\lambda_1)^{r+1}} \right\} \right]_{s=0}^y \qquad y > 0,$$

which completes the proof. □

The APPL procedure Cov(a, b, n) applies Theorem 13.2 to calculate the covariance between the sojourn times of customers a and b ($a < b$) in a system of n customers. For computational considerations (i.e., evaluating the fewest cases necessary for a given n), setting the number of customers $n = b$ provides the fastest result. Additionally, calling Cov(a, b, n) where $n > b$ produces a result identical to $n = b$ because customers arriving after customer b do not affect the covariance of previous customers.

Rewriting the integral as a sum via Theorem 13.2 avoids the calls to Convolution(X, Y) in APPL as well as integrating for each case and piece. One can always use this approach, even when the independent part of a particular customer's sojourn time contains many independent distribution segments. The times for these segments can only be exponential($\lambda+\mu$) distributed or exponential(μ) distributed, implying their sum can always be written as the sum of two independent Erlang random variables. This approach speeds computation time considerably. The symmetric variance–covariance matrix for $n = 10$ customers with parameters $\lambda = 1$, $\mu = 2$, and $\rho = 1/2$ is shown in Table 13.4 providing the exact values.

CPU time is a factor in these computations. Each element in the tenth column of the variance–covariance matrix is calculated from a joint PDF which is a mixture of $C_{10} = 20!/(10!11!) = 16{,}796$ component distributions, each corresponding to a unique ordering of arrivals and departures.

Table 13.4. Sojourn time variance–covariance matrix for the first $n = 10$ customers in an $M/M/1$ queue with $\lambda = 1$, $\mu = 2$

1/4	5/36	29/324	181/2916	1181/26244	2647/78732	18191/708588	127111/6377292	2699837/172186884	19319845/1549681956
•	7/18	13/54	239/1458	1543/13122	10303/118098	23485/354294	163493/3188646	3462503/86093442	24719519/774840978
•	•	1451/2916	8531/26244	53995/236196	356291/2125764	805705/6377292	5576849/57395628	39197977/516560652	836647331/13947137604
•	•	•	34514/59049	209794/531441	1357010/4782969	3031606/14348907	20810726/129140163	145390102/1162261467	3088887890/31381059609
•	•	•	•	12525605/19131876	77889229/172186884	170586983/516560652	1156711327/4649045868	8013045911/41841412812	169183999981/1129718145924
•	•	•	•	•	551583889/774840978	1162296371/2324522934	7727099083/20920706406	52871149859/188286357654	1106749378225/5083731656658
•	•	•	•	•	•	10582107143/13947137604	67728246079/125524238436	454382575415/1129718145924	9394007745229/30502389939948
•	•	•	•	•	•	•	225196533287/282429536481	1455144635743/2541865828329	29498588275973/68630377364883
•	•	•	•	•	•	•	•	75890492486993/91507169819844	1482244865480580/2470693585135780
•	•	•	•	•	•	•	•	•	28549065408995300/33354363399333100

Because these values are difficult to compare in fractional form, the same matrix is provided again, with matrix elements rounded to four decimal places:

$$
\begin{bmatrix}
0.2500 & 0.1389 & 0.0895 & 0.0621 & 0.0450 & 0.0336 & 0.0257 & 0.0199 & 0.0157 & 0.0125 \\
\bullet & 0.3889 & 0.2407 & 0.1639 & 0.1176 & 0.0872 & 0.0663 & 0.0513 & 0.0402 & 0.0319 \\
\bullet & \bullet & 0.4976 & 0.3251 & 0.2286 & 0.1676 & 0.1263 & 0.0972 & 0.0759 & 0.0600 \\
\bullet & \bullet & \bullet & 0.5845 & 0.3948 & 0.2837 & 0.2113 & 0.1611 & 0.1251 & 0.0984 \\
\bullet & \bullet & \bullet & \bullet & 0.6547 & 0.4524 & 0.3302 & 0.2488 & 0.1915 & 0.1498 \\
\bullet & \bullet & \bullet & \bullet & \bullet & 0.7119 & 0.5000 & 0.3694 & 0.2808 & 0.2177 \\
\bullet & \bullet & \bullet & \bullet & \bullet & \bullet & 0.7587 & 0.5396 & 0.4022 & 0.3080 \\
\bullet & \bullet & \bullet & \bullet & \bullet & \bullet & \bullet & 0.7974 & 0.5725 & 0.4298 \\
\bullet & \bullet & \bullet & \bullet & \bullet & \bullet & \bullet & \bullet & 0.8293 & 0.5999 \\
\bullet & \bullet & \bullet & \bullet & \bullet & \bullet & \bullet & \bullet & \bullet & 0.8559
\end{bmatrix}
$$

As the traffic intensity increases, so do the values in the variance–covariance matrix. To illustrate, the same matrix is provided for the increased traffic intensity parameters $\lambda = 1$, $\mu = 10/9$, and $\rho = 9/10$. The increasing sojourn-time variance along the diagonal is expected with the increasing traffic intensity. In addition, the rate that covariance between customers decreases as customer separation increases is less pronounced, as indicated in the variance–covariance matrix

$$
\begin{bmatrix}
0.8100 & 0.5856 & 0.4737 & 0.4040 & 0.3553 & 0.3189 & 0.2904 & 0.2673 & 0.2481 & 0.2318 \\
\bullet & 1.3956 & 1.1097 & 0.9393 & 0.8226 & 0.7363 & 0.6692 & 0.6150 & 0.5702 & 0.5323 \\
\bullet & \bullet & 1.9561 & 1.6298 & 1.4167 & 1.2626 & 1.1441 & 1.0494 & 0.9714 & 0.9057 \\
\bullet & \bullet & \bullet & 2.5021 & 2.1458 & 1.8995 & 1.7142 & 1.5679 & 1.4484 & 1.3485 \\
\bullet & \bullet & \bullet & \bullet & 3.0364 & 2.6565 & 2.3831 & 2.1715 & 2.0009 & 1.8593 \\
\bullet & \bullet & \bullet & \bullet & \bullet & 3.5605 & 3.1614 & 2.8652 & 2.6310 & 2.4389 \\
\bullet & \bullet & \bullet & \bullet & \bullet & \bullet & 4.0754 & 3.6600 & 3.3444 & 3.0904 \\
\bullet & \bullet & \bullet & \bullet & \bullet & \bullet & \bullet & 4.5818 & 4.1524 & 3.8199 \\
\bullet & \bullet & \bullet & \bullet & \bullet & \bullet & \bullet & \bullet & 5.0803 & 4.6386 \\
\bullet & \bullet & \bullet & \bullet & \bullet & \bullet & \bullet & \bullet & \bullet & 5.5713
\end{bmatrix}
$$

Using this variance–covariance matrix for traffic intensity $\rho = 9/10$, consider the following example.

Example 13.5. Let T_i, $i = 1, 2, \ldots, 10$, be the sojourn times for the first $n = 10$ customers in an $M/M/1$ queue with arrival rate $\lambda = 1$ and service rate $\mu = 10/9$ that is initially empty and idle. Find the variance of the average sojourn time for the first ten customers.

Define the average sojourn time as

$$
\bar{T} = \frac{1}{10} \sum_{i=1}^{10} T_i.
$$

Since the sojourn times are not independent random variables, the variance of the average sojourn time is

$$
\begin{aligned}
\mathrm{Var}\left[\bar{T}\right] &= \mathrm{Var}\left[\frac{1}{10} \sum_{i=1}^{10} T_i\right] \\
&= \frac{1}{100} \mathrm{Var}\left[\sum_{i=1}^{10} T_i\right] \\
&= \frac{1}{100}\left[\sum_{i=1}^{10} \mathrm{Var}\left[T_i\right] + 2 \sum \sum_{i<j} \mathrm{Cov}\left(T_i, T_j\right)\right].
\end{aligned}
$$

The result is the sum of all elements in the variance–covariance matrix multiplied by the constant $1/100$. The sum of the variance–covariance matrix rounded to four significant digits is 177.6642; therefore the variance of \bar{T} is

$$
\mathrm{Var}\left[\bar{T}\right] \cong 1.7766.
$$

To verify the calculation, a discrete-event simulation was conducted five times, each using one million replications. The resulting 95 % confidence interval for the variance of \bar{T} was $\bar{T} \in (1.773, 1.781)$, which agrees with the analytic result.

Traditional steady-state queueing theory and analysis lacks the insight provided in these transient variance–covariance matrices. For businesses in which the number of customers in a day is so small that true steady state is never achieved, routine queueing measures of performance are not representative of reality. Additionally, consider a system in which the traffic intensity exceeds one. For a such a system, an analyst might be interested in customer covariance. Increasing the traffic intensity so that $\rho > 1$ does not preclude covariance calculations using this method, and therefore allows transient analysis of such systems. A variance–covariance matrix for $\rho = 3/2$ is

$$
\begin{bmatrix}
2.2500 & 1.8900 & 1.7172 & 1.6135 & 1.5438 & 1.4937 & 1.4558 & 1.4263 & 1.4027 & 1.3835 \\
\bullet & 4.1400 & 3.7368 & 3.5018 & 3.3459 & 3.2344 & 3.1507 & 3.0856 & 3.0337 & 2.9913 \\
\bullet & \bullet & 6.0957 & 5.6825 & 5.4166 & 5.2292 & 5.0896 & 4.9817 & 4.8958 & 4.8261 \\
\bullet & \bullet & \bullet & 8.1312 & 7.7208 & 7.4397 & 7.2332 & 7.0747 & 6.9493 & 6.8479 \\
\bullet & \bullet & \bullet & \bullet & 10.2424 & 9.8410 & 9.5538 & 9.3361 & 9.1652 & 9.0276 \\
\bullet & \bullet & \bullet & \bullet & \bullet & 12.4235 & 12.0342 & 11.7463 & 11.5230 & 11.3444 \\
\bullet & \bullet & \bullet & \bullet & \bullet & \bullet & 14.6687 & 14.2931 & 14.0081 & 13.7828 \\
\bullet & \bullet & \bullet & \bullet & \bullet & \bullet & \bullet & 16.9727 & 16.6115 & 16.3319 \\
\bullet & \bullet & \bullet & \bullet & \bullet & \bullet & \bullet & \bullet & 19.3310 & 18.9846 \\
\bullet & \bullet & \bullet & \bullet & \bullet & \bullet & \bullet & \bullet & \bullet & 21.7397
\end{bmatrix}.
$$

Given this traffic intensity, the system is unstable and the expected sojourn times for successive customers increase without bound. Along the main diagonal the customer variance is clearly increasing, and the covariance decreases as the separation occurs between customers. This decrease is monotonic, and though not studied in detail here, it appears that the rate of covariance decrease might be of interest for an unstable traffic intensity.

13.7 Sojourn Time Covariance with k Customers Initially Present

When k customers are present in the $M/M/1$ queue at time zero, the approach used to compute sojourn-time covariance between customers becomes more difficult. When the two customers of interest possess indices larger than k (i.e., T_i where $i > k$), then the approach is similar to that derived in Sect. 13.6. However, there are two other possibilities. The first possibility is that the first customer has an index of k or less, and the second customer has an index larger than k. In this instance, the only difference in deriving the joint CDF is that the lower indexed customer begins his sojourn time at time zero. In the second possibility, both customers have an index of k or below. If these indices are i and j, where $i < j \le k$, the time intervals for sojourn times T_i and T_j begin at zero. It is obvious that $T_i \le T_j$, because the completion time for customer i must occur prior to the completion time for customer j. For each of the possibilities above the covariance derivation that follows will mirror the empty and idle covariance derivation in Sect. 13.6.

To illustrate the calculations, consider an $M/M/1$ queue with $k = 2$ customers initially present at time zero and a single additional customer, $n = 1$. The transition diagram where the first event (not including the k customers initially present at time zero) is an arrival, which is analogous to Figs. 13.9 and 13.10, is given in Fig. 13.12. The total number of customers passing

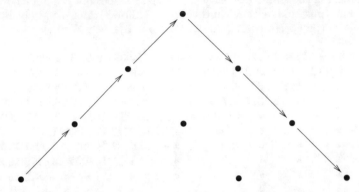

Fig. 13.12. Transition diagram for $n+k = 1+2 = 3$ customers when the first event is an arrival

through the system is $n + k = 3$. Using 1 to denote an arrival and -1 to denote a departure, each arrival/departure ordering instance for $n + k = 3$ customers must contain exactly three -1's (completions of service) and a single 1 (arrival). The algorithm presented by Ruskey and Williams [107] does not facilitate listing all orderings for an unbalanced system, where the number of departures is greater than the number of arrivals (as opposed to an empty and idle queue at time zero). However, we can produce all possible arrival–departure sequences with a simple manipulation of the algorithm, as well as count the number of possible sequences. The number of possible orderings, denoted by $C(n|k)$, follows, where n represents the number of customers passing through the system that arrive after time zero and k is the number of customers present at time zero:

$$C(n|k) = \sum_{j=0}^{\lfloor k/2 \rfloor} (-1)^j \binom{k-j}{j} C_{n+k-j}$$

for $k = 0, 1, 2, \ldots$ and $n = 1, 2, \ldots$, where $\lfloor \cdot \rfloor$ denotes the greatest integer function. The case matrix C is found by applying the Ruskey and Williams [107] algorithm for $n+k$ customers, then deleting the instances where the first k events do not correspond to arrivals. As seen previously, the case matrix for $n + k = 1 + 2 = 3$ customers is

$$C = \begin{bmatrix} 1 & -1 & 1 & 1 & -1 & -1 \\ 1 & 1 & -1 & 1 & -1 & -1 \\ 1 & -1 & 1 & -1 & 1 & -1 \\ 1 & 1 & -1 & -1 & 1 & -1 \\ 1 & 1 & 1 & -1 & -1 & -1 \end{bmatrix}.$$

Rows 2, 4, and 5 correspond to the first $k = 2$ events being arrivals. Rows 1 and 3 must be deleted from the case matrix, since for each row, a completion of service occurs prior to the first two arrivals. Deleting these rows results in the case matrix

$$C = \begin{bmatrix} 1 & 1 & -1 & 1 & -1 & -1 \\ 1 & 1 & -1 & -1 & 1 & -1 \\ 1 & 1 & 1 & -1 & -1 & -1 \end{bmatrix}$$

with the remaining rows representing all possible arrival–departure sequences. We can further simplify the case matrix by deleting the first k columns, resulting in

$$C = \begin{bmatrix} -1 & 1 & -1 & -1 \\ -1 & -1 & 1 & -1 \\ 1 & -1 & -1 & -1 \end{bmatrix}.$$

The rows of the case matrix correspond to the three cases shown in Fig. 13.13.

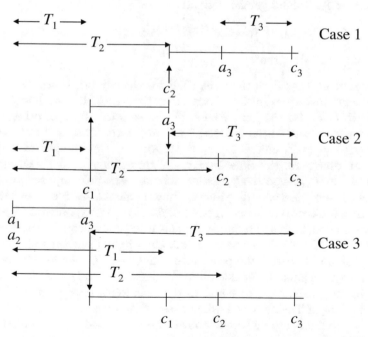

Fig. 13.13. Three cases for $k = 2$ initial customers and a single $n = 1$ additional customer in an $M/M/1$ queue

The algorithm for computing the joint PDF, and subsequently the covariance, of the sojourn times of any two customers does not differ significantly from the algorithm presented in Sect. 13.6. However, for the sojourn times T_1 and T_2 in Fig. 13.13, a new theorem is introduced.

Theorem 13.3. Let $X \sim$ exponential(λ_1) and $Y \sim$ exponential(λ_2) be independent random variables. The joint PDF of $(T_1, T_2) = (X, X + Y)$ is

$$f_{T_1,T_2}(t_1, t_2) = \lambda_1 \lambda_2 e^{-\lambda_2 t_1 - \lambda_1 t_2 + \lambda_1 t_1} \qquad 0 < t_1 < t_2.$$

Proof. The joint CDF of T_1 and T_2 is

$$
\begin{aligned}
F_{T_1,T_2}(t_1, t_2) &= \Pr\left(T_1 \leq t_1, T_2 \leq t_2\right) \\
&= \Pr\left(X \leq t_1, X + Y \leq t_2\right) \\
&= \Pr\left(X \leq t_1, Y \leq t_2 - X\right) \\
&= \int_0^{t_1} \int_0^{t_2-x} f_X(x) \cdot f_Y(y) \, dy \, dx \\
&= \int_0^{t_1} \int_0^{t_2-x} \left(\lambda_1 e^{-\lambda_1 x}\right) \cdot \left(\lambda_2 e^{-\lambda_2 y}\right) dy \, dx \\
&= \frac{\lambda_1 - \lambda_2 + \lambda_2 e^{-\lambda_1 t_2} + \lambda_2 e^{-\lambda_2 t_1} - \lambda_1 e^{-\lambda_2 t_1} - \lambda_2 e^{-\lambda_2 t_1 - \lambda_1 t_2 + \lambda_1 t_1}}{\lambda_1 - \lambda_2},
\end{aligned}
$$

for $0 < t_1 < t_2$. Taking partial derivatives, $f_{T_1,T_2}(t_1, t_2)$ is

$$f_{T_1,T_2}(t_1, t_2) = \lambda_1 \lambda_2 e^{-\lambda_2 t_1 - \lambda_1 t_2 + \lambda_1 t_1} \qquad 0 < t_1 < t_2,$$

which completes the proof. □

Theorem 13.3 provides the joint PDF for the sojourn times T_1 and T_2 of the first two customers initially present at time zero. It may be more complicated to calculate the joint PDFs for the sojourn times of other pairs of customers who were initially present at time zero. This is due to the fact that if $(i, j) \neq (1, 2)$ and $i < j \leq k$, where k is the number of customers present at time zero, the time intervals of duration X and Y during which customers i and j, respectively, are served may each be composed of multiple independent exponentially distributed time segments. Each of these multiple segments is limited to only one of two possibilities, an exponential($\lambda + \mu$) segment or an exponential(μ) segment. In this more complicated situation we let $(T_i, T_j) = (X, X + Y)$ as in Theorem 13.3 and apply Theorem 13.2 to find the PDF's of X and Y (using the procedure conv(m, n)), then let Maple handle the sojourn time joint PDF calculation. When the second customer of interest has an index greater than or equal to k, the sojourn time joint PDF follows an application of Theorem 13.1 when cases exist with dependence.

Using the final case matrix C above, the associated segment distribution matrix C' is

$$C' = \begin{bmatrix} 1 & 1 & 0 & 2 \\ 1 & 1 & 2 & 2 \\ 1 & 2 & 2 & 2 \end{bmatrix},$$

where the possible elements are the same as defined in Sect. 13.6. The probability vector associated with the case matrix C is

$$\begin{bmatrix} \dfrac{4}{9} & \dfrac{2}{9} & \dfrac{1}{3} \end{bmatrix}.$$

Using the case matrix C and the segment distribution matrix C', the joint PDFs for each case are created by selecting the appropriate segments for a given pair of customers, where the segments are identified by the R_l matrix discussed in Sect. 13.6. Once the joint PDF's are created for each case, they are mixed with the probability vector to determine the sojourn time joint PDF for covariance calculations. These calculations are coded in Maple as the procedure kCov(X, Y, a, b, n, k). The first two arguments X and Y are the distribution of time between arrivals, exponential(λ), and the service time distribution, exponential(μ), respectively. They are entered in the APPL list-of-lists format. The arguments a and b are the customers of interest for the covariance calculation, where $a < b$. The argument n is the number of customers processing through the system not including those present at time zero, which is indicated by the last argument, k. Therefore, the total number of customers processing through the system is $n + k$, and a covariance calculation between any two of these customers can be achieved with the appropriate function call. For example, the function call kCov(ExponentialRV(1), ExponentialRV(2), 1, 2, 1, 3) calculates the covariance between customers 1 and 2 in an $M/M/1$ queue with an arrival rate $\lambda = 1$, with service time rate $\mu = 2$, with three customers present at time zero, and a single additional customer processes through the system. The variance–covariance matrix for an $M/M/1$ queue with arrival rate $\lambda = 1$, and service rate $\mu = 2$, where $k = 4$ customers are present at time zero and an additional $n = 6$ customers process through the system is

$$\begin{bmatrix}
\frac{1}{4} & \frac{1}{4} & \frac{1}{4} & \frac{1}{4} & \frac{211}{972} & \frac{1579}{8748} & \frac{11651}{78732} & \frac{28553}{236196} & \frac{630131}{6377292} & \frac{4646155}{57395628} \\
\bullet & \frac{1}{2} & \frac{1}{2} & \frac{1}{2} & \frac{211}{486} & \frac{1579}{4374} & \frac{11651}{39366} & \frac{28553}{118098} & \frac{630131}{3188646} & \frac{4646155}{28697814} \\
\bullet & \bullet & \frac{3}{4} & \frac{3}{4} & \frac{211}{324} & \frac{1579}{2916} & \frac{11651}{26244} & \frac{28553}{78732} & \frac{630131}{2125764} & \frac{4646155}{19131876} \\
\bullet & \bullet & \bullet & 1 & \frac{211}{243} & \frac{1579}{2187} & \frac{11651}{19683} & \frac{28553}{59049} & \frac{630131}{1594323} & \frac{4646155}{14348907} \\
\bullet & \bullet & \bullet & \bullet & \frac{37289}{26244} & \frac{271153}{236196} & \frac{1966777}{2125764} & \frac{1588153}{2125764} & \frac{34755203}{57395628} & \frac{763875281}{1549681956} \\
\bullet & \bullet & \bullet & \bullet & \bullet & \frac{1629655}{1062882} & \frac{11663887}{9565938} & \frac{9353743}{9565938} & \frac{203800469}{258280326} & \frac{4465399991}{6973568802} \\
\bullet & \bullet & \bullet & \bullet & \bullet & \bullet & \frac{263490131}{172186884} & \frac{208262483}{172186884} & \frac{4506205633}{4649045868} & \frac{98323535707}{125524238436} \\
\bullet & \bullet & \bullet & \bullet & \bullet & \bullet & \bullet & \frac{63939878}{43046721} & \frac{1359189250}{1162261467} & \frac{29402061622}{31381059609} \\
\bullet & \bullet & \bullet & \bullet & \bullet & \bullet & \bullet & \bullet & \frac{179260456277}{125524238436} & \frac{379721786263}{3389154437772} \\
\bullet & \bullet & \bullet & \bullet & \bullet & \bullet & \bullet & \bullet & \bullet & \frac{62708955663745}{45753584909922}
\end{bmatrix}.$$

Unlike the previous variance–covariance matrices, some row elements, in particular those elements associated with customers that are initially present, do not decrease monotonically. To explain these entries, consider Theorem 13.4.

Theorem 13.4. If X_1, X_2, \ldots, X_n are iid exponential(μ) random variables and

$$T_s = \sum_{r=1}^{s} X_r \qquad s = 1, 2, \ldots, n,$$

then $\mathrm{Var}[T_i] = \mathrm{Cov}(T_i, T_l), 0 < i < l \leq n$.

Proof. Note that $E[T_k] = k/\mu$ for $k = 1, 2, \ldots, n$ and that T_i and X_r are independent for $1 \leq i < r \leq n$.

$$
\begin{aligned}
\mathrm{Cov}(T_i, T_l) &= E\left[\left(T_i - \frac{i}{\mu}\right)\left(T_l - \frac{l}{\mu}\right)\right] \\
&= E\left[\left(T_i - \frac{i}{\mu}\right)\left\{\left(T_i - \frac{i}{\mu}\right) + \sum_{r=i+1}^{l}\left(X_r - \frac{1}{\mu}\right)\right\}\right] \\
&= E\left[\left(T_i - \frac{i}{\mu}\right)^2\right] + E\left[\sum_{r=i+1}^{l}\left\{\left(T_i - \frac{i}{\mu}\right)\left(X_r - \frac{1}{\mu}\right)\right\}\right] \\
&= \mathrm{Var}[T_i] + \sum_{r=i+1}^{l} E\left[T_i - \frac{i}{\mu}\right] E\left[X_r - \frac{1}{\mu}\right] \\
&= \mathrm{Var}[T_i],
\end{aligned}
$$

which completes the proof. □

We can apply Theorem 13.4 to those customer pairs where both indices $i, j \leq k$. Therefore, the entries in the variance–covariance matrix for customer pairs $(1, 2)$, $(1, 3)$, and $(1, 4)$ are

$$\mathrm{Var}[T_1] = \mathrm{Cov}(T_1, T_2) = \mathrm{Cov}(T_1, T_3) = \mathrm{Cov}(T_1, T_4) = \frac{1}{4}.$$

Likewise, for the customer pairs $(2, 3)$ and $(2, 4)$

$$\mathrm{Var}[T_2] = \mathrm{Cov}(T_2, T_3) = \mathrm{Cov}(T_2, T_4) = \frac{1}{2}.$$

Furthermore, it can be shown in general that

$$\mathrm{Var}[T_i] = \mathrm{Cov}(T_i, T_j) = \frac{i}{\mu^2}$$

for $i < j \leq k$, where k customers are present at time zero. For example, consider a single-server box office with exponential(μ) service times which will be offering tickets to a popular concert the next day. If 1000 customers, each buying one ticket, camp out the night before determined to get the best seats for the concert, these $k = 1000$ customers are present at time zero and therefore we can pre-determine the covariance between any two of the

customers. Additionally, Theorem 13.4 presents the counter-intuitive result that $\text{Cov}(T_1, T_2) = \text{Cov}(T_1, T_{1000})$. The correlation decreases with increasing lag, however, as expected due to the diminishing effect of the intermediate customer sojourn times reflected in the denominator of the defining formula of the correlation.

Previous transient analysis results for the $M/M/1$ and $M/M/s$ queues have been combined with the functionality of the Maple computational engine (and subsequently APPL) to develop both symbolic and numeric exact sojourn time PDFs that can be manipulated to compute and study various measures of performance. A complete variance–covariance matrix for the first $n = 10$ customers and varying traffic intensity is calculated, illustrating this approach's ability to determine the joint PDF between two customer sojourn times. The results offer a framework for describing how the well-known $M/M/s$ queue steady-state results occur as the queue progresses toward steady-state. When possible, results are checked against corresponding discrete-event simulation and/or previous literature. The first principles derivation suggests a viable alternative for future research would be to apply the approaches provided in this work to a $G/G/1$ queue. Computational considerations take priority as n increases. Making use of other computational formulae (such as Hagwood [43]) may offer significant time savings and is another interesting avenue for future work.

14

Bayesian Applications

This chapter considers Bayesian applications of APPL. Section 14.1 introduces Bayesian statistics and motivates the use of a computer algebra system to derive posterior distributions. Section 14.2 develops algorithms in the case of a single unknown parameter. Section 14.3 develops algorithms in the case of multiple unknown parameters.

14.1 Introduction

Interest in Bayesian statistics has increased rapidly over the past few decades. In response to the increased interest, a variety of software packages have been developed in order to facilitate applied statistical analysis. These software packages, however, use numerical techniques implemented in languages such as R, Stan, and WinBUGS to produce discrete approximations of posterior distributions. Currently, no program automates the derivation of exact closed-form posterior distributions. These derivations must either be accomplished by hand, which is often intractable, or approximated through simulation. The computer implementations of Bayesian statistics have made it possible for practitioners of statistics to apply Bayesian methods in more complex settings. The growing popularity of Bayesian statistics has led to a proliferation of software packages capable of performing applied statistical analysis within the Bayesian framework.

In this chapter, we present software capable of automating the derivation of exact posterior distributions. Specifically, we extend APPL to include a number of procedures that integrate Bayesian methods into APPL. The Bayesian procedures developed here allow users to conduct inference on single- and multi-parameter distributions, derive posterior predictive distributions and derive Jeffrey's priors. By automating the application of Bayes' rule, APPL both serves as a time saving device and a tool for theoretical exploration.

© Springer International Publishing AG 2017 277
J.H. Drew et al., *Computational Probability*, International Series
in Operations Research & Management Science 246,
DOI 10.1007/978-3-319-43323-3_14

Bayesian computer applications have largely remained confined to numerical tools written in languages, such as R, Stan, and WinBUGS. While ideal for performing one-time computations within the context of applied statistical analysis, these software packages lack the symbolic manipulation capabilities present in computer algebra systems, such as Maple and Mathematica. Due to underutilization of computer algebra systems, a gap exists within the field of computational Bayesian statistics. Specifically, no software package currently has the ability to automate Bayesian inference on random variables, with the goal of producing the entire probability density function as its output, especially in the case of multi-parameter distributions.

This chapter presents a remedy for this problem in the form of software that produces posterior PDFs in closed-form. In particular, we outline the software suite APPLBayes, a Bayesian extension to APPL. Working in conjunction with existing APPL procedures, APPLBayes automates the application of Bayes' Theorem to random variables. At a practical level, these automated manipulations serve two primary purposes. First, they save time by eliminating the need to undergo mathematically tedious derivations. This saves time and allows for checking hand-derived results against the results produced by APPLBayes. Second, APPLBayes provides a tool for experimentation to solve time-consuming problems quickly. Moreover, it eliminates the need to rely on conjugate prior relationships. Even within the context of an introductory statistics course, APPLBayes gives students the ability to experiment with ad hoc prior distributions, thus allowing students to develop a deeper appreciation of theoretical concepts.

The procedures in APPLBayes fall broadly into two categories: Bayesian inference on single and multi-parameter models. This chapter will treat each of these categories in turn, and then present some avenues for further theoretical exploration that APPLBayes has already opened.

The literature contains only a few attempts to apply computational methods in order to derive closed-form posterior PDFs. Cook and Broehmling [18] present several Bayesian applications using Mathematica. Their code, however, is not reusable. They make no attempt to automate the ad hoc Bayesian procedures that they present in their article. APPLBayes, on the other hand, consists of automated procedures that can be readily applied to complex models without abstruse coding. Jaakkola and Jordan [54], on the other hand, present reusable code capable of producing closed-form posterior PDFs for logistic regression. Their algorithm, however, depends on producing normal approximations to the likelihood function. As such, they remain limited to the use of normal prior distributions. APPLBayes allows for greater flexibility than Jaakkola and Jordan's approximations because it produces exact closed-form PDFs without remaining constrained to conjugate prior relationships. Given these improvements over current systems, APPLBayes represents an advance in the field of computational Bayesian statistics.

14.2 Bayesian Inference on Single-Parameter Distributions

Bayesian inference on single-parameter distributions represents the simplest class of problems within Bayesian statistics. Due to their relative simplicity, introductory Bayesian courses typically use problems of inference on one variable as a means to motivate the theoretical basis for Bayesian statistics. A typical problem that would appear in a Bayesian statistics textbook would generally take the following form: given a likelihood function, $f(x|\theta)$, a prior distribution of a single unknown parameter θ, $g(\theta)$, and a data set, derive the posterior distribution $g(\theta|\mathbf{x})$. APPLBayes has the capability of automating the solution to these simple problems. Consider the following example.

Example 14.1. Suppose we conduct a series of experiments, each consisting of 12 Bernoulli trials (where 1 denotes a success and 0 denotes a failure). Further suppose that the likelihood function takes the form of a binomial$(12, \theta)$ random variable, where θ is an unknown parameter satisfying $0 < \theta < 1$. Assuming a beta$(\frac{1}{2}, \frac{1}{2})$ prior distribution on the parameter θ, and an observed data vector $\mathbf{x} = [5, 6, 3, 2, 5]$, derive the posterior distribution of the parameter θ.

APPLBayes can accomplish this derivation using the following commands.

```
> X := BinomialRV(12, theta);
> Y := BetaRV(1 / 2, 1 / 2);
> data := [5, 6, 3, 2, 5];
> P := Posterior(X, Y, data, theta);
```

This APPLBayes code returns a beta$(39, 21)$ random variable as the distribution of the random variable P. Notice that the posterior distribution in this case is in the same class of distributions as the prior distribution. Bayesians call this relationship between the prior and posterior distribution a *conjugate prior* relationship. Problems that use conjugate prior relationships abound in introductory textbooks because of their mathematical simplicity. To understand why, we employ Bayes' Theorem, which is typically written as

$$g(\theta|x) = \frac{f(x|\theta) \cdot g(\theta)}{\int_{-\infty}^{\infty} f(x|\theta) \cdot g(\theta) \, d\theta}.$$

Notice that the numerator requires just multiplication. Evaluating the integral in the denominator, however, can become rather complicated. Using conjugate prior relationships eases the burden of deriving posterior distributions because the posterior distribution falls in the same class of distributions as the prior. As such, the normalizing constant for the posterior distribution can be easily found through pattern

matching. In this example, for instance, we recognize the normalizing constant of a beta$(39, 21)$ random variable as the number $\frac{\Gamma(39+21)}{\Gamma(39)\Gamma(21)}$. Thus, introductory Bayesian statistics students can derive a posterior distribution without ever evaluating a complex integral.

Not everyone, however, will choose the beta distribution as a prior for the binomial distribution. Any probability distribution with support $(0, 1)$, in fact, could classify our prior beliefs about the parameter. This disillusionment with an arbitrary choice of prior distributions may even lead to preferring classical methods, without ever giving Bayesian methods a fair chance. APPLBayes helps by eliminating the need to rely on conjugate priors. APPLBayes, in fact, makes deriving posterior distributions with ad hoc priors as simple as deriving posteriors with conjugate priors. Consider Example 14.1. Instead of a beta$(\frac{1}{2}, \frac{1}{2})$ prior, suppose that we place a triangular$(0, \frac{1}{2}, 1)$ prior on the parameter θ. Such a problem would never appear in an introductory Bayesian text. Not only does the posterior not fall into a known conjugate prior relationship, but obtaining the normalizing constant also requires integration over a piecewise function. Nevertheless, APPLBayes can solve this problem with the same amount of effort as in the conjugate prior case. The following APPLBayes commands can be used to derive the posterior distribution.

```
> X := BinomialRV(12, theta);
> Y := TriangularRV(0, 1 / 2, 1);
> data := [5, 6, 3, 2, 5];
> P := Posterior(X, Y, data, theta);
```

The resulting posterior distribution is a piece-wise distribution, which appears as

$$
g(\theta|\mathbf{x}) = \begin{cases} -\frac{543985284099935107948831513667174400}{395821499369962291} \theta^{22} (\theta - 1)^{39} & 0 \leq \theta \leq \frac{1}{2} \\ \frac{543985284099935107948831513667174400}{395821499369962291} \theta^{21} (\theta - 1)^{40} & \frac{1}{2} \leq \theta \leq 1. \end{cases}
$$

As illustrated, the APPLBayes software allows experimentation with a wider range of possibilities within Bayesian statistics than is currently possible.

Along with the derivation of posterior distributions, additional functionalities provided by APPLBayes allows solution of follow-on problems related to the posterior distribution. Consider our modification to Example 14.1 presented above. Given a data set, a binomial likelihood function, and a triangular prior, APPLBayes computed a posterior distribution describing our updated belief about the parameter θ. Another typical problem in a Bayesian statistics textbook may ask a student to conduct statistical inference on the parameter, given this posterior distribution. For instance, suppose we wish to test the null hypothesis $H_0 : \theta = 1/2$ versus the alternative hypothesis $H_1 : \theta > 1/2$. The APPL command SF(X, x) computes the survivor function of a random variable, and allows us to calculate the probability that θ is greater than $1/2$ based on its posterior distribution. APPL can likewise compute the left-tail probability using an analogous command, CDF(X,x).

The *additional* APPLBayes statement that calculates the floating-point representation of the probability that θ exceeds $1/2$ is

```
> prob := evalf(SF(P, 1 / 2));
```

The result of this code is $\Pr(\theta > \frac{1}{2}|\mathbf{x}) = 0.013775$, which allows us to reject or fail to reject the null hypothesis, depending on the confidence level α.

Similarly, a Bayesian credible set is calculated with the APPLBayes `CredibleSet(X, alpha)` procedure, which constructs a Bayesian credible set with confidence level α. APPLBayes can construct the Bayesian credible set with the additional statement

```
> CS := CredibleSet(P, 0.05);
```

The APPLBayes output gives the 95% Bayesian credible set $[0.2517, 0.3149]$. These examples show how APPLBayes can be used to conduct inference on a parameter using an ad hoc prior, using only a few simple lines of code. In so doing, APPLBayes allows solution of a much broader range of problems than was previously possible.

Another typical problem presented to introductory students in Bayesian statistics involves the derivation of posterior predictive distributions, which shows the distribution of the $(n+1)$th observation, given a vector of n observations. Again, the problems within grasp of introductory students remain confined to distributions that demonstrate a conjugate prior relationship. In order to see why, consider the following example.

Example 14.2. Suppose we are observing a process that is described by a Poisson(θ) likelihood, a gamma($2, 5$) prior distribution on the parameter θ, and a single observation: $x = 5$. Derive the posterior predictive distribution of y, the next observation.

APPLBayes can compute the posterior predictive distribution with the following commands.

```
> X := PoissonRV(theta);
> Y := GammaRV(2, 5);
> data := [5];
> PP := PosteriorPredictive(X, Y, data, theta);
```

The resulting APPLBayes output shows that the posterior predictive distribution is

$$f(y|x) = \frac{729}{4697620490} 4^{-y}(y+1)(y+2)\ldots(y+9) \qquad y > 0.$$

In the case of posterior predictive distributions, it is less clear that this solution relies on the conjugate prior relationship. In order to clarify the nature of this dependence, it is useful to derive the posterior predictive distribution. In this case, the posterior predictive distribution

can be derived as follows from the definition of the posterior predictive, which states that $f(y|x) = \int_{-\infty}^{\infty} g(\theta|x) \cdot f(y|\theta) \, d\theta$. The likelihood function is

$$f(x = 5|\theta) = \frac{\theta^5 e^{-\theta}}{5!} \qquad \theta > 0,$$

and the prior distribution is

$$g(\theta) = \frac{2^5}{\Gamma(5)} \theta^4 e^{-2\theta} = \frac{4}{3} \theta^4 e^{-2\theta} \qquad \theta > 0.$$

The posterior distribution is derived as

$$f(x = 5|\theta)g(\theta) = \left(\frac{\theta^5 e^{-\theta}}{5!} \cdot \frac{4}{3} \theta^4 e^{-2\theta}\right) \propto e^{-3\theta} \theta^9$$

$$g(\theta|x = 5) = \frac{f(x = 5|\theta)g(\theta)}{\int_0^\infty f(x = 5|\theta)g(\theta)d\theta} = \frac{3^{10}}{\Gamma(10)} e^{-3\theta} \theta^9 \qquad \theta > 0.$$

The posterior predictive distribution is

$$f(y|x = 5) = \int_0^\infty g(\theta|x = 5) \cdot f(y|\theta)d\theta$$

$$= \int_0^\infty \frac{3^{10}}{\Gamma(10)} e^{-3\theta} \theta^9 \cdot \frac{\theta^y e^{-\theta}}{y!} \, d\theta$$

$$= \frac{3^{10}}{\Gamma(10)y!} \int_0^\infty e^{-3\theta} \theta^9 \theta^y e^{-\theta} \, d\theta$$

$$= \frac{3^{10}}{\Gamma(10)y!} \int_0^\infty \theta^{9+y} e^{-4\theta} \, d\theta$$

$$= \frac{3^{10}}{\Gamma(10)y!} \cdot \frac{\Gamma(10 + y)}{4^{10+y}} \int_0^\infty \frac{4^{10+y}}{\Gamma(10 + y)} e^{-4\theta} \theta^{9+y} \, d\theta$$

$$= \frac{3^{10}}{\Gamma(10)y!} \cdot \frac{\Gamma(10 + y)}{4^{10+y}} \qquad y = 0, 1, 2, \ldots .$$

Note that, although the posterior predictive distribution derived above takes a different form than the posterior predictive derived by APPLBayes, they are in fact the same distribution. A simple check in Maple shows that $\frac{3^{10}}{\Gamma(10) \cdot 4^{10}} = \frac{729}{4697620490}$, the constant produced by APPLBayes, while the identity $\Gamma(10+y) = (y+9)(y+8)\ldots 1$ proceeds directly from the definition of the gamma function. Notice that, unlike the posterior distribution, the resulting posterior predictive distribution does not fall in the same class of distributions as the prior. However, the conjugate prior relationship also made the derivation of the posterior predictive easier by eliminating the integral $\int_0^\infty e^{-4\theta} \theta^{9+y}$. Since this takes the form of the gamma distribution, we can avoid evaluating the integral by multiplying the inside of the integral by $\frac{4^{10+y}}{\Gamma(10+y)}$, the constant

that normalizes the gamma distribution, and the outside of the integral by its inverse. By doing this, the integral simply evaluates to one and the function remains unchanged, because multiplying the function by $\frac{4^{10+y}}{\Gamma(10+y)} \cdot \frac{\Gamma(10+y)}{4^{10+y}}$ is equivalent to multiplying the function by one. As with posterior distributions, introductory Bayesian statistics textbooks remain confined to addressing the derivation of posterior predictive distributions in which conjugate prior relationships holds. Again, with APPLBayes we can eliminate the need to use conjugate prior relationships with no additional effort. Considering Example 14.2, assume that, instead of a gamma prior, we instead wish to place a $N\left(1/2, (1/2)^2\right)$ prior distribution on the parameter θ. APPLBayes can accomplish this derivation using the following code:

```
> X := PoissonRV(theta);
> N := NormalRV(1 / 2, 1 / 2);
> data := [5];
> PP := PosteriorPredictive(X, N, data, theta);
```

This APPL code calculates the posterior predictive distribution as

$$f(y|x = 5) = \frac{-16\left(e^{-1/8}(y+4)(y+2)\sqrt{2}((-1)^y - 1)2^{-3y/2}\right)}{281\Gamma(y/2 + 1/2)} \qquad y > 0.$$

Deriving this distribution by hand would be very tedious due to the lack of a conjugate prior relationship. Nevertheless APPLBayes has the ability to produce a closed-form posterior predictive distribution, even in this relatively complex case, with no more effort than it requires for the posterior predictive distribution in the simple case using conjugate priors. This presents an opportunity for fruitful theoretical exploration. Specifically, posterior predictive distributions using ad hoc distributions can be compared to their conjugate prior counterparts in order to determine which makes better predictions regarding the next observation in the data set. If the ad hoc prior performs better, then it makes little sense to remain constrained by the use of conjugate prior relationships.

In the case of single-parameter distributions, APPLBayes also has a function that can automatically derive Jeffrey's priors, which are uninformed priors that are invariate under reparameterizations of the distribution. For distributions with one parameter, Jeffrey's priors can be found using

$$g(\theta) = \sqrt{E\left[\frac{d \ln(f(x|\theta))}{d\theta}\right]}.$$

This function is useful because it allows users to check posterior distributions produced using informed priors against a result using a non-informed prior. APPLBayes has the ability to produce Jeffrey's priors for both likelihood functions with well-known Jeffrey's priors, as well as ad hoc likelihood functions. An an example, consider a Poisson likelihood function. The Jeffrey's prior

for this likelihood function can be found by using the following APPLBayes commands.

```
> X := PoissonRV(lambda);
> J := Jeffreys(X, 0, infinity, lambda);
```

The output of these APPLBayes commands shows that the Jeffrey's prior for the Poisson distribution is $g(\lambda) = 1/\sqrt{\lambda}$ for $\lambda > 0$, consistent with the well-known analytic result.

Using these Bayesian procedures, APPLBayes provides greater flexibility in using ad hoc prior distributions for inference on a single parameter. This has the potential not only to aid students in the exploration of Bayesian concepts, but also to open the door to theoretical experimentation.

Theoretical Explorations Using APPLBayes. So far, it may appear as though APPLBayes is only useful in solving simple, introductory level problems in the single-parameter case. However, the computer algebra capabilities inherent in APPLBayes have already facilitated important theoretical exploration.

APPLBayes, for instance, can be used to compare the effect of using different prior distributions. Using previous software, such comparisons would have required an extraordinary amount of coding. The reusable APPLBayes code makes such comparisons simple to implement. For example, suppose we have observed a data vector $\mathbf{x} = [2, 3, 2, 5, 6]$. We posit that this data has been drawn from a $N\left(\mu, 2^2\right)$ distribution, and would like to estimate the unknown parameter μ. Under the typical Bayesian paradigm, the choice of prior distributions would be limited to either conjugate or uninformed priors. Here we will evaluate the effect of using two ad hoc, but nevertheless reasonable prior distributions, the Laplace and triangular distributions. Of these two distributions, the triangular is particularly interesting because it provides a compromise between an uninformed and a well informed prior. In spite of its utility, the use of the triangular distribution as a prior is notably absent from the literature.

We can set up our experiment using the following APPLBayes commands for each prior distribution.

```
> X := NormalRV(mu, 2);
> data := [2, 3, 2, 5, 6];
> Y1 := NormalRV(2, 2);
> B1 := BayesUpdate(X, Y1, data, mu);
```

These commands produce the following posterior distributions as output. For the normal prior distribution, the posterior distribution is

$$g(\mu|\mathbf{x}) = \frac{\sqrt{3}\, e^{-3\mu^2/4 + 5\mu - 47/5}}{2\sqrt{\pi}\, e^{-16/15}} \qquad -\infty < \mu < \infty.$$

For the uniform prior distribution, the posterior distribution is

$$g(\mu|\mathbf{x}) = \frac{\sqrt{10}\,e^{-5\mu^2/8+9\mu/2-9}}{\sqrt{\pi}e^{-9/10}\left(\mathrm{erf}\left(17\sqrt{10}/5\right)+\mathrm{erf}\left(8\sqrt{10}/5\right)\right)} \quad -10 < \mu < 10.$$

For the Laplace prior distribution, the posterior distribution is

$$g(\mu|\mathbf{x}) = \frac{e^{-\frac{1}{2}|\mu-2|-5\mu^2/8+9\mu/2-163\sqrt{10}/16}}{\sqrt{\pi}\left(e^{-\frac{19}{16}}+e^{-\frac{223}{80}}-\mathrm{erf}\left(\sqrt{10}/2\right)e^{-\frac{19}{16}}+\mathrm{erf}\left(3\sqrt{10}/10\right)e^{-\frac{223}{80}}\right)}$$

on the support $-\infty < \mu < \infty$. For the triangular prior distribution, the posterior distribution is

$$g(\mu|\mathbf{x}) = \begin{cases} 14.83\,(0.0164\mu + 0.164)\,e^{-0.625\mu^2+4.5\mu-9.75} & -10 \le \mu \le 2 \\ 14.83\,(2.456 - 0.02456\mu)\,e^{-0.625\mu^2+4.5\mu-9.75} & 2 \le \mu \le 10. \end{cases}$$

Note that the posterior distribution for the triangular prior is displayed using floating point rather than exact numbers. The PDF using exact values covers several pages. With these functions in hand, we can plot the posterior distributions in order to visualize how the choice of prior distribution affects the posterior. A plot of the various prior distributions appears in Fig. 14.1. A plot of the corresponding posterior distributions appears in Fig. 14.2. Using AP-PLBayes, it is not only possible to use ad hoc priors such as the Laplace and

Fig. 14.1. Prior distributions for four parametric distributions

Fig. 14.2. Posterior distributions for four parametric prior distributions

triangular distributions. It is also possible to show visually how the posterior distribution changes with the choice of priors. Thus, APPLBayes serves as an excellent tool for investigating the impact of using various prior distributions.

Along with comparisons, APPLBayes also simplifies theoretical explorations. For example, we can use APPLBayes to confirm that the posterior distribution of the μ parameter in a normal distribution with a uniform prior follows a truncated normal distribution. Using the following APPL commands, we can derive the PDF of a $N\left(18/5, (2\sqrt{5}/5)^2\right)$ distribution truncated between -10 and 10:

```
> N := NormalRV(18 / 5, (2 / 5) * sqrt(5));
> N2 := Truncate(N, -10, 10);
```

The output of these APPLBayes statements shows that this truncated normal has PDF $c \cdot \exp(-(5x - 18)^2/40)$ for $-10 < x < 10$, where c is a normalizing constant. Completing the square shows that $\exp(-(5x - 18)^2/40) = \exp(-5x^2/8 + 9x/2 - 9)$, or the same distribution that we found as the posterior for μ using the uniform prior. As such, we can use APPLBayes to show that the posterior distribution for μ with a uniform prior follows a truncated normal distribution.

Leaving the parameters of the Laplace distribution unspecified, we can use APPLBayes to derive a closed-form solution to the posterior distribution for μ with a Laplace prior. This derivation, which otherwise is quite complicated, can be accomplished using three lines of code in APPLBayes.

```
> X := NormalRV(mu, sigma);
> Y := LaPlaceRV(a, b);
> B := BayesUpdate(X, Y, [x], mu);
```

The APPLBayes output shows that if we have a $N(\mu, \sigma^2)$ likelihood function, a Laplace(a, b) prior distribution, and a data observation x, then the posterior distribution takes the form

$$g(\mu|x) = 2e^{-\left(|-\mu+b|\sigma^2 + ax^2 - 2ax\mu + a\mu^2\right)/(a\sigma^2)} / \left[\sqrt{\pi}\sigma e^{\left(-\sigma^2 + 4ab + 4xa\right)/(4a^2)}\right.$$

$$\left. \left(-e^{2x/a} - e^{2b/a} - \mathrm{erf}\left(\frac{2ab - 2xa - \sigma^2}{2a\sigma}\right)e^{2x/a} + \mathrm{erf}\left(\frac{2ab - 2xa - \sigma^2}{2a\sigma}\right)e^{2b/a}\right)\right]$$

for $\sigma > 0$, $b > 0$, and $-\infty < \mu < \infty$. This closed-form solution, which does not appear to exist anywhere in the literature despite the potential usefulness of the Laplace as a prior distribution for μ, can be calculated using APPLBayes with minimal effort. Similar code shows if we have a Binomial(n, p) likelihood function—where p is the unknown parameter and n is a positive integer—and a Triangular$(0, b, 1)$ prior distribution with $0 < b < 1$, then the posterior distribution for p is

$$g(p|x) = \begin{cases} \dfrac{\frac{2p(n!)(p^x)(1-p)^{n-x}}{b(n-x)!x!}\left(\frac{n!bp^x(1-p)^{n-x}}{(n-x)!x!} + \frac{p^x(1-p)^{n-x}n!(1-b^2)}{(b-1)(n-x)!x!} - \frac{2p^x(1-p)^{n-x}n!(1-b)}{(b-1)(n-x)!x!}\right)}{} & 0 \le p \le b \\[2em] \dfrac{\frac{2(p-1)p^x(1-p)^{n-x}}{b(n-x)!x!}\left(\frac{n!bp^x(1-p)^{n-x}}{(n-x)!x!} + \frac{p^x(1-p)^{n-x}n!(1-b^2)}{(b-1)(n-x)!x!} - \frac{2p^x(1-p)^{n-x}n!(1-b)}{(b-1)(n-x)!x!}\right)}{} & b \le p \le 1. \end{cases}$$

The ability of APPLBayes to produce generalized closed-form posterior distributions with ad hoc priors highlights the program's theoretical usefulness. These examples represent a small portion of what has been accomplished so far using APPLBayes. As these examples demonstrate, leveraging the computer algebra capabilities of APPLBayes makes fruitful theoretical exploration a relatively simple matter.

14.3 Bayesian Inference on Multiple-Parameter Distributions

As demonstrated in the previous section, deriving posterior distributions using Bayesian methods quickly becomes intractable when we deviate from the use of conjugate prior distributions, even in the single-parameter case. This problem compounds rapidly when we enter the realm of distributions with multiple parameters. To understand why, we can begin by discussing two-parameter distributions. For a distribution with two parameters, Bayes' Theorem becomes

$$g(\theta_1, \theta_2|\mathbf{x}) = \frac{f(x|\theta_1, \theta_2) \cdot g(\theta_1, \theta_2)}{\int_{-\infty}^{\infty} \int_{-\infty}^{\infty} f(x|\theta_1, \theta_2) \cdot g(\theta_1, \theta_2) \, d\theta_1 \, d\theta_2}.$$

This relationship differs from Bayes' Theorem for single-parameter distributions in two important ways. First, the multiple-parameter case requires a joint prior distribution, $g(\theta_1, \theta_2)$. Second, it requires the evaluation of a double integral. The multiple-parameter case, in fact, requires one iterated integration for each parameter in the distribution. As we saw in the previous sections, these integrals are typically rather difficult to evaluate. Thus, for Bayesians seeking a closed-form posterior distribution, the dependence on conjugate priors becomes even stronger in the multiple-parameter case.

Bayesians have several methods for dealing with this complication. When conducting applied statistical analysis, Bayesians depend on Markov Chain Monte Carlo (MCMC) methods. Without delving into too much detail, MCMC methods essentially simulate the prior distribution and the likelihood function, and use the data points gathered from this simulation in order to construct a discrete approximation to the posterior distribution. While ideal for one-time use in statistical analysis, since the simulations depend on pseudo-random number generators, the posterior distributions that result from MCMC are different after each run. As such, it is difficult to use these results for theoretical experimentation. Moreover, MCMC simulations fail to produce closed-form posterior distributions; the output of MCMC is essentially a matrix of probabilities.

Several attempts have been made to devise methods capable of producing closed-form posterior distributions in the multiple-parameter case. Jaakkola and Jordan [54] present a method capable of producing approximate closed-form distributions of the posterior for logistic regression. In order to accomplishes this, Jaakkola and Jordan utilize variational methods. Variational methods attempt to find a normal approximation to the posterior distribution using computational optimization techniques to minimize the difference between the true posterior distribution and the Gaussian approximation. This method, however, has limitations. First, in some cases, normal distributions provide a poor approximation to the logit model. Second, the method only applies to logistic regression, whereas a more general method would be preferable. Finally, because of the reliance on a normal approximation to the logit curve, closed-form posterior distributions can only be found when normal prior distributions are used.

APPLBayes, on the other hand, contains an algorithm capable of producing approximate closed-form marginal distributions for each parameter in a multiple-parameter distribution without depending on conjugate priors. The algorithm works by holding all but one parameter constant, conducting Bayesian inference on that parameter, and then doing the same for the remaining parameters. We then iterate this process until the posterior distributions converge to a final solution. This process results in increasingly better approximations of the marginal distribution for this parameter. Under the assumption of independence, these approximations converge to the true marginal posterior distribution for each parameter. This section will describe this algorithm, the cases under which the algorithm produces good

approximations to the marginal distributions for each parameter, potential issues with the algorithm, as well as its some interesting avenues of theoretical exploration that it has already opened.

Assumptions. In order for the algorithm to produce marginal posterior distributions for the parameters in a multiple-parameter likelihood function, we must make several assumptions concerning the parameters, their interdependence, and the shape of the posterior distribution. In order to simplify the notation, we will present these assumptions here in the two-parameter case. First, we assume that in the joint prior distribution, $g(\theta_1, \theta_2)$, the parameters θ_1 and θ_2 are independent. By the definition of independence, this allows us to write the prior distribution as $g(\theta_1, \theta_2) = g_1(\theta_1) \cdot g_2(\theta_2)$. This assumption serves two purposes. First, APPL's data structure currently only treats single-variable distributions. As such, including joint prior distributions would require significant modifications to the underlying structure of APPL. Second, since the posterior distributions must be independent for the conditional distributions that the algorithm produces to equal the marginal distributions for each parameter, the prior distributions must be independent as well. Second, we must assume that the parameters are independent a posteriori. That is, the value of each parameter cannot have any effect on the value of the other parameters. This allows us to use our method to compute marginal distributions, rather than simply computing conditional approximations to the marginal distributions. Under the independence assumption, the conditional distribution $g(\theta_1|x, \hat{\theta}_2)$ is the same as the marginal distribution $g(\theta_1|x)$. This follows directly from the definition of statistical independence.

Along with these assumptions regarding the independence of the parameters, we must also assume that the resulting posterior distributions are convex. This is necessary in order for the iterating mechanism of the algorithm to converge on the correct conditional distribution. This assumption should hold in the vast majority of cases, especially given a large data vector. Well-known results in Bayesian theory show that the posterior distribution is asymptotically normal under a relatively broad class of conditions (see, for example, Ghosal et al. [36]). This is a stricter condition than convexity, which tends to hold even for small data vectors.

Before proceeding to describe the algorithm, it is useful to see how these assumptions allow us to produce an exact marginal distribution. We begin by repeating Bayes' Theorem in the two-parameter case:

$$g(\theta_1, \theta_2|\mathbf{x}) = \frac{f(x|\theta_1, \theta_2) \cdot g(\theta_1, \theta_2)}{\int_{-\infty}^{\infty} \int_{-\infty}^{\infty} f(x|\theta_1, \theta_2) \cdot g(\theta_1, \theta_2) \, d\theta_1 \, d\theta_2}.$$

Assuming independence of the parameters in the prior distributions, we can rewrite Bayes' Theorem as

$$g(\theta_1, \theta_2|\mathbf{x}) = \frac{f(x|\theta_1, \theta_2) \cdot g_1(\theta_1) \cdot g_2(\theta_2)}{\int_{-\infty}^{\infty} \int_{-\infty}^{\infty} f(x|\theta_1, \theta_2) \cdot g_1(\theta_1) \cdot g_2(\theta_2) \, d\theta_1 \, d\theta_2}.$$

Convexity guarantees that the algorithm will find the most likely conditional distribution, given a likelihood function, a data vector, and a set of prior distributions. Suppose that we have used our algorithm to find such a distribution for the first parameter. The relationship now becomes

$$g(\theta_1|\mathbf{x}, \hat{\theta}_2) = \frac{f(x|\theta_1, \hat{\theta}_2) \cdot g_1(\theta_1)}{\int_{-\infty}^{\infty} f(x|\theta_1, \hat{\theta}_2) \cdot g_1(\theta_1) \, d\theta_1}.$$

The integral in the denominator in this case reduces to a single integral. This can either be evaluated symbolically or numerically. In either case, finding the conditional distribution $g(\theta_1|\mathbf{x}, \hat{\theta}_2)$ now reduces to a computational problem. Now, under the assumption of independence in the parameters in the posterior distribution, it follows from the definition of independence that $g(\theta_1|\mathbf{x}, \hat{\theta}_2) = g(\theta_1|\mathbf{x})$, the marginal distribution of the parameter θ_1. In this manner, the algorithm is capable of producing marginal posterior distributions for each parameter in a multiple-parameter likelihood function.

The Algorithm. With these assumptions in place, we can now proceed to describe the steps that the algorithm uses to produce marginal distributions for each parameter. First, the algorithm requires as arguments the likelihood function, a list of marginal prior distributions, a list of unknown parameters, a data vector, a list of variables in the likelihood function, a list of initial guess values for the parameters and a level of precision. Written in APPL, the call sequence for the procedure is

```
> PosteriorMV(X, data, PriorList, paramList,
>              variables, guess, precision);
```

The likelihood function is input as a random variable using APPL's random variable data structure. Data is entered as a list-of-sublists. One data vector is required for each variable in the likelihood function. While typically a likelihood function may only contain one variable, for instance an x variable, this becomes important when the algorithm is used for regression. In this case, the likelihood function may contain several variables, such as y, x_1, x_2, etc. The list of marginal priors is entered as a list of random variables in the APPL random variable format. The list of parameters, variables, and the initial guess for each parameter are entered as a list. The number of elements in the initial guess list, however, must equal the number of elements in the unknown parameter list. In other words, there must be a single guess for each parameter. Precision is a floating point number on $[0, 1]$.

The Likelihood Function. The `PosteriorMV` algorithm begins by deriving the likelihood function using the input data set. Here we introduce the following notation: $f(\mathbf{x}|\boldsymbol{\theta})$ denotes the likelihood function of the data set, given a vector of unknown parameters, \mathbf{x} denotes the data set, and $f(x|\boldsymbol{\theta})$ refers to the input likelihood function for one piece of data. This procedure requires multiplication, using Maple's built-in symbolic manipulator:

$$f(\mathbf{x}|\boldsymbol{\theta}) \mapsto \prod_{i=1}^{n} f(x_i|\boldsymbol{\theta})$$

where n is equal to the number of observations in the data set.

Initial Approximations. After deriving the likelihood function, PosteriorMV produces initial approximations for the marginal distribution of each parameter, using the initial guess values provided in the call sequence. In order to accomplish this, the algorithm substitutes the initial guess for all but one parameter, and then conducts Bayesian updating on the resulting approximate marginal distributions. Mathematically, this step of the algorithm appears as $f_{\mathrm{approx}} \mapsto g(\hat{\theta}_i|\mathbf{x}, \boldsymbol{\theta}\backslash\theta_i)$ for each θ_i in the parameter vector $\boldsymbol{\theta}$. These approximations to the marginal distributions serve as the initial guesses supplied to the algorithm's iteration procedure. Note that, at this point in the algorithm, the approximations remain un-normalized. Since normalizing the marginal distributions is computationally expensive, it occurs only once, at the end of the algorithm.

Iteration. After finding an initial set of approximations, the algorithm proceeds to find an increasingly better set of approximations. It accomplishes this by repeating the fix-then-update process until the resulting distributions cease to change. In order to determine the degree to which the approximate distributions change after each iteration, however, we require a measure that allows us to track how the distribution changes. The ideal measure would be $\left(\int f_{\mathrm{approx}}^{n+1}\right)^2 - \left(\int f_{\mathrm{approx}}^{n}\right)^2$ which would measure the area between the approximation on step $n+1$ and step n. However, integrals are computationally expensive, making this approach impractical.

An alternative approach considers using the mean of the distribution. The mean of the approximate marginal can be computed as

$$E\left[\hat{\theta}_i\right] = \int_{-\infty}^{\infty} f_{\mathrm{approx}}^{n+1}(\hat{\theta}_i|\mathbf{x}, \boldsymbol{\theta}\backslash\theta_i)\, d\theta_i.$$

Again, since computing the mean also requires integration, using the mean to measure the difference between successive approximations is not ideal.

A more fruitful option uses the mode of the distribution. Using the mode is useful because, owing to the convexity of the marginal approximations, all possible marginal approximations have a unique mode. The mode has the added advantage that it avoids the need to evaluate an integral on each step of the iteration. Several issues nevertheless arose in the initial attempts to use the mode as the basis for measuring changes in the approximate marginal distribution. In particular, the usual method used to compute the mode involves taking the first derivative of the PDF, and then solving for its critical point. This requires us to solve $\frac{df_{\mathrm{approx}}^{n+1}(\hat{\theta}_i)}{d\theta_i} = 0$ for the critical $\hat{\theta}_i$ via Maple's symbolic solve function. This approach proved unsatisfactory because the marginal distributions are only defined within the limits of the support of the

parameter. Consider, for instance, an approximate marginal distribution for a parameter θ that is defined for $\theta \geq 0$. In this case, the function describing the approximate marginal may contain critical values for $\theta < 0$. Using the traditional differentiation approach, Maple sometimes returns these incorrect critical values. Because of this, numerical methods that take an initial guess as an input provide a better alternative. Again, the most natural choice involves differentiation, and then using a numerical method, perhaps a bisection search, to solve for $\hat{\theta}_i$ in $\frac{df^{n+1}_{\text{approx}}(\hat{\theta}_i)}{d\theta_i} = 0$. The limitations of Maple's floating point number evaluator, however, caused this approach to fail as well. Specifically, in the tails of these approximate marginal distributions, solving for $\hat{\theta}_i$ in $\frac{df^{n+1}_{\text{approx}}(\hat{\theta}_i)}{d\theta_i} = 0$ evaluates to a number very close to 0. Even adjusting Maple to track floating point numbers to 50 digits, Maple has difficulty distinguishing these numbers from 0, and as a result frequently returns values in the tail as critical values. As such, an ideal method computes the mode without having to differentiate the approximate marginal distribution. This can be accomplished using a golden section search, which works by evaluating the function itself until it converges to the functions highest value. Since the approximate marginal distribution are convex functions, the golden section search is guaranteed to accurately compute the mode of the distribution, to a pre-specified level of precision. In consequence of these advantages, the `PosteriorMV` algorithms employs a Maple implementation of the golden section search to compute the mode of the approximate marginal distributions. Using the golden section search has the added advantage that it requires no calculus. This leaves open the possibility that the algorithm could be implemented in a lower level language, such as C, which would substantially reduce the algorithm's run-time.

Using the mode as a number that uniquely characterizes each approximate marginal distribution, the algorithm produces increasingly accurate approximations to the marginal distribution using the following steps, beginning with a set of guesses, $\hat{\boldsymbol{\theta}}$:

1. Fix all parameters except for θ_i.
2. Perform Bayesian updating to find the approximate marginal distribution $g(\theta_i | \mathbf{x}, \hat{\boldsymbol{\theta}})$.
3. Replace $\hat{\theta}_i \in \hat{\boldsymbol{\theta}}$ with $\text{argmax}_{\theta_i} g(\theta_i | \mathbf{x}, \hat{\boldsymbol{\theta}})$.
4. Repeat for all $\theta_i \in \boldsymbol{\theta}$.

The algorithm continues this iteration process until the guesses for each parameter cease to change, within a pre-set level of precision. In order to determine when to stop, the algorithm tracks the percent change in the guesses for each parameter on each step of the iteration. The iteration process then stops after the condition $\left| \frac{\hat{\theta}_{i2} - \hat{\theta}_{i1}}{\hat{\theta}_{i1}} \right| < \rho$ for all $\theta_i \in \boldsymbol{\theta}$ is met, where ρ is the precision variable described in the call sequence.

The set of marginal distributions that the algorithm converges to does not depend on the initial conditions provided in the algorithm's call sequence.

However, the iteration process does end considerably sooner if good guesses are supplied by the user. So far, the maximum likelihood estimates for the parameters have provided the best performance with regard to the initial guesses. With independent or weakly dependent parameters the algorithm tends to converge after only a handful of iterations when maximum likelihood estimates are used as initial guesses.

Normalizing the Marginals. After completing the iteration process, the algorithm proceeds to normalize the resulting marginal distributions. Due to the assumption that the parameters are independent a posteriori, this can be accomplished with a single integral. Experience has shown this integral to be problematic, even given Maple's symbolic manipulation capabilities, due to the complexity of the functions that result when ad hoc, non-conjugate prior distributions are used for the parameters. As such, numerical methods are used to obtain the normalizing constant. This is disadvantageous insofar as it eliminates the possibility of producing exact closed-form marginal distributions. Nevertheless, given sufficiently accurate numerical techniques, the resulting approximate marginal will be very close to the true distribution.

Problems with this approach arise because it is difficult to perform numerical integration on functions that have infinite support. This problem can be remedied using one of two methods. First, truncated prior distributions can be used. The use of truncated prior distributions is convenient because, in the majority of cases, the probability density contained in the tails of the marginal posterior distributions is negligible. As such, little information is lost when a truncated prior is used, as long as the truncation occurs in the tail. Actually, this highlights the usefulness of APPLBayes in this respect. Only with the aid of software such as APPLBayes does the use of ad hoc, truncated priors become feasible. Second, the limits of the numerical integration can be modified when the support extends to infinity. Presently, in cases in which infinite support poses a problem, the algorithm changes the support of the numerical integration to a point where Maple's floating point evaluator fails to distinguish the value of the function $g(\hat{\theta}_i|\mathbf{x}, \hat{\boldsymbol{\theta}})$ from zero. Since only a negligible amount of the area beneath the curve falls beyond this point, omitting it from the integration has little effect on the result. In cases where the numerical integrator fails to normalize the area under the posterior distribution to 1, the algorithm repeats the normalization process until the area under the curve falls between 0.999999 and 1.000001. As a practical matter, experience has shown that using Gaussian quadrature with a sufficiently high precision setting succeeds in producing a normalizing constant in most cases.

Output. The output of the algorithm is a list of random variables, describing a closed-form approximation to the marginal posterior distribution for each unknown parameter in the likelihood function. Standard APPL procedures can be applied to these random variables in order to construct percentiles, confidence intervals, and other statistical measures. The algorithm is

capable of producing closed-form approximations to the marginal distributions for each parameter in all cases where the problem can be re-stated as performing inference on a distribution with n unknown parameters. This not only includes inference on distributions with unknown parameters, but also linear regression. As such, the range of problems to which the algorithm applies is much broader than the set of APPL procedures that operates on single-variable distributions.

Issues with the Algorithm and Theoretical Possibilities. A variety of issues may limit the usefulness of the `PosteriorMV` algorithm under certain conditions. These issues fall broadly into two categories, computational and theoretical.

- **Computational Issues:** A number of computational issues have arisen during experimentation with the algorithm. These issues relate to the manner in which Maple handles floating point numbers. The first issue results when an insufficiently high level of precision is used in the implementation of the golden section search for finding the mode. In this case, the golden section search finds a mode that closely approximates, but is not exactly equal to, the mode of the intermediate approximation to the marginal distribution. When the golden section search does not find the mode to a sufficiently high degree of precision, the algorithm has in some cases, failed to converge. Instead, the algorithm bifurcates between two sets of approximations, each very close to the set of true marginal distributions. Under ideal circumstances, the algorithm ought to converge to somewhere between these two sets of approximations. This issue can be addressed by increasing the precision setting on the golden section search algorithm. Increasing the precision setting on the mode-finding procedure appears to solve this problem in the majority of cases. An alternative approach may be to modify the algorithm so that it recognizes when these oscillations take place. The algorithm could then produce a final approximation that lies in between the two sets of looser approximations. This method, however, may fail if the algorithm instead oscillates between three or more intermediate approximations.

 Second, when working with posterior distributions whose support extends to infinity, our method of modifying the limits of integration sometimes fails to produce an area of one under the approximate marginal distributions. As outlined above, this problem has been addressed by feeding the resulting posteriors into a loop that re-normalizes them until the area under the distributions becomes one. In practice, using this looped re-normalization technique, this particular computational problem has not proven to be particularly inhibiting.

- **Theoretical Issues:** A host of additional theoretical issues also accompany the algorithm, especially when the assumption of parameter independence is violated. In particular, when the parameters are dependent, even weakly dependent, the algorithm does not truly produce marginal

distributions. Rather, it provides a heuristic technique that produces the conditional distribution for each parameter that most closely matches the marginal distribution. When parameter dependence is weak, the difference between the distributions that the algorithm produces and the true marginal distributions is likely to be extremely small. It is extenuated, however, when the parameters are strongly dependent, for instance between the parameters in a logistic regression. In the case of logistic regression, the algorithm produces credible sets that center around the same values found using MCMC methods, as well as classical maximum likelihood estimation. However, the Bayesian credible sets in these cases tend to be somewhat narrower. In other words, the variance of the distributions produced by the algorithm may tend to underestimate the variance of the true marginal distributions. This is difficult to establish, however, because the exact marginal distributions are not available, making comparison impossible. Efforts are underway to analyze the nature of this discrepancy. It may be possible to transform the closed-form approximations produced by the algorithm in such a way that it better captures the variance of the true marginal distribution.

In addition to this, an analytical proof that the algorithm converges to the marginal distribution in cases in which parameter independence holds has yet to be written. This area of theoretical endeavor is currently under exploration. In spite of the lack of an analytical proof of correctness, however, in practice the algorithm appears to produce solid results for independent parameters.

Regardless of these difficulties, the `PosteriorMV` represents an advance in the field of computational Bayesian statistics, for several reasons. First, it has produced the first set of closed-form posterior distributions for several outstanding problems in Bayesian statistics. This includes a logistic regression on the failure data in the O-rings on the Challenger space shuttle. Second, because the algorithm is completely analytic, as opposed to MCMC methods, it serves as a convenient tool for theoretical experimentation, especially on distributions with independent parameters. For instance, with the algorithm, it is possible to repeatedly update using the same data set. After each run of the algorithm, the resulting posteriors can be used as the prior distributions for the next run. Repeating this process a sufficient number of times results in a set of degenerate distributions equal to the mean of each of the initial posterior distributions. While it is unclear whether this is of any theoretical interest, such experimentation would not be possible outside of the framework of the APPLBayes software.

Example 14.3. This example illustrates the APPLBayes `PosteriorMV` procedure. First, we will generate a data of $n = 20$ observations set by drawing randomly from a $N\left(4, 2^2\right)$ distribution. The following Maple/APPLBayes code accomplishes this task.

```
> xdata:=[];
> for i from 1 to 20 do
>    xdata:=[op(xdata), NormalVariate(4, 2)];
> end do;
```

This produces the data vector

```
xdata = [4.95237218, 5.70944159, 5.4330855, 6.66171129,
         3.05933919, 3.82564198, 0.6325586, 4.16359636,
         2.54410626, 3.24794389, 5.1276851, 2.08433906,
         2.82013433, 7.04576974, 2.4937987, 2.65766740,
         4.71412016, 3.48119893, 2.5813510, 3.97188172]
```

Next, we assume that the likelihood function takes the form of a normal distribution, which, omitting the normalizing constant, is

$$f(x|\mu,\sigma) = \frac{1}{\sigma} e^{-(x-\mu)^2/(2\sigma^2)} \qquad -\infty \leq x \leq \infty,$$

for μ, and σ in the parameter space $\{(\mu,\sigma)| -\infty < \mu < \infty, \sigma > 0\}$. (Omitting the normalizing constant is acceptable because multiplying the likelihood by a constant has no effect on the posterior distribution.) In order to apply the algorithm, we must verify that our assumption of independent parameters is valid. The independence assumption can be supported with a contour plot of the two parameters in the likelihood function, as shown in Fig. 14.3. This can be accomplished using the following APPLBayes code.

```
> X := NormalRV(mu, sigma);
> likeFunc := simplify(Likelihood(X, [xdata], [x]));
> with(plots);
> likeContour := contourplot(likefunc, mu = 0 .. 10,
                    sigma = 0 .. 10, numpoints = 20000);
```

The lack of slope of the contour plot supports the independence assumption on the parameters. Since it appears that our assumptions hold, we can now proceed to apply prior distributions to the two unknown parameters, μ and σ. In our example, we will use an exponential($2/7$) distribution truncated between 0 and 20 as the prior for μ and a $N(2.3, 1^2)$ distribution truncated between 0 and 12 as the prior for σ. These distributions appear in APPLBayes as

$$g(\mu) = 0.2867 \cdot e^{-.2857\mu} \qquad\qquad 0 \leq \mu \leq 20$$
$$g(\sigma) = 0.4033 \cdot e^{-.0050(10\sigma-23)^2} \qquad 0 \leq \sigma \leq 12.$$

In APPLBayes, we set up the priors using the following code.

```
> Y1 := Truncate(ExponentialRV(2 / 7), 0, 20);
> Y2 := Truncate(NormalRV(2.3, 1), 0, 12);
> PriorList := [Y1, Y2];
```

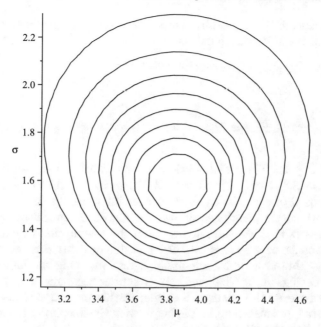

Fig. 14.3. Likelihood function contours for the simulated data values

We must now produce sensible initial guesses for each parameter. Typically, using the maximum likelihood estimators for the likelihood function works quite effectively. In this example, we can use $\hat{\mu} = \bar{x}$ and $\hat{\sigma} = \bar{s}$ as sensible initial guesses for the parameters. Note that the output of the algorithm is independent of the initial guesses. Effective initial guesses simply serve to limit the amount of time the algorithm requires to converge. We can establish our initial guesses using the following APPL commands.

```
> B := BootstrapRV(xdata);
> muHat := Mean(B);
> sigHat := sqrt(Variance(B));
> guess := [muHat, sigHat];
```

We are now ready to apply the algorithm.

```
> paramList := [mu, sigma];
> variables := [x];
> precision := 0.001;
> Params := PosteriorMV(X, data, PriorList, paramList,
>                          variables, guess, precision);
```

The algorithm converges after three iterations, and produces the two marginal posterior distributions

$$g(\mu|\mathbf{x}) \cong 74494.74 \cdot e^{30.5864\mu - 3.9986\mu^2 - 69.5891\mu^3} \qquad 0 \le \mu \le 20$$

and

$$g(\sigma|\mathbf{x}) \cong \frac{3.9847 \cdot 10^8}{\sigma^{20}} e^{-10^{-28}(2.5001\cdot10^{29}\sigma^{-2} + 5.000\cdot10^{27}\sigma^2 - 2.300\cdot10^{28}\sigma + 2.645\cdot10^{28})}$$

for $0 \le \sigma \le 12$. These distributions have conditional expected values $E[\mu|\mathbf{x}] = 3.825$ and $E[\sigma|\mathbf{x}] = 1.000$. Additional APPL commands can be used in order to conduct further inference on these posterior distributions. As mentioned above, the algorithm produces Bayesian credible sets that are too narrow when the assumption of parameter independence does not hold. In order to demonstrate this, we will use the algorithm to conduct a logistic regression, using the failure data for the O-rings in the Challenger space shuttle accident. Since the code used to generate this example is similar to that presented above, it will be omitted for this example. We begin with the likelihood function for the logistic regression, which is

$$f(x|\beta_0, \beta_1) = \left(\frac{e^{\beta_0 + \beta_1 x}}{1 + e^{\beta_0 + \beta_1 x}} \right)^y \cdot \left(1 - \frac{e^{\beta_0 + \beta_1 x}}{1 + e^{\beta_0 + \beta_1 x}} \right)^{1-y} \qquad 0 \le x \le 1.$$

A contour plot of the two parameters in the likelihood function is used to determine whether or our assumption of independence is valid. Using the O-ring data set, this likelihood function appears in Fig. 14.4. The pronounced slope in the contour plot indicates that the two parameters are dependent. Thus, the output of the algorithm will be a conditional slice of the multi-parameter distribution that is narrower than the true marginal distribution. In order to determine an approximation to the marginal distribution, we will begin by applying prior distributions to each parameter. In this case, we will use an exponential$(1/8.9)$ distribution as a prior for the parameter β_0 and a $U(-1, 1)$ distribution as a prior for the parameter β_1. The prior distributions are

$$g(\beta_0) = (1/8.9) \cdot e^{-\beta_0/8.9} \qquad \beta_0 > 0$$

and

$$g(\beta_1) = \frac{1}{2} \qquad -1 < \beta_1 < 1.$$

Applying these prior distributions to the likelihood function, the algorithm converges after five iterations and produces closed-form approximations to the marginal distributions for each parameter. The resulting approximate marginal distributions take the form

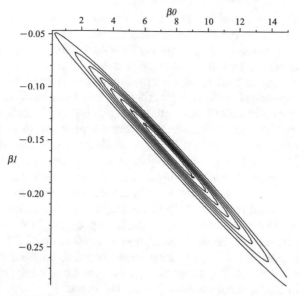

Fig. 14.4. Likelihood function contours for the O-ring data set

$$g(\beta_0|\mathbf{x}) \cong \left(1.02 \cdot 10^{60} e^{6.89\beta_0 - 76.64}\right) / \left[(1 + e^{\beta_0 - 11.93})^6 (1 + e^{\beta_0 - 12.65})^{24}\right.$$
$$(1 + e^{\beta_0 - 12.47})^6 (1 + e^{\beta_0 - 14.46})^6 (1 + e^{\beta_0 - 12.29})^6 (1 + e^{\beta_0 - 12.11})^{18}$$
$$(1 + e^{\beta_0 - 13.01})^6 (1 + e^{\beta_0 - 13.20})^6 (1 + e^{\beta_0 - 10.30})^6 (1 + e^{\beta_0 - 11.39})^6$$
$$(1 + e^{\beta_0 - 14.09})^6 (1 + e^{\beta_0 - 9.58})^6 (1 + e^{\beta_0 - 13.56})^{12} (1 + e^{\beta_0 - 14.64})^6$$
$$\left.(1 + e^{\beta_0 - 13.74})^{12} (1 + e^{\beta_0 - 14.28})^6 (1 + e^{\beta_0 - 10.48})^6\right] \qquad \beta_0 > 0$$

and

$$g(\beta_1|\mathbf{x}) \cong 2.39 \cdot 10^{11} e^{62.23 + 424\beta_1} / \left[(1 + e^{8.89 + 66\beta_1})^6 (1 + e^{8.89 + 70\beta_1})^{24}\right.$$
$$(1 + e^{8.89 + 69\beta_1})^6 (1 + e^{8.89 + 80\beta_1})^6 (1 + e^{8.89 + 68\beta_1})^6$$
$$(1 + e^{8.89 + 67\beta_1})^{18} (1 + e^{8.89 + 72\beta_1})^6 (1 + e^{8.89 + 73\beta_1})^6$$
$$(1 + e^{8.89 + 57\beta_1})^6 (1 + e^{8.89 + 63\beta_1})^6 (1 + e^{8.89 + 78\beta_1})^6$$
$$(1 + e^{8.89 + 53\beta_1})^6 (1 + e^{8.89 + 75\beta_1})^{12} (1 + e^{8.89 + 81\beta_1})^6$$
$$\left.(1 + e^{8.89 + 76\beta_1})^{12} (1 + e^{8.89 + 79\beta_1})^6 (1 + e^{8.89 + 58\beta_1})^6\right]$$

for $-1 < \beta_1 < 1$. From these distributions, we can construct 95% Bayesian credible sets. For the intercept parameter, β_0, APPL produces a 95% Bayesian credible set

$$8.4 < \beta_0 < 9.2$$

and for the slope parameter, β_1, APPL produces a 95% Bayesian credible set

$$-0.19 < \beta_1 < -0.18.$$

The applications demonstrated in this chapter show how APPLBayes can expand the current state of understanding in Bayesian statistics. Concerning single-parameter models, it might appear that APPLBayes only has the ability to solve elementary problems. In a sense this is true: in this case APPLBayes primarily serves as a tool to automate the application of Bayes' Theorem and ease mathematical tedium. Nevertheless, APPLBayes allows users to experiment with non-standard prior distributions using Maple's symbolic manipulation capability and offer a mechanism whereby Bayesians can test the efficacy of different prior distributions. In the realm of multiple-variable distributions, APPLBayes has proven capable of solving problems that Bayesians would have previously considered intractable. On the one hand, having closed-form solutions to these problems may prove little more useful than discrete approximations produced by MCMC methods. On the other hand, closed-form solutions may provide an unparalleled opportunity for new theoretical insights. Regardless of their usefulness, however, without APPL's unique random variable data structure, deriving these closed-form distributions would be tedious. As such, APPLBayes has opened up an entirely new avenue of computational exploration for Bayesian statisticians.

15

Other Applications

This chapter contains miscellaneous computational probability applications. Section 15.1 concerns algorithms for calculating the probability distribution of the longest path of a series-parallel stochastic activity network with continuous activity durations. Section 15.2 concerns the use of APPL in determining whether a continuous random variable obeys Benford's law. Finally, Sect. 15.3 contains miscellaneous computational probability applications that are not covered elsewhere in the monograph.

15.1 Stochastic Activity Networks

Project planning is typically aided by the use of *activity networks*, which show precedence relationships between the various activities that constitute the project (see Adlakha and Kulkarni [2]). An activity network is a special case of a directed graph in which the nodes (or vertices) represent points in time and the arcs (or edges) represent activities with time values modeling activity

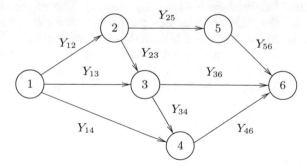

Fig. 15.1. A six-node, nine-arc stochastic activity network

© Springer International Publishing AG 2017
J.H. Drew et al., *Computational Probability*, International Series
in Operations Research & Management Science 246,
DOI 10.1007/978-3-319-43323-3_15

durations. Figure 15.1 shows an example of a *stochastic* activity network in which the positive random variable Y_{ij} denotes the time to complete the activity associated with the arc from node i to node j. The Y_{ij}'s are assumed to be mutually independent. Common-cause delays, e.g., rain delays on an outdoor project, are not considered here. Activity start times are constrained in that no activity emanating from a given node can start until all activities that enter that node have been completed.

Notation and Assumptions. Activity networks have n nodes and m arcs, with a single source node (labeled node 1), a single terminal node (labeled node n), and no loops. Nodes are labeled $1, 2, \ldots, n$ in topological order as defined by Laplante [67] so that if there is an arc a_{ij} from node i to node j, then i is less than j. Thus the nodes along each path occur in ascending order.

For each node j, the "backward" set $\mathcal{B}(j)$ is the set of all nodes that are immediately *before* node j on some path, that is, $\mathcal{B}(j)$ is the set of all nodes i for which there is an arc from i to j. Similarly, for each node j, the "forward" set $\mathcal{A}(j)$ is the set of all nodes that are immediately *after* node j on some path, that is, $\mathcal{A}(j)$ is the set of all nodes k for which there is an arc from j to k.

For each arc a_{ij}, there is a continuous random activity duration Y_{ij} with positive support. The distribution of Y_{ij} is represented by its CDF $F_{Y_{ij}}(t)$ or its PDF $f_{Y_{ij}}(t)$. For each node j there is a random time value T_j, which is the time of completion of all activities entering node j. The quantity T_n is therefore the time of completion of the entire network. Shier [108, pages 122–123] gives an example of an activity network with discrete activity durations.

For each network, there will typically be several paths leading from node 1 to node n. Let M be the set of all paths, with the paths labeled $\pi_1, \pi_2, \ldots, \pi_r$, where $r = |M|$ is the number of paths. A path may be viewed as an ordered set of arcs leading in succession from node 1 to node n. The *length* of path π_k, denoted L_k, is the sum of all Y_{ij} corresponding to the arcs $a_{ij} \in \pi_k$.

For each realization of a stochastic network, the *critical path* π_c is the path with the greatest length, $L_c = \max\{L_1, L_2, \ldots, L_r\}$. The length of the critical path determines the time to complete the entire network. For a stochastic network, a path in M is the critical path with some probability $p(\pi_k) = \Pr(L_k = L_c)$, $k = 1, 2, \ldots, r$. Some arcs may be along more than one path. The probability that arc a_{ij} is along the critical path, also called the arc's *criticality*, denoted by ρ_{ij}, is the sum of all $p(\pi_k)$ where $a_{ij} \in \pi_k$. The criticality of arc a_{ij} can be calculated by

$$\rho_{ij} = \sum_{k=1}^{r} p(\pi_k)\delta_{ij}^k,$$

where δ_{ij}^k is 1 if $a_{ij} \in \pi_k$ and 0 otherwise.

Matrix Representation of a Network. The first step in building a model is defining a mathematical representation of a network. Matrices are well suited for this task because (a) each node and arc in the network can be

designated by using the rows and columns of a matrix, and (b) matrices are easily instantiated in most computer languages. A node-arc incidence matrix was chosen due to its suitability for the algorithms developed here.

The node-arc incidence matrix associated with a network is an $n \times m$ matrix N, where each row represents a node and each column represents an arc. Let

$$N[i,k] = \begin{cases} 1 & \text{if arc } k \text{ leaves node } i \\ -1 & \text{if arc } k \text{ enters node } i \\ 0 & \text{otherwise.} \end{cases}$$

A node-arc incidence matrix that describes the network in Fig. 15.1 is

$$N = \begin{bmatrix} 1 & 1 & 1 & 0 & 0 & 0 & 0 & 0 & 0 \\ -1 & 0 & 0 & 1 & 1 & 0 & 0 & 0 & 0 \\ 0 & -1 & 0 & 0 & -1 & 1 & 1 & 0 & 0 \\ 0 & 0 & -1 & 0 & 0 & 0 & -1 & 0 & 1 \\ 0 & 0 & 0 & -1 & 0 & 0 & 0 & 1 & 0 \\ 0 & 0 & 0 & 0 & 0 & -1 & 0 & -1 & -1 \end{bmatrix}.$$

The only limitation of this representation is that it cannot show an arc that leaves and enters the same node. This ability will not be necessary since feedback is not allowed by our definition of a stochastic activity network.

Series-Parallel Networks. The traditional approach to analyzing stochastic activity networks is simulation. While straightforward to apply, simulation has a distinct drawback. Each additional digit of accuracy in the estimate of $E[T_n]$, for example, requires approximately a 100-fold increase in replications due to the square root in the denominator of the confidence interval formulas for $E[T_n]$. We now derive the various performance measures analytically— eliminating the need for simulation—for a special case of a stochastic activity network known as a *series-parallel network*.

A series-parallel activity network is a special case of an activity network that can be simplified by a sequence of reductions to a simple network consisting of just two nodes and one arc. Reductions consist of taking two arcs, either in series or in parallel, and replacing them with a single arc that has a random duration whose distribution is calculated in the algorithm. This is a good match for APPL's symbolic capabilities because each reduction can be preformed using either the Maximum or Convolution procedures. Once the network is completely reduced, the distribution of the single remaining arc is the distribution of the time T_n to complete the original network. The recursive algorithm then reconstructs the network to determine the critical path probabilities and arc criticalities. Our algorithm will use these operations to simplify a series-parallel network, and then, by determining the distribution of T_n and reconstructing the network, determine the critical path probabilities and the criticalities. The earliest work on the exact analysis of series-parallel networks is Martin [84]. The next paragraphs describe the parallel and series reductions and reconstructions.

Parallel Reduction. The algorithm described subsequently is capable of processing two arcs in *parallel*, as illustrated in Fig. 15.2. A parallel reduction is the process of combining these two arcs into a single arc. Let X_{ij} and Y_{ij} denote the random arc durations. Without loss of generality, if $T_i = 0$, then $T_j = \max\{X_{ij}, Y_{ij}\}$. So

$$F_{T_j}(t) = \Pr(T_j \le t) = \Pr(X_{ij} \le t \text{ and } Y_{ij} \le t) = F_{X_{ij}}(t)F_{Y_{ij}}(t)$$

on the support of T_j. The portion of the algorithm prior to the second recursive call uses this formula to reduce two arcs in parallel to a single arc.

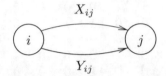

Fig. 15.2. Two arcs in parallel

Parallel Reconstruction. Consider the single arc a_{ij} resulting from a parallel reduction. If the probability ρ_{ij} that this arc is on the critical path is known, then this probability can be allocated to the two original parallel arcs x and y based on each arc's activity duration CDF:

$$\rho_x = \Pr(X_{ij} > Y_{ij}) \cdot \rho_{ij} = \Pr(X_{ij} - Y_{ij} > 0) \cdot \rho_{ij}$$

and

$$\rho_y = \rho_{ij} - \rho_x.$$

This calculation involves determining the distribution of the difference between the two random variables, which can be calculated using the APPL `Difference` procedure. The portion of the algorithm after the second recursive call calculates the distribution of $X_{ij} - Y_{ij}$ and the probability that this random variable is positive.

Series Reduction. The algorithm finds the CDF of the time to complete two arcs in *series*, as illustrated in Fig. 15.3. Let Y_{ij} and Y_{jk} denote the durations of the two arcs with CDFs $F_{Y_{ij}}(t)$ and $F_{Y_{jk}}(t)$. Without loss of generality, if $T_i = 0$, then $T_k = Y_{ij} + Y_{jk}$. The CDF of T_k on its support is (see Casella and Berger [16, page 215])

$$F_{T_k}(t) = \Pr[T_k \le t] = \int_0^t F_{Y_{ij}}(t - y_{jk}) f_{Y_{jk}}(y_{jk}) \, dy_{jk}.$$

Series Reconstruction. Consider the single arc a_{ik} resulting from a series reduction. For any two arcs a_{ij} and a_{jk} that have been reduced into a single

Fig. 15.3. Two arcs in series

arc a_{ik}, $\rho_{ij} = \rho_{jk} = \rho_{ik}$. If two arcs in series are along the critical path, then so is the respective arc resulting from the reduction.

Algorithm. Two arcs in series are detected by searching the rows of the node-arc incidence matrix N for the first row containing exactly two nonzero elements whose sum is 0. Such a row represents a single arc entering and a single arc exiting the given node. To reduce a series of two arcs to a single arc, the algorithm

- zeros out the row of N corresponding to the deleted middle node,
- links the first node of the series directly to the last node of the series, and
- computes the distribution of the sum of the durations of the two arcs.

Two arcs in parallel are identified by two identical nonzero columns of N, which represent two arcs linking the same two nodes. The algorithm iterates through each column and if parallel arcs are found, the algorithm

- computes the distribution of the maximum of the two arc durations, and
- zeros out the column of N corresponding to the eliminated arc.

The following recursive algorithm reduces a series-parallel network, where N is the node-arc incidence matrix of the network. The PDFs for the activity durations of each arc are stored in a vector f (indexed in the same way that the arcs are indexed to correspond to the columns of the matrix N), and the criticalities of each arc are stored in a vector C. This algorithm recursively performs one arc reduction at a time, arbitrarily giving priority to series reductions over parallel reductions, until the network consists of just two nodes connected by a single arc. At this point, the value of C for that one arc is set to one and then the network is reconstructed on the return calls in reverse order, determining values of C for each pair of series or parallel arcs based on the value of C for the reduced arc. The algorithm returns the distribution of the time to complete the network and the final value of C, which contains ρ_{ij} for each arc a_{ij} in the network. Indentation denotes nesting.

Procedure `GetCriticalities`: Find the distribution of the time to complete a series-parallel stochastic activity network and the arc criticalities.

Input: The $n \times m$ node-arc incidence matrix N, the PDFs of m activity durations Y_{ij} stored in the vector f, the number of nodes n, the number of arcs m, the number of arcs to be reduced *arcsremaining* (initialized to m).

Output: Criticalities $C[m]$

Algorithm:
local C, i, j, *rowsum*, *absrowsum*, *negidx*, *posidx*, c, d, l, k, *empty*, *numdiff*
if ($arcsremaining = 1$) if network reduced to one arc
 for $i \leftarrow 1$ to n loop through rows of N
 for $k \leftarrow 1$ to m loop through columns of N
 if ($N[i,k] \neq 0$) if true then arc k is the last existing arc
 $C[k] \leftarrow 1$ set criticality to 1
 print(CDF($f[k]$)) print CDF of project completion time
 return(C) return criticalities
for $i \leftarrow 1$ to n **begin series reduction**
 rowsum $\leftarrow 0$ initialize *rowsum*
 absrowsum $\leftarrow 0$ initialize sum of absolute values in a row
 for $k \leftarrow 1$ to m loop through columns of N
 rowsum \leftarrow *rowsum* $+ N[i,k]$ sum values in the ith row of N
 absrowsum \leftarrow *absrowsum* $+ |N[i,k]|$ sum the absolute values
 if ($N[i,k] = -1$) if arc k enters node i
 negidx $\leftarrow k$ store the index of arc k
 if ($N[i,k] = 1$) if arc k leaves from node i
 posidx $\leftarrow k$ store the index of arc k
 if (*rowsum* $= 0$ and *absrowsum* $= 2$) only one arc enters & leaves node i
 $c \leftarrow f[negidx]$ save the PDF of first arc for reconstruction
 $d \leftarrow f[posidx]$ save the PDF of second arc for reconstruction
 $f[negidx] \leftarrow$ Convolution(c,d) combine the arcs into a single arc
 for $j \leftarrow 1$ to n loop through rows of N
 $N[j, negidx] \leftarrow N[j, negidx] + N[j, posidx]$ change entering node
 $N[j, posidx] \leftarrow 0$ delete middle node
 arcsremaining \leftarrow *arcsremaining* $- 1$ reduce # of arcs in network
 $C \leftarrow$ GetCriticalities($N, f, n, m, arcsremaining$) recursive call
 for $j \leftarrow 1$ to n **begin series reconstruction**
 if ($N[j, negidx] = -1$) if arc *negidx* enters node j
 $N[j, negidx] \leftarrow 0$ delete arc made previously by reduction
 $N[j, posidx] \leftarrow -1$ delete arc made previously by reduction
 $N[i, negidx] \leftarrow -1$ reconstruct node deleted previously by reduction
 $N[i, posidx] \leftarrow 1$ reconstruct node deleted previously by reduction
 $f[negidx] \leftarrow c$ reconstruct arc duration
 $f[posidx] \leftarrow d$ reconstruct arc duration
 $C[posidx] \leftarrow C[negidx]$ reconstruct arc criticality
 return(C) return criticalities
for $l \leftarrow 1$ to m **begin parallel reduction**
 for $k \leftarrow l + 1$ to m loop through columns of N
 empty $\leftarrow 0$ initialize *empty*
 numdiff $\leftarrow 0$ initialize *numdiff*
 for $i \leftarrow 1$ to n loop through rows of N

if $(N[i,l] \neq N[i,k])$	if column difference found
$\quad numdiff \leftarrow 1$	set boolean variable $numdiff$
if $(N[i,l] \neq 0)$	if a column is not empty
$\quad empty \leftarrow 1$	set boolean variable $empty$
if $(numdiff = 0$ and $empty \neq 0)$	columns l & k identical & nonzero
$\quad c \leftarrow f[l]$	save the PDF of arc l for reconstruction
$\quad d \leftarrow f[k]$	save the PDF of arc k for reconstruction
$\quad f[l] \leftarrow \text{Maximum}(c, d)$	store PDF of the maximum
\quad for $i \leftarrow 1$ n	loop through rows of N
$\quad\quad N[i, k] \leftarrow 0$	zero out column of removed arc
$\quad arcsremaining \leftarrow arcsremaining - 1$	reduce # network arcs
$\quad C \leftarrow \text{GetCriticalities}(N, f, n, m, arcsremaining)$	recursive call
\quad for $i \leftarrow 1$ to n	**begin parallel reconstruction**
$\quad\quad N[i, k] \leftarrow N[i, l]$	reinsert arc k by copying column l
$\quad f[l] \leftarrow c$	rebuild distribution
$\quad f[k] \leftarrow d$	rebuild distribution
$\quad C[k] \leftarrow \text{CDF}(\text{Difference}(f[l], f[k]), 0) \cdot C[l]$	$\Pr(X_{il} - Y_{il} > 0) \cdot \rho_{il}$
$\quad C[l] \leftarrow C[l] - C[k]$	calculate $\rho_y = \rho_{il} - \rho_x$
\quad return(C)	return criticalities
return(C)	return criticalities

This algorithm requires symbolic processing capability in order to calculate the distribution of the maximum of two independent random variables for parallel reduction and the distribution of the sum of two independent random variables for series reduction. The APPL procedures Maximum and Convolution are used for these operations.

Example 15.1. Figure 15.4 shows an example of a series-parallel network from Elmaghraby [27, page 261]. This network can be described by the node-arc incidence matrix

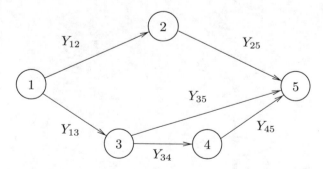

Fig. 15.4. Series-parallel network from Elmaghraby [27, page 261]

$$N = \begin{bmatrix} 1 & 1 & 0 & 0 & 0 & 0 \\ -1 & 0 & 1 & 0 & 0 & 0 \\ 0 & -1 & 0 & 1 & 1 & 0 \\ 0 & 0 & 0 & -1 & 0 & 1 \\ 0 & 0 & -1 & 0 & -1 & -1 \end{bmatrix}.$$

The network can be reduced and reconstructed as illustrated in the steps separated by arrows in Fig. 15.5. The algorithm has been applied to this stochastic activity network in which the duration of each arc $Y_{ij} \sim$ exponential(b), where $1/b$ is the mean activity duration. The CDF of the time to complete the network (T_5) is returned by the algorithm as

$$F_{T_5}(t) = 1 - 3bte^{-bt} - \frac{b^2t^2}{2}e^{-bt} - 3e^{-2bt} + \frac{5b^2t^2}{2}e^{-2bt} + \frac{b^3t^3}{2}e^{-2bt} +$$

$$2e^{-3bt} + 3bte^{-3bt} + b^2t^2e^{-3bt} \qquad\qquad t > 0.$$

The $r = 3$ paths in the network are listed in Table 15.1, along with the critical path probabilities. Table 15.2 contains the criticalities for the $m = 6$ arcs when $b = 2$. (Fisher et al. [32] determine criticalities when activity durations are iid gamma random variables.) This example has a small number of nodes and arcs along with simple distributions for the arc durations so as to make the final CDF for the project duration simple. If, however, the arc durations are each independent triangular$(0, 1, 2)$ random variables, then the problem of determining the distribution of T_5 becomes much more tedious. The arc criticalities returned by the algorithm for this example with triangular$(0, 1, 2)$ arc

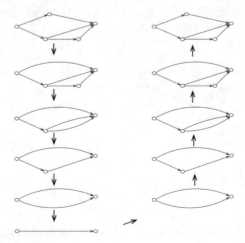

Fig. 15.5. Reduction and reconstruction of the series-parallel network from Elmaghraby [27, page 261]

Table 15.1. Paths π_k and critical path probabilities $p(\pi_k)$ for exponentially distributed arc durations with $b = 2$

k	Node sequence	π_k	$p(\pi_k)$
1	$1 \to 2 \to 5$	$\{a_{12}, a_{25}\}$	$115/432 \cong 0.266$
2	$1 \to 3 \to 5$	$\{a_{13}, a_{35}\}$	$317/1728 \cong 0.183$
3	$1 \to 3 \to 4 \to 5$	$\{a_{13}, a_{34}, a_{45}\}$	$317/576 \cong 0.550$

Table 15.2. Criticalities ρ_{ij} for the exponentially distributed arc durations with $b = 2$

Arc	Paths	ρ_{ij}
a_{12}	π_1	$115/432 \cong 0.266$
a_{13}	π_2, π_3	$317/432 \cong 0.734$
a_{25}	π_1	$115/432 \cong 0.266$
a_{35}	π_2	$317/1728 \cong 0.183$
a_{34}	π_3	$317/576 \cong 0.550$
a_{45}	π_3	$317/576 \cong 0.550$

durations are shown in Table 15.3. The CDF of the project completion time is returned as

$$
F_{T_5}(t) =
\begin{cases}
0 & t \le 0 \\[4pt]
\frac{1}{64512} t^{12} & 0 < t \le 1 \\[4pt]
\begin{aligned}
&-\frac{1}{21504} t^{12} + \frac{5}{4032} t^{11} - \frac{1357}{120960} t^{10} + \frac{1553}{30240} t^9 \\
&-\frac{2431}{17280} t^8 + \frac{15389}{60480} t^7 - \frac{5531}{17280} t^6 + \frac{1237}{4320} t^5 \\
&-\frac{44039}{241920} t^4 + \frac{607}{7560} t^3 - \frac{13}{560} t^2 + \frac{59}{15120} t - \frac{17}{60480}
\end{aligned} & 1 < t \le 2 \\[4pt]
\begin{aligned}
&-\frac{5}{21504} t^{12} + \frac{5}{672} t^{11} - \frac{1813}{17280} t^{10} + \frac{8693}{10080} t^9 \\
&-\frac{556873}{120960} t^8 + \frac{1018271}{60480} t^7 - \frac{749309}{17280} t^6 + \frac{89059}{1120} t^5 \\
&-\frac{8332907}{80640} t^4 + \frac{156139}{1680} t^3 - \frac{414913}{7560} t^2 + \frac{57587}{3024} t - \frac{176249}{60480}
\end{aligned} & 2 < t \le 3 \\[4pt]
\begin{aligned}
&-\frac{1}{21504} t^{12} + \frac{25}{12096} t^{11} - \frac{499}{12096} t^{10} + \frac{703}{1440} t^9 \\
&-\frac{556873}{120960} t^8 + \frac{1018271}{60480} t^7 - \frac{672253}{8640} t^6 + \frac{699347}{3360} t^5 \\
&-\frac{2587009}{6720} t^4 + \frac{197879}{420} t^3 - \frac{376261}{1080} t^2 + \frac{94931}{756} t - \frac{68353}{7560}
\end{aligned} & 3 < t \le 4 \\[4pt]
\frac{1}{144} t^6 - \frac{1}{5} t^5 + \frac{19}{8} t^4 - \frac{89}{6} t^3 + \frac{409}{8} t^2 - \frac{1829}{20} t + \frac{7969}{120} & 4 < t \le 5 \\[4pt]
-\frac{1}{720} t^6 + \frac{1}{20} t^5 - \frac{3}{4} t^4 + 6 t^3 - 27 t^2 + \frac{324}{5} t - \frac{319}{5} & 5 < t \le 6 \\[4pt]
1 & t > 6.
\end{cases}
$$

Activity networks are useful when modeling a project having distinct activities that exhibit precedence relationships. When the durations of the activities are deterministic, it is easy to find the time needed to complete a project by using a simple longest path algorithm. However, when only random activity durations are given, the analysis is much more complicated. Monte Carlo

Table 15.3. Criticalities ρ_{ij} for triangular(0, 1, 2) arc durations

Arc	Paths	ρ_{ij}
a_{12}	π_1	$30233/237600 \cong 0.127$
a_{13}	π_2, π_3	$207367/237600 \cong 0.873$
a_{25}	π_1	$30233/237600 \cong 0.127$
a_{35}	π_2	$6013643/85536000 \cong 0.070$
a_{34}	π_3	$68638477/85536000 \cong 0.802$
a_{45}	π_3	$68638477/85536000 \cong 0.802$

simulation can provide an approximate solution in this case, but the simulation must be written specifically for the network and could require millions of replications for a reliable estimate. In some networks, series-parallel reduction can supply an exact solution analytically, but very few networks can be reduced in this manner. Leemis et al. [75] give an algorithm that can be applied to a more general class of networks.

15.2 Benford's Law

This section states Benford's law and shows how APPL can be used to check a probability model for conformity. The traditional analysis of Benford's law considers its applicability to *data sets*, but we consider whether *probability distributions* obey Benford's law. The material in this section is taken from Leemis et al. [74], where a more thorough treatment of the topic can be found.

Benford's law was first encountered by astronomer and mathematician Simon Newcomb, who noticed "how much faster the first pages (of tables of logarithms) wear out than the last ones" leading to the counter-intuitive conclusion that the first significant digit in the values in a logarithm table is not uniformly distributed between 1 and 9. Using a heuristic argument, he found that ones occur most often (more than 30% of the time) and nines least often (less than 5% of the time), and

$$\Pr(X = x) = \log_{10}(1 + 1/x) \qquad x = 1, 2, \ldots, 9,$$

where X is a random variable denoting the first significant digit. He published this result in 1881 as the "logarithm law" in the *American Journal of Mathematics*.

General Electric physicist Frank Benford [9] apparently independently arrived at the same conclusion as Newcomb concerning logarithm tables in 1938. He proceeded to "collect data from as many fields as possible" to see if natural and sociological data sets would also obey the logarithm law. He often found good agreement between the logarithm law for his 20,229 total observations, including data sets as diverse as the areas of rivers, American League baseball

statistics, atomic weights of elements, death rates, and numbers appearing in *Reader's Digest*. What has become known as "Benford's law" has found applications in the distribution of the one-day return on stock market indices (see Ley [78]), the distribution of the populations of 3141 counties in the 1990 U.S. Census, and the detection of accounting fraud (see Nigrini [93]). Recent books on Benford's law are Berger and Hill [10], Kossovsky [64], and Miller [88].

A mathematically rigorous proof of Benford's law has proven elusive. This is in part due to the fact that certain data sets (e.g., random numbers) do not follow Benford's law. Recent attempts have considered the effect of scale invariance (e.g., dollars vs. yen), base invariance (e.g., octal vs. base ten), and mixtures (i.e., sample data drawn from several population distributions that are selected at random), as indicated in Hill [47, 48].

The purpose of this section is to switch the emphasis from the examination of *data sets* that obey Benford's law to *probability distributions* that obey Benford's law. The emphasis here is on survival distributions (i.e., distributions with positive support), although more general distributions can be examined in the same fashion. Hill [47, pages 361–362] states that "An interesting open problem is to determine which common distributions (or mixtures thereof) satisfy Benford's law ...". General conditions associated with the distribution of the random variable T will now be derived in order to determine when Benford's law applies.

Conditions for Conformance to Benford's Law. We refer to a random variable X with PDF

$$f_X(x) = \Pr(X = x) = \log_{10}(1 + 1/x) \qquad x = 1, 2, \ldots, 9$$

as a "Benford random variable." This random variable is pre-defined in APPL as BenfordRV. The associated CDF is

$$F_X(x) = \Pr(X \le x) = \log_{10}(1 + x) \qquad x = 1, 2, \ldots, 9.$$

Inverting the CDF, a Benford random variate X can be generated by

$$X \leftarrow \lceil 10^U - 1 \rceil,$$

or, equivalently, saving a subtraction, by

$$X \leftarrow \lfloor 10^U \rfloor,$$

where $U \sim U(0, 1)$.

Let T be a nonnegative random variable whose first significant digit is of interest. Let the integer-valued random variable D satisfy

$$10^D \le T < 10^{D+1}$$

(e.g., $T = 365 \Rightarrow D = 2$ and $T = 1/10 \Rightarrow D = -1$). This definition of D allows the first significant digit Y of T to be written in terms of T and D as

$$Y = \lfloor T \cdot 10^{-D} \rfloor = \lfloor 10^{\log_{10} T - D} \rfloor$$

(e.g., $T = 365 \Rightarrow Y = \lfloor 365 \cdot 10^{-2} \rfloor = \lfloor 3.65 \rfloor = 3$).

Referring back to the variate generation algorithm, it is clear that if the random variable $Z = \log_{10} T - D \sim U(0,1)$, which represents the result from the logarithm table, then the first significant digit Y has the Benford distribution. Using conditioning, the CDF of Z is given by

$$F_Z(z) = \Pr(Z \le z)$$
$$= \sum_{d=-\infty}^{\infty} \Pr\left(10^d \le T < 10^{d+1}\right) \cdot \Pr\left(\log_{10} T - d \le z \mid 10^d \le T < 10^{d+1}\right),$$

for $0 < z < 1$. Thus conformance to Benford's law implies that the weights (the first factor in the product) associated with each order of magnitude (indexed by d) and the distribution of $Z = \log_{10} T - D$ (the second factor in the product) are such that the infinite sum produces a linear function in z.

Three examples of distributions providing a perfect fit for Benford's law are given below.

Example 15.2. Let $W \sim U(0,2)$. Let $T = 10^W$. The PDF of T is

$$f_T(t) = \frac{1}{2t \ln 10} \qquad 1 < t < 100.$$

The PDF of D is

$$f_D(0) = \Pr(D = 0) = \Pr(1 < T < 10) = \int_1^{10} f_T(t)dt = \frac{1}{2}$$

and

$$f_D(1) = \Pr(D = 1) = \Pr(10 < T < 100) = \int_{10}^{100} f_T(t)dt = \frac{1}{2}.$$

The PDF of the leading digit Y is

$$f_Y(y) = \Pr(Y = y)$$
$$= \Pr(y < T < y+1) + \Pr\left(10y < T < 10(y+1)\right)$$
$$= \int_y^{y+1} f_T(t)\, dt + \int_{10y}^{10(y+1)} f_T(t)\, dt$$
$$= \int_y^{y+1} \frac{1}{2t \ln 10}\, dt + \int_{10y}^{10(y+1)} \frac{1}{2t \ln 10}\, dt$$
$$= \log_{10}\left(\frac{y+1}{y}\right) \qquad y = 1, 2, \ldots, 9.$$

This PDF matches Benford's distribution exactly. Alternatively, one can proceed by determining the distribution of $Z = \log_{10} T - D$ as follows, where $W = \log_{10} T$:

$$F_Z(z) = \Pr(Z \le z)$$

$$= \sum_{d=-\infty}^{\infty} \Pr\left(10^d \le T < 10^{d+1}\right) \cdot \Pr\left(\log_{10} T - d \le z \mid 10^d \le T < 10^{d+1}\right)$$

$$= \sum_{d=0}^{1} \Pr\left(d \le W < d+1\right) \cdot \Pr\left(W - d \le z \mid d \le W < d+1\right)$$

$$= \Pr\left(0 \le W < 1\right) \cdot \Pr\left(W \le z \mid 0 \le W < 1\right) +$$
$$\Pr\left(1 \le W < 2\right) \cdot \Pr\left(W - 1 \le z \mid 1 \le W < 2\right)$$

$$= \frac{1}{2}z + \frac{1}{2}z$$

$$= z \qquad 0 < z < 1.$$

Since this is the CDF for a $U(0,1)$ random variable, Benford's law is satisfied exactly.

This example can be generalized as follows. Let $W \sim U(a,b)$, where a and b are real numbers satisfying $a < b$. If the interval $\left(10^a, 10^b\right)$ covers an integer number of orders of magnitude, then the first significant digit of the random variable $T = 10^W$ satisfies Benford's law exactly. Equivalently, if $b - a$ is a positive integer, then the first significant digit of $T = 10^W$ satisfies Benford's law. Examples include $a = -2$, $b = 1$ and $a = \log_{10}(3/2)$, $b = \log_{10}(150)$, with $\left(10^a, 10^b\right)$ equal to $(0.01, 10)$ and $(1.5, 150)$ respectively.

There is no need for the support of the distribution of $T = 10^W$ to span several orders of magnitude as is the case for many of the data sets that conform to Benford's law. A single order of magnitude (e.g., $a = 5$, $b = 6$) is sufficient.

The next example considers a non-uniform distribution for W that satisfies Benford's law exactly.

Example 15.3. Let $W \sim \text{triangular}(0, 1, 2)$. The PDF for W is

$$f_W(w) = \begin{cases} w & 0 < w < 1 \\ 2 - w & 1 \le w < 2. \end{cases}$$

As before, let $T = 10^W$ and $Z = W - D$. The CDF of Z is

$$F_Z(z) = \sum_{d=0}^{1} \Pr\left(d \le W < d+1\right) \cdot \Pr\left(W - d \le z \mid d \le W < d+1\right)$$

$$= \frac{1}{2}z^2 + \frac{1}{2}\left(2z - z^2\right)$$

$$= z \qquad 0 < z < 1.$$

Thus the first significant digit of T satisfies Benford's law exactly.

This example can also be generalized. Let $W \sim \text{triangular}(a, b, c)$, where a, b, and c are real numbers satisfying $a < b < c$. The first significant digit of the random variable $T = 10^W$ satisfies Benford's law exactly if a, b, and c are integers. The next example considers a non-symmetric distribution for W.

Example 15.4. Let W have PDF

$$f_W(w) = \begin{cases} 1 - w^2 & -1 < w < 0 \\ (w-1)^2 & 0 \le w < 1. \end{cases}$$

As before, let $T = 10^W$ and $Z = W - D$. The CDF of Z is

$$F_Z(z) = \sum_{d=-1}^{0} \Pr(d \le W < d+1) \cdot \Pr(W - d \le z \,|\, d \le W < d+1)$$

$$= \frac{2}{3}\left[-\frac{z^3}{2} + \frac{3z^2}{2}\right] + \frac{1}{3}\left[1 + (z-1)^3\right]$$

$$= z \qquad 0 < z < 1.$$

Thus the first significant digit of T satisfies Benford's law exactly.

This example can be generalized for W with PDF

$$f_W(w) = \begin{cases} 1 - w^n & -1 < w < 0 \\ (w-1)^n & 0 \le w < 1, \end{cases}$$

where n is a positive, even integer.

In order to test other probability distributions for conformance to Benford's law, we automated the process of deriving the distribution of Z by writing the APPL procedure `Benford`, whose argument is a random variable representing W and whose returned value is the distribution of Z. The algorithm is shown below.

Procedure `Benford`: Compute the distribution of Z to check conformity to Benford's law.

Input: A random variable W formatted in the usual APPL list-of-sublists format.

Output: The PDF of Z.

Algorithm:

$\Omega \leftarrow Support(W)$ the set Ω is the support of the random variable W
$Lo \leftarrow \lfloor \Omega \rfloor$ lower loop limit
$Hi \leftarrow \lceil \Omega \rceil - 1$ upper loop limit
Dimension $W[Hi - Lo + 1]$ W holds the mixture probabilities
Dimension $TranW[Hi - Lo + 1]$ $TranW$ holds transformed segments of W
For $d \leftarrow Lo$ to Hi loop through the support of W
 $W[d] \leftarrow F_W(d+1) - F_W(d)$ calculate weights for the mixture
 $TruncW[d] \leftarrow \texttt{Truncate}(W, d, d+1)$ truncate W between d and $d+1$
 $TranW[d] \leftarrow \texttt{Transform}(TruncW, w - d)$ shift W by d units
$Z \leftarrow \texttt{Mixture}(W, TranW)$ compute the PDF of the mixture

The APPL code required to return the distribution of Z for Example 15.2 is

```
> W := UniformRV(0, 2);
> Z := Benford(W);
```

The APPL code required to return the distribution of Z for Example 15.3 is

```
> W := TriangularRV(0, 1, 2);
> Z := Benford(W);
```

The APPL code required to return the distribution of Z for Example 15.4 is

```
> W := [[w -> 1 - w ^ 2, w -> (1 - w) ^ 2],
        [-1, 0, 1], ["Continuous", "PDF"]];
> Z := Benford(W);
```

All three code segments return $Z \sim U(0,1)$.

15.3 Miscellaneous Applications

This final section contains a series of short examples that are appropriate for
the symbolic capabilities of APPL that were not covered in previous chapters.

Example 15.5. The central limit theorem can be used to facilitate in-
ference about the population mean. The central limit theorem implies
that $\bar{X} = \frac{1}{n} \sum_{i=1}^{n} X_i$ is asymptotically normally distributed when the
X_i's are iid random variables with finite mean and variance. Appli-
cation of the central limit theorem becomes restricted, however, with
small sample sizes, with skewed data, or in the presence of outliers. In
these cases, an alternative method is needed to determine the exact
distribution of \bar{X}. Once the distribution \bar{X} is found, one can use it
to make statistical inferences about the population mean. Consider
the case of the population associated with $X_i \sim$ exponential(1). By
the central limit theorem, the distribution of $Y = \bar{X}$ is approximately
$N(1, \frac{1}{n})$ for large n. For the case $n = 3$, one can compute the exact
distribution of Y with the following APPL statements.

```
> X := ExponentialRV(1);
> n := 3;
> Y := ConvolutionIID(X, n);
> Y := Transform(Y, [[x -> x / n], [0, infinity]]);
```

The resulting Erlang PDF is

$$f_Y(y) = \frac{27}{2} y^2 e^{-3y} \qquad y > 0,$$

which is predictably non-normal in shape since n is too small for the
normality of the central limit theorem to apply to such a skewed under-
lying distribution. Now consider the case of $n = 30$, often considered

a sufficiently large sample size for an adequate central limit theorem approximation. Here a computation of the error associated with the approximation is provided. To obtain the PDF of \bar{X}, simply change the second line of code to n := 30; The result is the following Erlang PDF:

$$f_Y(y) = \frac{2463153133392333398437500000}{10577732774609} y^{29} e^{-30y} \qquad y > 0.$$

Table 15.4 contains selected fractiles for both the exact distribution of \bar{X} and its central limit theorem approximation, the $N(1, \frac{1}{30})$ distribution. For the selected fractiles, the approximation is only good to about one digit beyond the decimal point. The additional APPL statements that generate the central limit theorem values in Table 15.4 are

```
> Z := NormalRV(1, 1 / sqrt(n));
> alpha := [0.9, 0.95, 0.975, 0.99];
> for i from 1 to 4 do
>   evalf(IDF(Z, alpha[i]));
> od;
```

Table 15.4. Fractiles for exact and approximate distributions of \bar{X} for $n = 30$

		Fractile			
Approach	Distribution	0.9	0.95	0.975	0.99
Exact	\bar{X}	1.240	1.318	1.388	1.473
Central limit theorem	$N\left(1, \frac{1}{30}\right)$	1.234	1.300	1.358	1.425

Example 15.6. As a second application related to the central limit theorem, consider an old, approximate method of generating random variates that have a standard normal distribution:

$$Z^* = U_1 + U_2 + \cdots + U_{12} - 6,$$

where the U_i's are iid $U(0,1)$ random variables. The following APPL statements determine the exact distribution of this 12-fold sum:

```
> U := UniformRV(0, 1);
> U12 := ConvolutionIID(U, 12);
> Zstar := Transform(U12, [[x -> x - 6], [0, 12]]);
```

which returns the PDF of Z^* as

$$
f_{Z^*}(x) = \begin{cases}
\frac{1}{39916800}(x+6)^{11} & -6 < x < -5 \\[2mm]
\begin{aligned}
&-\frac{1}{3628800}x^{11} - \frac{1}{67200}x^{10} - \frac{11}{30240}x^9 - \frac{107}{20160}x^8 - \frac{517}{10080}x^7 - \\
&\frac{2477}{7200}x^6 - \frac{11737}{7200}x^5 - \frac{54797}{10080}x^4 - \frac{250657}{20160}x^3 - \frac{1113317}{60480}x^2 - \\
&\frac{4726777}{302400}x - \frac{18595037}{3326400}
\end{aligned} & -5 < x < -4 \\[2mm]
\begin{aligned}
&\frac{1}{725760}x^{11} + \frac{1}{17280}x^{10} + \frac{11}{10080}x^9 + \frac{7}{576}x^8 + \frac{99}{1120}x^7 + \\
&\frac{631}{1440}x^6 + \frac{1199}{800}x^5 + \frac{1009}{288}x^4 + \frac{12199}{2240}x^3 + \frac{9385}{1728}x^2 + \\
&\frac{38533}{11200}x + \frac{894727}{665280}
\end{aligned} & -4 < x < -3 \\[2mm]
\begin{aligned}
&-\frac{1}{241920}x^{11} - \frac{1}{8064}x^{10} - \frac{11}{6720}x^9 - \frac{25}{2016}x^8 - \frac{33}{560}x^7 - \\
&\frac{13}{72}x^6 - \frac{143}{400}x^5 - \frac{239}{504}x^4 - \frac{583}{1120}x^3 - \frac{1619}{3024}x^2 - \\
&\frac{781}{5600}x + \frac{61297}{166320}
\end{aligned} & -3 < x < -2 \\[2mm]
\begin{aligned}
&\frac{1}{120960}x^{11} + \frac{1}{6720}x^{10} + \frac{11}{10080}x^9 + \frac{1}{252}x^8 + \frac{11}{1680}x^7 + \\
&\frac{1}{360}x^6 + \frac{11}{1200}x^5 + \frac{25}{504}x^4 + \frac{11}{3360}x^3 - \frac{563}{3024}x^2 + \\
&\frac{11}{50400}x + \frac{65521}{166320}
\end{aligned} & -2 < x < -1 \\[2mm]
\begin{aligned}
&-\frac{1}{86400}x^{11} - \frac{1}{14400}x^{10} + \frac{1}{1440}x^8 - \frac{23}{3600}x^6 + \frac{31}{720}x^4 - \\
&\frac{809}{4320}x^2 + \frac{655177}{1663200}
\end{aligned} & -1 < x < 0 \\[2mm]
\begin{aligned}
&\frac{1}{86400}x^{11} - \frac{1}{14400}x^{10} + \frac{1}{1440}x^8 - \frac{23}{3600}x^6 + \frac{31}{720}x^4 - \\
&\frac{809}{4320}x^2 + \frac{655177}{1663200}
\end{aligned} & 0 < x < 1 \\[2mm]
\begin{aligned}
&-\frac{1}{120960}x^{11} + \frac{1}{6720}x^{10} - \frac{11}{10080}x^9 + \frac{1}{252}x^8 - \frac{11}{1680}x^7 + \\
&\frac{1}{360}x^6 - \frac{11}{1200}x^5 + \frac{25}{504}x^4 - \\
&\frac{11}{3360}x^3 - \frac{563}{3024}x^2 - \frac{11}{50400}x + \frac{65521}{166320}
\end{aligned} & 1 < x < 2 \\[2mm]
\begin{aligned}
&\frac{1}{241920}x^{11} - \frac{1}{8064}x^{10} + \frac{11}{6720}x^9 - \frac{25}{2016}x^8 + \frac{33}{560}x^7 - \\
&\frac{13}{72}x^6 + \frac{143}{400}x^5 - \frac{239}{504}x^4 + \frac{583}{1120}x^3 - \frac{1619}{3024}x^2 + \\
&\frac{781}{5600}x + \frac{61297}{166320}
\end{aligned} & 2 < x < 3 \\[2mm]
\begin{aligned}
&-\frac{1}{725760}x^{11} + \frac{1}{17280}x^{10} - \frac{11}{10080}x^9 + \frac{7}{576}x^8 - \frac{99}{1120}x^7 + \\
&\frac{631}{1440}x^6 - \frac{1199}{800}x^5 + \frac{1009}{288}x^4 - \frac{12199}{2240}x^3 + \frac{9385}{1728}x^2 - \\
&\frac{38533}{11200}x + \frac{894727}{665280}
\end{aligned} & 3 < x < 4 \\[2mm]
\begin{aligned}
&\frac{1}{3628800}x^{11} - \frac{1}{67200}x^{10} + \frac{11}{30240}x^9 - \frac{107}{20160}x^8 + \frac{517}{10080}x^7 - \\
&\frac{2477}{7200}x^6 + \frac{11737}{7200}x^5 - \frac{54797}{10080}x^4 + \frac{250657}{20160}x^3 - \frac{1113317}{60480}x^2 + \\
&\frac{4726777}{302400}x - \frac{18595037}{3326400}
\end{aligned} & 4 < x < 5 \\[2mm]
\begin{aligned}
&-\frac{1}{39916800}x^{11} + \frac{1}{604800}x^{10} - \frac{1}{20160}x^9 + \frac{1}{1120}x^8 - \frac{3}{280}x^7 + \\
&\frac{9}{100}x^6 - \frac{27}{50}x^5 + \frac{81}{35}x^4 - \frac{243}{35}x^3 + \frac{486}{35}x^2 - \\
&\frac{2916}{175}x + \frac{17496}{1925}
\end{aligned} & 5 < x < 6.
\end{cases}
$$

The following additional APPL statements calculate the kurtosis of Z^* and the kurtosis of a standard normal random variable Z:

```
> Z := NormalRV(0, 1);
> Kurtosis(Zstar);
> Kurtosis(Z);
```

The distribution of Z^* is useful because, like the standard normal distribution, its mean is zero, its variance is one, and its skewness is zero. However, the fourth moments of Z^* and Z differ, with

$$E\left[\left(\frac{Z^* - E\left[Z^*\right]}{\sqrt{V\left[Z^*\right]}}\right)^4\right] = E\left[(Z^*)^4\right] = 2.9,$$

while the kurtosis of Z is 3. An overlaid Maple plot of the standard normal PDF and $f_{Z^*}(x)$ reveals that this approximation does quite well. It is apparent that the PDFs are very close except around $x = 0$ [where $f_Z(0) = \frac{1}{\sqrt{2\pi}} \cong 0.399$ and $f_{Z^*}(0) = \frac{655177}{1663200} \cong 0.394$] and, of course, in the tails, because the approximate distribution only has support on $(-6, 6)$. To obtain overlaid plots of these two distributions, use the PlotDist procedure. First create a plot of the PDF of Z^* and store it as Plot1. Next, create a plot of the PDF of Z and store it as Plot2. Finally, use the plots[display] procedure to combine the two plots. This is summarized in the statements

```
> Plot1 := PlotDist(Zstar, -6, 6):
> Plot2 := PlotDist(Z, -6, 6):
> plots[display]({Plot1, Plot2}, scaling = unconstrained);
```

Dozens of tables of mathematically intractable statistical values have been published over the years. Some of these constants can be calculated in APPL, as illustrated in the next example.

Example 15.7. Balakrishnan and Chen's [6] text is a valuable resource for comparing various measures associated with probability distributions. Here we use APPL's capabilities to produce similar measures. This book begins with 39 pages of theory and references to the inverse Gaussian (IG) distribution and its order statistics, followed by 645 pages of PDF plots, expected value tables, covariance tables, estimation tables, and so forth. APPL is able to create a number of these figures and tables, and in some cases goes beyond those presented in the text. Page 50, for example, shows the $N(0,1)$ and the standardized IG(0.8) PDFs overlaid in a single plot. The PDF of a standardized IG random variable X [which is listed as Eq. (2.14) on page 7 of the text] is

$$f_X(x) = \frac{1}{\sqrt{2\pi}} \left(\frac{3}{3 + kx}\right)^{3/2} e^{-3x^2/(6+2kx)} \qquad -\frac{3}{k} < x < \infty.$$

The two plots may be overlaid and animated for k increasing from zero to one as follows:

```
> Z := NormalRV(0, 1);
> X := [[x -> (3 / (3 + k * x)) ^ (3 / 2) *
              exp(-3 * x ^ 2 / (6 + 2 * k * x)) /
              sqrt(2 * Pi)],
        [-3 / k, infinity], ["Continuous", "PDF"]];
> NormalExpr := op(unapply(Z[1](x))(x));
> InvGaussExpr := op(unapply(X[1](x))(x));
> unassign('k');
> plots[animate]({NormalExpr, InvGaussExpr},
              x = -4 .. 4, k = 0 .. 1);
```

The plot is shown in Fig. 15.6 for $k = 0.8$. To execute the animation, first select the plot by clicking on it. Then choose "Play" from the "Animation" menu. Balakrishnan and Chen use such plots to show how the two distributions separate as k increases from zero. Clearly APPL is not limited to plotting the standardized version of the IG PDF, but can be used to plot IG PDF, CDF, HF, and so on. Their text also gives a listing of selected means of order statistics for the standardized IG distribution Z. Using APPL one can calculate, for example, the mean of the order statistic $Z_{(2)}$ for $n = 4$ (when $k = 0.7$, as on page 74 of their text) as follows:

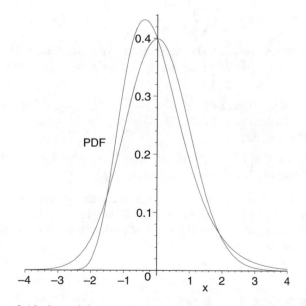

Fig. 15.6. Overlaid plots of the standard normal and standard IG(0.8) distributions

```
> k := 0.7;
> Z := [[z -> 1 / sqrt(2 * Pi) * (3 / (3 + k * z)) ^
                (3 / 2) * exp(-3 * z ^ 2 /
                (2 * (3 + k * z)))],
            [-3 / k, infinity], ["Continuous", "PDF"]];
> Zos := OrderStat(Z, 4, 2);
> avg := Mean(Zos);
```

which returns the value -0.35202 in agreement with the authors. In addition to giving the mean, note that the variable Zos above is in fact the list-of-sublists representation of the distribution of that order statistic. Thus one may use any number of functions on that variable, in order to find critical points, significance levels, and using that distribution in determining other distributions, e.g., by invoking Product or Convolution.

Another resource, such as Harter and Balakrishnan's [45] handbook of tables for the use of order statistics in estimation also gives extensive charts for various applications of order statistics. APPL can be used in a similar fashion to duplicate many of their tabled values, and, in some cases, go beyond the tabled values. See Glen et al. [38] for a specific example.

Example 15.8. Define a geometric random variable X with parameter $p = 1/4$ to model the number of trials up to and including the first success, i.e.,

$$f_X(x) = \frac{1}{4} \cdot \left(\frac{3}{4}\right)^{x-1} \qquad x = 1, 2, \dots .$$

Calculate the median of the maximum order statistic when $n = 5$ items are sampled *with replacement* from this geometric distribution. The APPL statements

```
> X := GeometricRV(1 / 4);
> Y := OrderStat(X, 5, 5);
> IDF(Y, 0.5);
```

return the median of the distribution as 8. The final statement could have equivalently been issued as Median(Y);

Example 15.9. A sequence of order statistics drawn from a unit exponential distribution was considered in Example 3.12. Five plots were generated corresponding to the PDF of each of the order statistics via the APPL code

```
> X := ExponentialRV(1);
> n := 5;
> for i from 1 to n do
>   PlotDist(OrderStat(X, n, i), 0, 6);
> od;
```

But what if one wanted to see all five of the order statistics on a single graph? This can be done by modifying the code to

```
> X := ExponentialRV(1);
> n := 5;
> OSgraphs := [];
> for i from 1 to n do
>    OSgraphs := [op(OSgraphs),
                    PlotDist(OrderStat(X, n, i), 0, 6)]:
> od:
> with(plots);
> display(OSgraphs);
```

This code initializes **OSgraphs** to an empty list, and concatenates the various graphs in the loop. The **with(plots)** command is used to include additional Maple plotting options. Finally, **display** is used to plot multiple graphs on one plot. The graph is shown in Fig. 15.7.

Fig. 15.7. Overlaid plots of the PDFs of order statistics drawn from an exponential population

References

1. Abate J, Whitt W (1988) Transient behavior of the $M/M/1$ queue via Laplace transforms. Adv Appl Probab 20:145–178
2. Adlakha VG, Kulkarni VG (1989) A classified bibliography of research on stochastic PERT networks: 1966–1987. INFOR 27:272–296
3. Andrews DWK, Buchinsky M (2000) A three-step method for choosing the number of bootstrap repetitions. Econometrica 68:23–51
4. Andrews DWK, Buchinsky M (2002) On the number of bootstrap repetitions for BC_a confidence intervals. Econometric Theory 18:962–984
5. Arnold BC, Balakrishnan N, Nagaraja HN (1992) A first course in order statistics. SIAM, Philadelphia
6. Balakrishnan N, Chen WWS (1997) CRC handbook of tables for order statistics from inverse Gaussian distributions with applications. CRC Press, Boca Raton
7. Banks J, Carson JS, Nelson BL, Nicol DM (2005) Discrete-event system simulation, 4th edn. Prentice-Hall, Upper Saddle River, New Jersey
8. Barr D, Zehna PW (1971) Probability. Brooks/Cole
9. Benford F (1938) The law of anomalous numbers. Proc Am Philos Soc 78: 551–572
10. Berger A, Hill, TP (2015) An introduction to Benford's law. Princeton University Press, Princeton
11. Billingsley P (1995) Probability and measure, 3rd edn. Wiley, New York
12. Birnbaum ZW (1952) Numerical tabulation of the distribution of Kolomogorov's statistic for finite sample size. J Am Stat Assoc 47:425–441
13. Box GEP, Jenkins GM (1994) Time series analysis: forecasting & control, 3rd edn. Prentice-Hall
14. Burr IW (1955) Calculation of exact sampling distribution of ranges from a discrete population. Ann Math Stat 26:530–532 (correction, volume 38, 280)
15. Carrano FM, Helman P, Veroff R (1998) Data abstraction and problem solving with C++: walls and mirrors, 2nd edn. Addison-Wesley Longman, Reading, Massachusetts
16. Casella G, Berger R (2002) Statistical inference, 2nd edn. Duxbury, Pacific Grove, California

© Springer International Publishing AG 2017
J.H. Drew et al., *Computational Probability*, International Series
in Operations Research & Management Science 246,
DOI 10.1007/978-3-319-43323-3

17. Ciardo G, Leemis LM, Nicol D (1995) On the minimum of independent geometrically distributed random variables. Stat Probab Lett 23:313–326

18. Cook P, Broemeling LD (1995) Bayesian statistics using mathematica. Am Stat 49:70–76

19. D'Agostino RB, Stephens MA (1986) Goodness-of-fit techniques. Marcel Dekker, New York

20. David HA, Nagaraja HN (2003) Order statistics, 3rd edn. Wiley, New York

21. Devroye L (1996) Random variate generation in one line of code. In: Charnes J, Morrice D, Brunner D, Swain J (eds) Proceedings of the 1996 winter simulation conference. Institute of Electrical and Electronics Engineers, Coronado, CA, pp 265–272

22. Doss H, Chiang Y (1994) Choosing the resampling scheme when bootstrapping: a case study in reliability. J Am Stat Assoc 89:298–308

23. Drew JH, Glen AG, Leemis LM (2000) Computing the cumulative distribution function of the Kolmogorov–Smirnov statistic. Comput Stat Data Anal 34: 1–15

24. Duggan MJ, Drew JH, Leemis LM (2005) A test of randomness based on the distance between consecutive random number pairs. In: Kuhl ME, Steiger NM, Armstrong FB, Joines JA (eds) Proceedings of the 2005 winter simulation conference. Institute of Electrical and Electronics Engineers, Piscataway, New Jersey, pp 741–748

25. e Silva ES, Gail HR, Campos RV (1995) Calculating transient distributions of cumulative reward. In: Proceedings of the 1995 ACM SIGMETRICS joint international conference on measurement and modeling of computer systems, pp 231–240

26. Efron B, Tibshirani RJ (1993) An introduction to the bootstrap. Chapman & Hall, New York

27. Elmaghraby SE (1977) Activity networks: project planning and control by network models. Wiley, New York

28. Evans DL, Leemis LM (2000) Input modeling using a computer algebra system. In: Joines J, Barton R, Fishwick P, Kang K. (eds) Proceedings of the 2000 winter simulation conference. Institute of Electrical and Electronics Engineers, Piscataway, New Jersey, pp 577–586

29. Evans DL, Leemis LM (2004) Algorithms for determining the distributions of sums of discrete random variables. Math Comput Modell 40:1429–1452

30. Evans DL, Leemis LM, Drew JH (2006) The distribution of order statistics for discrete random variables with applications to bootstrapping. INFORMS J Comput 18:19–30

31. Evans DL, Drew JH, Leemis, LM (2008) The distribution of the Kolmogorov–Smirnov, Cramer–von Mises, and Anderson–Darling test statistics for exponential populations with estimated parameters. Commun Stat–Simul Comput 37:1396–1421

32. Fisher DL, Saisi D, Goldstein WM (1985) Stochastic PERT networks: OP diagrams, critical paths and the project completion time. Comput Oper Res 12:471–482

33. Fishman GS (2001) Discrete-event simulation: modeling, programming, and analysis. Springer, Berlin

34. Gafarian AV, Ancker CJ Jr, Morisaku T (1976) The problem of the initial transient in digital computer simulation. In: Proceedings of the 76 bicentennial conference on winter simulation, pp 49–51

35. Gehan EA (1965) A generalized Wilcoxon test for comparing arbitrarily singly-censored samples. Biometrika 52(parts 1 and 2):203–223

36. Ghosal S, Ghosh JK, Ramamoorthi RV (1999) Consistency issues in Bayesian nonparametrics. In: Ghosh S (ed) Asymptotics, nonparametrics and time series: a tribute to Madan Lal Puri. Statistics textbooks and monographs, vol 158. Marcel Dekker, New York, pp 639–667

37. Glen AG, Drew JH, Leemis LM (1997) A generalized univariate change-of-variable transformation technique. INFORMS J Comput 9:288–295

38. Glen AG, Evans DL, Leemis LM (2001) APPL: a probability programming language. Am Stat 55:156–166

39. Glen AG, Leemis LM, Drew JH (2004) Computing the distribution of the product of two continuous random variables. Comput Stat Data Anal 44: 451–464

40. Grassmann WK (1977) Transient solutions in Markovian queueing systems. Comput Oper Res 4:47–53

41. Grassmann WK (2008) Warm-up periods in simulation can be detrimental. Probab Eng Inform Sci 22:415–429

42. Grinstead CM, Snell JL (1997) Introduction to probability, 2nd rev. edn. American Mathematical Society, Providence, Rhode Island

43. Hagwood C (2009) An application of the residue calculus: the distribution of the sum of nonhomogeneous gamma variates. Am Stat 63:37–39

44. Hamilton JD (1994) Time series analysis. Princeton University Press, Princeton

45. Harter HL, Balakrishnan N (1996) CRC handbook of tables for the use of order statistics in estimation. CRC Press, Boca Raton

46. Hasting KJ (2006) Introduction to the mathematics of operations research with mathematica, 2nd edn. CRC Press, Boca Raton

47. Hill TP (1995) A statistical derivation of the significant-digit law. Stat Sci 86:354–363

48. Hill TP (1998) The first digit phenomenon. Am Sci 86:358–363

49. Hillier FS, Lieberman GJ (2010) Introduction to operations research, 9th edn. McGraw–Hill, New York

50. Hogg RV, Craig AT (1995) Introduction to the mathematical statistics, 5th edn. Prentice-Hall, Upper Saddle River, New Jersey

51. Hogg RV, Tanis EA (2001) Probability and statistical inference, 6th edn. Prentice-Hall, Upper Saddle River, New Jersey

52. Hogg RV, McKean JW, Craig AT (2005) Introduction to the mathematical statistics, 6th edn. Prentice-Hall, Upper Saddle River, New Jersey

53. Hutson AD, Ernst MD (2000) The exact bootstrap mean and variance of an L-estimator. J R Stat Soc Ser B 62:89–94

54. Jaakkola TS, Jordan MI (2000) Bayesian parameter estimation via variational methods. Stat Comput 10:25–37

55. Johnson NL, Kotz S, Balakrishnan N (1995) Continuous univariate distributions, vol 2, 2nd ed. Wiley, New York

56. Kaczynski WH, Leemis LM, Drew JH (2012) Transient queueing analysis. INFORMS J Comput 24:10–28

57. Kalbfleisch JD, Prentice RL (2002) The statistical analysis of failure time data, 2nd edn. Wiley, Hoboken, New Jersey

58. Karian ZA, Tanis EA (1999) Probability and statistics: explorations with Maple, 2nd edn. Prentice-Hall, Upper Saddle River, New Jersey

59. Kelton WD (1985) Transient exponential-Erlang queues and steady-state simulation. Commun ACM 28:741–749

60. Kelton WD, Law AM (1985) The transient behavior of the $M/M/s$ queue, with implications for steady-state simulation. Oper Res 33:378–396

61. Khoshkenara A, Mahloojia H. (2013) A new test of randomness for Lehmer generators based on the Manhattan Distance Between Pairs. Commun Stat–Simul Comput 42:202–214

62. Kleinrock L (1975) Queueing systems. Wiley, New York

63. Knuth DE (1998) The art of computer programming, volume 2: seminumerical algorithms, 3rd edn. Addison-Wesley, Reading, Massachusetts

64. Kossovsky AE (2015) Benford's law : theory, the general law of relative quantities, and forensic fraud detection applications. World Scientific, Singapore

65. L'Écuyer P, Cordeau J-F, Simard R (2000) Close-point spatial tests and their application to random number generators. Oper Res 48:308–317

66. Law AM (1975) A comparison of two techniques for determining the accuracy of simulation output. Technical Report 75–11, University of Wisconsin at Madison

67. Laplante PA (ed) (2001) Dictionary of computer science, engineering and technology. CRC Press

68. Larsen RJ, Marx ML (2001) An introduction to mathematical statistics and its applications, 3rd edn. Prentice-Hall, Upper Saddle River, New Jersey

69. Larsen RJ, Marx ML (2006) An introduction to mathematical statistics and its applications, 4th edn. Prentice-Hall, Upper Saddle River, New Jersey

70. Law AM (2015) Simulation modeling and analysis, 5th edn. McGraw-Hill, New York

71. Lawless JF (2003) Statistical models and methods for lifetime data, 2nd edn. Wiley, Hoboken, New Jersey

72. Leemis L (1995) Reliability: probabilistic models and statistical methods. Prentice-Hall, Upper Saddle River, New Jersey

73. Leemis L (2006) Lower system reliability bounds from binary failure data using bootstrapping. J Qual Technol 38:2–13

74. Leemis L, Schmeiser B, Evans D (2000) Survival distributions satisfying Benford's law. Am Stat 54:236–241

75. Leemis LM, Duggan MJ, Drew JH, Mallozzi JA, Connell KW (2006) Algorithms to calculate the distribution of the longest path length of a stochastic activity network with continuous activity durations. Networks 48:143–165

76. Leguesdron P, Pellaumail J, Rubino G, Sericola B (1993) Transient analysis of the $M/M/1$ queue. Adv Appl Probab 25:702–713

77. Lehmer DH (1951) Mathematical methods in large-scale computing units. In: Proceedings of the 2nd symposium on large-scale calculating machinery. Harvard University Press, pp 141–146

78. Ley E (1996) On the peculiar distribution of the U.S. stock indices digits. Am Stat 50:311–313

79. Lieblein J, Zelen M (1956) Statistical investigation of the fatigue life of deep-groove ball bearings. J Res Natl Bur Stand 57:273–316

80. Maplesoft (2013) Maple, Version 17. Waterloo

81. Margolin BH, Winokur HS (1967) Exact moments of the order statistics of the geometric distribution and their relation to inverse sampling and reliability of redundant systems. J Am Stat Assoc 62:915–925

82. Marks CE, Glen AG, Robinson MW, Leemis LM (2014) Applying bootstrap methods to system reliability. Am Stat 68:174–182

83. Marsaglia G (1968) Random numbers fall mainly in the planes. Proc Natl Acad Sci 61:25–28

84. Martin JJ (1965) Distribution of the time through a directed, acyclic network. Oper Res 13:44–66

85. Martin MA (1990) On bootstrap iteration for coverage correction in confidence intervals. J Am Stat Assoc 85:1105–1118

86. Meeker WQ, Escobar LA (1998) Statistical methods for reliability data. Wiley, New York

87. Miller LH (1956) Table of percentage points of Kolmogorov statistics. J Am Stat Assoc 51:111–121

88. Miller S (ed) (2015) Benford's law: theory & applications. Princeton University Press, Princeton

89. Miller I, Miller M (2004) John E. Freund's mathematical statistics, 7th edn. Prentice-Hall, Upper Saddle River, New Jersey

90. Morisaku T (1976) Techniques for data-truncation in digital computer simulation. Ph.D. thesis, University of Southern California, Los Angeles

91. Nelson BL, Yamnitsky M (1998) Input modeling tools for complex problems. In: Medeiros DJ, Watson EF, Carson JS, Manivannan, MS (eds) Proceedings of the 1998 winter simulation conference. Institute of Electrical and Electronics Engineers, pp 105–112

92. Nicol D (2000) Personal communication

93. Nigrini M (1996) A taxpayer compliance application of Benford's law. J Am Taxat Assoc 18:72–91

94. Odoni AR, Roth E (1983) Empirical investigation of the transient behavior of stationary queueing systems. Oper Res 31:432–455

95. Owen DB (1962) Handbook of statistical tables. Addison-Wesley, Reading, Massachusetts

96. Padgett WJ, Tomlinson MA (2003) Lower confidence bounds for percentiles of Weibull and Birnbaum–Saunders distributions. J Stat Comput Simul 73: 429–443

97. Park SK, Miller KW (1988) Random number generators: good ones are hard to find. Commun ACM 31:1192–1201

98. Parlar M (2000) Interactive operations research with Maple. Birkhäuser, Boston

99. Parthasarathy PR (1987) A transient solution to an $M/M/1$ queue: a simple approach. Adv Appl Probab 19:997–998

100. Parzen E (1960) Modern probability theory and its applications. Wiley, New York

101. Pegden CD, Rosenshine M (1982) Some new results for the $M/M/1$ queue. Manage Sci 28:821–828

102. Port SC (1994) Theoretical probability for applications. Wiley

103. Rice JA (2007) Mathematical statistics and data analysis, 3rd edn. Thompson and Brooks/Cole, Belmont, California

104. Rohatgi VK (1976) An introduction to probability theory and mathematical statistics. Wiley, New York

105. Rose C, Smith MD (2002) Mathematical statistics and mathematica. Springer, New York

106. Ross S (2006) A first course in probability, 7th edn. Prentice Hall, Upper Saddle River, New Jersey

107. Ruskey F, Williams A (2008) Generating balanced parentheses and binary trees by prefix shifts. In: Proceedings of the 12th computing: the Australasian theory symposium (CATS2008), CRPIT, vol 77, pp 107–115

108. Shier DR (1991) Network reliability and algebraic structures. Oxford University Press, New York
109. Springer MD (1979) The algebra of random variables. Wiley, New York
110. Srivastava RC (1974) Two characterizations of the geometric distribution. J Am Stat Assoc 69:267–269
111. Stanley RP (1999) Enumerative combinatorics. Volume 62 of Cambridge studies in advanced mathematics. Cambridge University Press, Cambridge
112. Thompson P (2000) Getting normal probability approximations without using normal tables. College Math J 31:51–54
113. Trosset M (2001) Personal communication
114. Vargo E, Pasupathy R, Leemis L (2010) Moment-ratio diagrams for univariate distributions. J Qual Technol 42:276–286
115. Webb KH, Leemis LM (2014) Symbolic ARMA model analysis. Comput Econ 43:313–330
116. Weiss MA (1994) Data structures and algorithm analysis in C++. Addison-Wesley Publishing Company, Menlo Park, California
117. Weisstein EW (2016) "Generalized Hypergeometric Function" from MathWorld–A Wolfram Web Resource. http://mathworld.wolfram.com/GeneralizedHypergeometricFunction.html
118. Winston WL (2004) Operations research: applications and algorithms, 4th edn. Thompson, Belmont, California
119. Wolfram Research, Inc. (2013) Mathematica, Version 10, Champaign, IL
120. Woodward WA, Gray HL (1981) On the relationship between the S array and the Box–Jenkins method of ARMA model identification. J Am Stat Assoc 76:579–587
121. Woodward JA, Palmer CGS (1997) On the exact convolution of discrete random variables. Appl Math Comput 83:69–77
122. Yang JX, Drew JH, Leemis LM (2012) Automating bivariate transformations. INFORMS J Comput 24:1–9
123. Young DH (1970) The order statistics of the negative binomial distribution. Biometrika 57:181–186

Index

© Springer International Publishing AG 2017 329
J.H. Drew et al., *Computational Probability*, International Series
in Operations Research & Management Science 246,
DOI 10.1007/978-3-319-43323-3

Printed in the United States
By Bookmasters